YOU'VE JUST PURCHASED MORE THAN A BOOK!

Resources for Thrall/Robertson: Atlas of Normal Radiographic Anatomy and Anatomic Variants in the Dog and Cat, Second Edition

A unique feature of the second edition of *Atlas of Radiographic Anatomy and Anatomic Variants in the Dog and Cat* is the new companion website.

The authors have collaborated to create assets they hope will make a valuable contribution to the reader's understanding and will provide a useful adjunct to formal instructional courses. The website features:

- A collection of more than 200 radiographic CT scans to supplement images in the atlas
- More than 100 questions with answers and rationales to help students assess their comprehension of the content in the atlas
- Questions that include exercises to aid in identifying clinically important parts of the anatomy

Activate the complete learning experience that comes with each book purchase by registering at

www.thrallrobertsonatlas.com

REGISTER TODAY!

You can now purchase Elsevier products on Elsevier Health!
Go to us.elsevierhealth.com to search and browse for products.

ATLAS OF Normal Radiographic Anatomy & Anatomic Variants IN THE Dog AND Cat

ATLAS OF Normal Radiographic Anatomy & Anatomic Variants IN THE Dog AND Cat

SECOND EDITION

DONALD E. THRALL, DVM, PhD, DACVR
(Radiology, Radiation Oncology)
Clinical Professor
Department of Molecular Biomedical Sciences
College of Veterinary Medicine
North Carolina State University
Raleigh, North Carolina

Quality Control, Radiologist
VDIC—IDEXX Telemedicine Consultants
IDEXX Laboratories, Inc.
Clackamas, Oregon

IAN D. ROBERTSON, BVSc, DACVR
Clinical Professor
Department of Molecular Biomedical Sciences
College of Veterinary Medicine
North Carolina State University
Raleigh, North Carolina

ELSEVIER

ELSEVIER

3251 Riverport Lane
St. Louis, Missouri 63043

ATLAS OF NORMAL RADIOGRAPHIC ANATOMY AND ANATOMIC VARIANTS IN THE DOG AND CAT, SECOND EDITION

ISBN: 978-0-323-31225-7

Copyright © 2016 by Elsevier, Inc. All rights reserved.

No part of this publication may be reproduced or transmitted in any form or by any means, electronic or mechanical, including photocopying, recording, or any information storage and retrieval system, without permission in writing from the publisher. Details on how to seek permission, further information about the Publisher's permissions policies and our arrangements with organizations such as the Copyright Clearance Center and the Copyright Licensing Agency, can be found at our website: www.elsevier.com/permissions.

This book and the individual contributions contained in it are protected under copyright by the Publisher (other than as may be noted herein).

Notices

Knowledge and best practice in this field are constantly changing. As new research and experience broaden our understanding, changes in research methods, professional practices, or medical treatment may become necessary.

Practitioners and researchers must always rely on their own experience and knowledge in evaluating and using any information, methods, compounds, or experiments described herein. In using such information or methods they should be mindful of their own safety and the safety of others, including parties for whom they have a professional responsibility.

With respect to any drug or pharmaceutical products identified, readers are advised to check the most current information provided (i) on procedures featured or (ii) by the manufacturer of each product to be administered, to verify the recommended dose or formula, the method and duration of administration, and contraindications. It is the responsibility of practitioners, relying on their own experience and knowledge of their patients, to make diagnoses, to determine dosages and the best treatment for each individual patient, and to take all appropriate safety precautions.

To the fullest extent of the law, neither the Publisher nor the authors, contributors, or editors, assume any liability for any injury and/or damage to persons or property as a matter of products liability, negligence or otherwise, or from any use or operation of any methods, products, instructions, or ideas contained in the material herein.

Previous edition copyrighted 2011.

Library of Congress Cataloging-in-Publication Data

Thrall, Donald E., author.
  [Atlas of normal radiographic anatomy & anatomic variants in the dog and cat]
  Atlas of normal radiographic anatomy and anatomic variants in the dog and cat / Donald E. Thrall, Ian D. Robertson.—Second edition.
      p. ; cm.
  Preceded by Atlas of normal radiographic anatomy & anatomic variants in the dog and cat / Donald E. Thrall, Ian D. Robertson. c2011.
  Includes bibliographical references and index.
  ISBN 978-0-323-31225-7 (hardcover : alk. paper)
  I. Robertson, Ian D. (Ian Douglas), 1958- , author. II. Title.
  [DNLM: 1. Cats—anatomy & histology—Atlases. 2. Dogs—anatomy & histology—Atlases. 3. Radiography—veterinary—Atlases. SF 767.D6]
  SF757.8
  636.089′607572—dc23

2015012557

Content Strategy Director: Penny Rudolph
Senior Content Development Specialist: Courtney Sprehe
Publishing Services Manager: Jeff Patterson
Project Manager: Lisa A. P. Bushey
Manager, Art & Design: Teresa McBryan

Printed in the United States of America.

Last digit is the print number: 9 8 7 6 5 4 3 2 1

Preface

Becoming a proficient diagnostic radiologist is a long journey. Specialty training leading to board certification entails at least 4 years of post-DVM structured learning, followed by a rigorous multistage examination. However, board-certified radiologists make up only a small fraction of all veterinarians who interpret radiographs each day. Most radiographic studies are interpreted by competent veterinarians whose training in image interpretation has been limited to relatively few contact hours of didactic instruction and supervised clinical training. All of these veterinarians, as well as students who are just beginning to develop their interpretive skills, must have a solid appreciation for normal radiographic anatomy, anatomic variants, and things that mimic disease, which are affectionately termed "fakeouts" by those of us who spend our lives interpreting images.

The vastness of normal variation within dogs and cats is staggering. Although the generic cat is relatively standard, dogs come in all shapes and sizes, with innumerable inherent variations that can be misinterpreted as disease unless recognized as normal. On top of this inherent variation is the variation introduced by radiographic positioning that can lead to countless variations in the appearance of a normal structure. During their training, specialists have this information drilled into them during many hours of mentored learning and brow-beating by experienced radiologists. Nonspecialists, on the other hand, may have had some introduction to normal radiographic anatomy during veterinary school, but the acuity of recall becomes dulled by the sheer volume of memory-bank information needed to be a competent, licensed, contemporary veterinarian. During one's education as a student, it is impossible to be exposed to the range of normal that is likely to be encountered in practice and then influenced by radiographic positioning. Therefore there is a real need for a reference source for practicing veterinarians and students to assist them in the daunting task of interpreting clinical radiographs competently. This need led to the development of this atlas.

In this book, we have not only pointed out the identity of essentially every clinically significant anatomic part of a dog or cat that can be seen radiographically, we have also included more than one example of those parts where normal inherent variation can confuse interpretation. Simply labeling structures in radiographs of a generic dog or cat is highly inadequate in addressing the mission of providing a clinically relevant resource. Additionally, this atlas includes context relevant to the description of normal anatomy that only a radiologist can provide. Normal is presented in the context of how it is modified by the procedure of making the radiograph. Although this is not a radiographic positioning guide, specific technical factors have been included to the extent that their influence on the image is so great that they must be understood for the image to be interpreted accurately.

Finally, this book is not simply a picture atlas. Every body part is put into context with a textual description. This provides a basis for the reader to understand why a structure appears as it does in radiographs, and it enables the reader to appreciate variations of normal that are not included based on an understanding of basic radiographic principles. This may require a bit of effort from the reader in comparison to a picture atlas, but this small investment of time has the potential for a big payoff in terms of interpretive ability.

Acknowledgments

We acknowledge the many dedicated, inquisitive, and intelligent veterinary students and radiology residents at North Carolina State University whose innumerable questions over the years helped us focus on clinically relevant radiographic anatomy and anatomic variants.

I wish to thank my wife and colleague, Debbie, for her support over the years and my children, Heather and Matt, for tolerating my time-consuming career.

Ian Robertson

I have enjoyed assisting with the compilation of this atlas, but I know it detracted from activities I could have shared with my children, Hilary and Tristan, of whom I am so proud. I appreciate their support of me and of my efforts in projects such as this book.

Don Thrall

Contents

1
Basic Imaging Principles and Physeal Closure Time, 1

2
The Skull, 20

3
The Spine, 49

4
The Thoracic Limb, 90

5
The Pelvic Limb, 136

6
The Thorax, 182

7
The Abdomen, 241

Index, 296

The Images

All images in this book were acquired using a commercially available indirect digital imaging plate. The images that were created using this technology have tremendous contrast resolution compared with images acquired using a film-screen system. What this means to the reader is that every image in this book is the highest quality possible and all regions of the part being displayed can be assessed. Both thick and thin parts are assessable, which is something that is impossible when using film screen–based images. Things that are described in the text and labeled in the image can be seen. Imagination is not needed to gain an appreciation for the message being delivered.

Over 95% of the images in this book were acquired on clinical patients. This introduces a level of relevance that is extremely valuable in terms of putting radiographic anatomy into perspective. Because the images were derived from clinical patients, there will be some minor disease that is visible in some images. This is pointed out where it is relevant to make sure that the reader does not misinterpret this as part of the normal variation process. Having absolutely no abnormality in any image could have been avoided by imaging cadavers, but the breadth of variation in patient size, age, and breed could not have been duplicated in that instance. The value gained by this variation far outweighs any minor disease that may be seen occasionally.

ATLAS OF Normal Radiographic Anatomy & Anatomic Variants IN THE Dog AND Cat

CHAPTER 1

Basic Imaging Principles and Physeal Closure Time

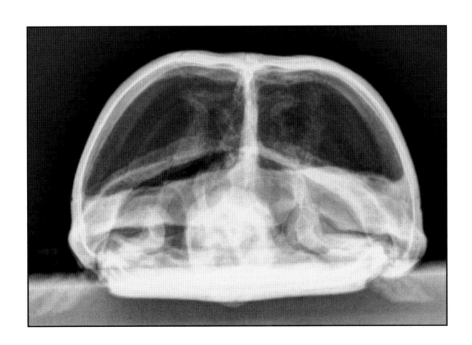

HOW TO USE THIS ATLAS

As described in the Preface, a radiographic atlas is intended to help decide whether any given radiographic finding is normal or abnormal. Determining normal from abnormal is one of the most difficult parts of interpreting a radiograph (if not the most difficult). No atlas will be able to provide a definitive answer to the "What is that, and is it normal?" question in every circumstance, but the information in this atlas can help guide the decision-making process.

The best way to use *this* atlas is to spend some time with it, and get to know it. Of course, labeled images are provided—every atlas needs these. But, contrary to a pure picture atlas, some of the most valuable information in this atlas is contained in the text. Being familiar with the text, which is brief and focused, and noting the important principles that have been augmented with illustrative examples can help formulate a basis for interpretation that extends beyond structure identification alone.

WHAT IS NORMAL?

Many dogs and cats have congenital or developmental changes that are apparent radiographically but insignificant clinically. Discussing some of these common variations in this atlas, along with normal structures, is justified even though they are not completely normal because they are often confused with disease. This book demonstrates much of the morphologic diversity currently present in domestic canine and feline companions that has come to be commonly accepted as normal or clinically insignificant.

WHY ARE COMPUTED TOMOGRAPHY IMAGES INCLUDED IN THIS ATLAS?

Because this is a radiographic atlas, the majority of images are radiographs. This book is not intended to be an atlas of normal computed tomography (CT) appearances, but some CT images are included to reinforce the appearance of selected structures in radiographs. A radiograph is a 2-dimensional image of a 3-dimensional (3-D) object, and, as a result, spatial localization of structures cannot always be done accurately, even with multiple views of the object or patient. CT images are tomographic, creating multiple image *slices* of the volume of interest. *Slicing* the patient solves the problem of spatial localization and can assist with understanding radiographic anatomy more thoroughly. Additionally, CT images, which are typically acquired in a transverse plane, can be reformatted into other planes, typically sagittal and dorsal planes, for clarification or to improve structure visualization. Transverse CT images can even be reformatted into volumetric images for illustration of complex structures. Therefore, a careful selection of planar and volumetric CT images has been included with this goal in mind.

RADIOGRAPHIC TERMINOLOGY

This book uses the standard method for naming radiographic projections approved by the American College of Veterinary Radiology.[1] In general, this naming method is based on anatomic directional terms (as defined by the *Nomina Anatomica Veterinaria*) combined with the point-of-entrance to point-of-exit of the primary x-ray beam. In other words, any radiograph can be named by knowing the point of entrance and point of exit of the primary x-ray beam. For example, a spinal radiograph made with a dog lying in dorsal recumbency would be called a *ventrodorsal view* because the x-ray beam strikes the ventral aspect of the dog and exits dorsally. To name radiographs correctly, the accepted anatomic directional terms must be known (Figure 1-1). Several important concepts are commonly violated, leading to improper image identification. In summary, these are:

- The terms *anterior* and *posterior* should not be used when describing a radiographic projection.
- In the head, the term *cranial* should not be used; *rostral* is substituted.
- In the forelimb, the terms *cranial* and *caudal* should not be used distal to the antebrachiocarpal joint; *dorsal* and *palmar*, respectively, are substituted.

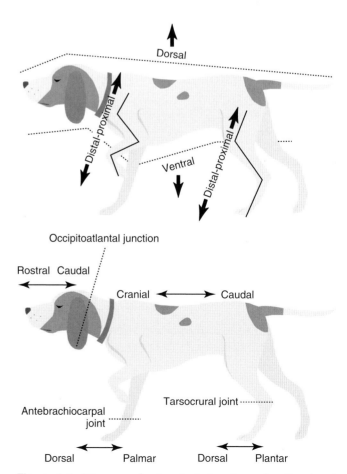

Figure 1-1. Diagram of a dog depicting the major directional anatomic terms, accepted by *Nominica Anatomica Veterinaria*.

- In the hindlimb, the terms *cranial* and *caudal* should not be used distal to the tarsocrural joint; *dorsal* and *plantar*, respectively, are substituted.
- For CT or magnetic resonance (MR) images, the terms *coronal* and *axial* should not be used; the correct terms are *dorsal* and *transverse*, respectively.

One common exception to this standard method needs clarification, that being the acquisition of *lateral* radiographs. For example, the correct name of a lateral thoracic radiograph of a dog made with the subject lying on the left side and the x-ray beam entering the right side is a *right-left lateral*. However, it has become commonplace to take a shortcut and name lateral views according to the side of the subject that is closest to the x-ray table (i.e., the side that the subject is lying on). Thus a *right-left lateral* is typically shortened to *left lateral* because the dog is lying on the left side. Similarly, the correct name of a lateral thoracic radiograph made with a dog lying on the right side and the x-ray beam entering from the left is a *left-right lateral*, but the shortcut convention is to call this view a *right lateral* because the dog is lying on the right side. This shortcut is usually applied to lateral views of the skull, spine, thorax, abdomen, and pelvis.

VIEWING IMAGES

When all radiographic images were recorded on film, a method for consistently hanging film on a viewbox was developed. Viewing radiographs in the same orientation for every subject reduces variation, and the brain becomes more familiar with the way a certain body part should appear in an image. The standard orientation of images is independent of the actual position of the patient during the making of the radiograph. The basic aspects of that radiograph-hanging system are:
- Lateral images of any body part should be oriented with the subject's head, or the cranial or rostral aspect of the body part, facing to the examiner's left.
- Ventrodorsal or dorsoventral images of the head, neck, or trunk should be oriented with the cranial or rostral aspect of the subject pointing up, toward the ceiling, and the left side of the subject positioned on the examiner's right.
- Lateromedial or mediolateral images of extremities should be oriented with the proximal aspect of the subject's limb pointing up, toward the ceiling, and the cranial or dorsal aspect of the subject's limb on the examiner's left.
- Caudocranial (palmarodorsal or plantarodorsal) or craniocaudal (dorsopalmar or dorsoplantar) images of an extremity should be oriented with the proximal end of the extremity pointing up, toward the ceiling, and the distal end pointing down, toward the floor. There is no convention with regard to whether the medial or lateral side of the extremity should be placed to the examiner's left.

Table 1-1 Common Orthogonal Views for Major Body Parts

Body Part	View	Orthogonal View
Skull	Left-right lateral or right-left lateral	Ventrodorsal or dorsoventral
Spine	Left-right lateral or right-left lateral	Ventrodorsal*
Thorax	Left-right lateral or right-left lateral	Ventrodorsal or dorsoventral
Abdomen	Left-right lateral or right-left lateral	Ventrodorsal*
Pelvis	Left-right lateral or right-left lateral	Ventrodorsal*
Brachium, antebrachium, thigh, crus	Lateral-medial or medial-lateral	Craniocaudal or caudocranial
Manus	Lateral-medial or medial-lateral	Dorsopalmar or palmarodorsal
Pes	Lateral-medial or medial-lateral	Dorsoplantar or plantarodorsal

*Dorsoventral views of the spine, abdomen or pelvis are rarely acquired.

Although these principles were developed to define how a film should be hung on a viewbox, they have carried over to the digital age and are used to direct how the digital image should be oriented on a monitor or in print.

STANDARD PROJECTIONS

Nearly every body part should be radiographed using multiple views with the beam entering multiple points on the surface of the body. Most commonly, two projections are acquired, made at 90 degrees to each other. Such views are termed *orthogonal views*. Table 1-1 lists the most common orthogonal views for the major body parts. As described, these views are named according to the point-of-entrance to point-of-exit of the primary x-ray beam method, as already described.

Making the same standard orthogonal views during routine radiographic examinations is very important. The repetitive aspect of looking at the same radiographic projections and orientations over and over makes it easier to recognize abnormal from normal. On the other hand, when an object or body region is radiographed using an uncommon or unfamiliar point of entrance to point of exit of the x-ray beam, the radiograph features become less recognizable and more difficult to interpret (Figure 1-2).

OBLIQUE PROJECTIONS

For anatomically complex regions, such as the manus and pes, two orthogonal radiographic views are not adequate

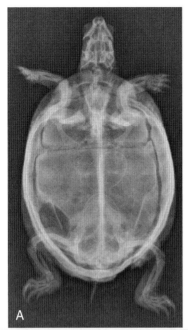
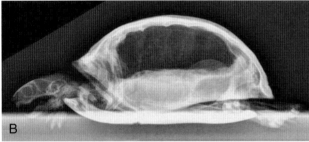
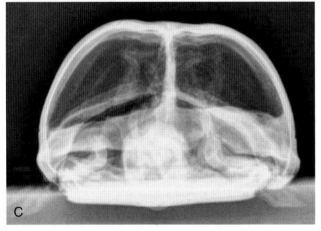

Figure 1-2. Dorsoventral (**A**), lateral (**B**), and rostocaudal (**C**) radiographs of a box turtle. That the subject is a turtle is easily recognizable in **A** and **B**, which are orthogonal radiographs. That the subject is a turtle is less obvious in **C**, which is also an orthogonal view with respect to both **A** and **B**. However, this view is acquired less frequently, making it unfamiliar. In addition, the eggs in the celom would not be identified if only view **C** is being evaluated. This example emphasizes the need for at least two standardized orthogonal views of any body part being radiographed and the need to use the same views in every subject.

to assess all aspects of the region. There is too much superimposition for all surfaces to be assessed completely in two orthogonal views, and important lesions can be missed. To overcome this, other projections are used along with the two standard orthogonal projections. The objective of radiographing complex structures using these multiple views is to project as many surfaces or edges in the most unobstructed manner possible. The internal structure of the region can sometimes still be assessed, even with overlapping, because of the penetrating nature of x-rays. However, the assessment of a complex structure is going to be most accurate when as many edges as possible are projected in an unobstructed manner.

The best solution to solving the problem of superimposition is to use a *tomographic imaging modality*. Tomographic imaging modalities display images in slices, thus avoiding the problem of superimposition completely. Ultrasound, computed tomography, and magnetic resonance imaging are all tomographic modalities. Because these modalities may not be available on a daily basis, the use of *oblique radiographic projections* is another method to assist in reducing the complexity created by superimposition of structures.

For oblique radiography, projections in addition to standard orthogonal projections are acquired. In other words, the angle of the primary x-ray beam with respect to the part being radiographed is somewhere between the angles used for the standard orthogonal projections. Typically, this angle is approximately 45 degrees, but other angles can be used depending on the circumstances. The radiographic naming concept previously described is crucial to understanding the terminology of oblique radiography. Remember, radiographic views are named according to the direction of the primary x-ray beam, from point-of-entrance to point-of-exit, using correct anatomic terminology.

The following example of oblique radiography is based on canine tarsal radiographs.[*] These principles apply to other regions, such as the manus. If these principles are applied to the forelimb, plantar is replaced by palmar.

Dorsoplantar View

The dorsoplantar view is one of the two basic orthogonal radiographic views of rear extremities, distal to the tarsocrural joint (Table 1-2). In a dorsoplantar view of a pes, for example, the x-ray beam strikes the dorsal surface of the pes with the image plate plantar to (behind) the pes oriented perpendicular to the primary x-ray beam (Figure 1-3). In this geometric arrangement, the medial and lateral aspects of the pes are visualized in an unobstructed manner (see Figures 1-3 and 1-4). This does not mean that only the medial and lateral edges of the pes can be evaluated because the infrastructure can be

---

[*]The colorized surface renderings in Figures 1-4, 1-6, 1-8, and 1-10 were graciously prepared by Sarena Sunico, DVM, DACVR.

Table 1-2	Correct Names for Radiographic Projections of a Limb Where the X-Ray Beam Strikes the Front Surface of the Limb and the Cassette or Imaging Plate Is Directly behind the Limb
Correct Name of View	**Orientation**
Dorsopalmar	Primary x-ray beam strikes front surface of *forelimb* at antebrachiocarpal joint or distal. Cassette or imaging plate is perpendicular to primary x-ray beam.
Dorsoplantar	Primary x-ray beam strikes front surface of *hindlimb* at tarsocrural joint or distal. Cassette or imaging plate is perpendicular to primary x-ray beam.
Craniocaudal	Primary x-ray beam strikes front surface of *forelimb or hindlimb* proximal to antebrachiocarpal joint or tarsocrural joint. Cassette or imaging plate is perpendicular to primary x-ray beam.

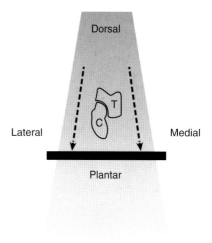

Figure 1-3. The level of the pes containing the calcaneus and talus was sliced transversely. The x-ray beam strikes the tarsal bones from the front. In this projection, the only surfaces at this level that will be projected in an unobstructed fashion are the medial side of the talus *(T)*, and the lateral side of the calcaneus *(C); dotted arrows* indicate these surfaces. These are the only locations that can be evaluated for surface lesions, such as periosteal reaction or cortical lysis. Other surfaces will be superimposed on another structure and cannot be assessed accurately.

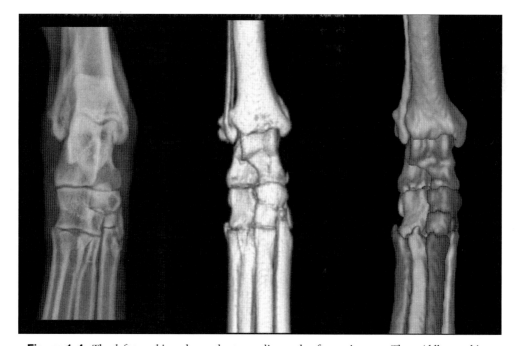

Figure 1-4. The *left panel* is a dorsoplantar radiograph of a canine pes. The *middle panel* is a three-dimensional rendering of a normal right canine pes as seen from the perspective of the x-ray beam when making a dorsoplantar radiograph. The *right panel* is also a three-dimensional rendering of a normal right canine pes, also as seen from the perspective of the x-ray beam when making a dorsoplantar radiograph, but where each bone has been colorized (see Color Plate 1). The colorized version makes it easier to comprehend the extent of overlap. Note in the radiograph how the only aspects of the tarsal bones that are projected in an unobstructed fashion where the surface can be evaluated are the medial and lateral aspects of the tarsus.

assessed but the lateral and medial surfaces are primarily where a periosteal reaction or cortical erosion can be identified.

Lateral View

The complementary orthogonal view to the dorsoplantar view is the lateral-medial or medial-lateral view. It is made when the x-ray beam strikes the side surface of a limb with the cassette or imaging plate on the opposite side of the limb, perpendicular to the primary x-ray beam (Figure 1-5). These views are most often referred to as *lateral views*, although *lateral-medial* or *medial-lateral* is more correct depending on whether the lateral or medial aspect of the limb, respectively, is struck by the primary x-ray beam.

In a medial-lateral view of a pes, for example, the x-ray beam strikes the medial surface of the pes with the image plate lateral to the pes, oriented perpendicularly to the primary x-ray beam (see Figure 1-5). In this geometric arrangement, the dorsal and palmar aspects of the pes are visualized in an unobstructed manner (see Figures 1-5 and 1-6). This does not mean that only the lateral and medial edges of the pes can be evaluated because the infrastructure can be assessed but the dorsal and plantar surfaces are the only surfaces where a surface change, such as a periosteal reaction or cortical erosion, can be identified.

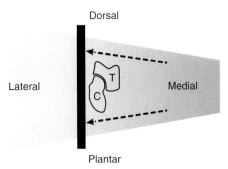

Figure 1-5. The level of the pes containing the calcaneus and talus was sliced transversely. The x-ray beam strikes the tarsal bones from the medial side, in this instance, a mediolateral view. As can be seen, the only surfaces that will be projected in an unobstructed fashion are the dorsal surface of the talus (*T*) and the plantar surface of the calcaneus (*C*); *dotted arrows* indicate these surfaces. These surfaces are the only ones that can be evaluated for surface lesions, such as periosteal reaction or cortical lysis.

Oblique Views

In oblique views of the pes, the entrance point of the primary x-ray beam is intentionally shifted to some location between dorsal and lateral or between dorsal and medial. Typically this position is approximately midway between dorsal and lateral, or midway between dorsal

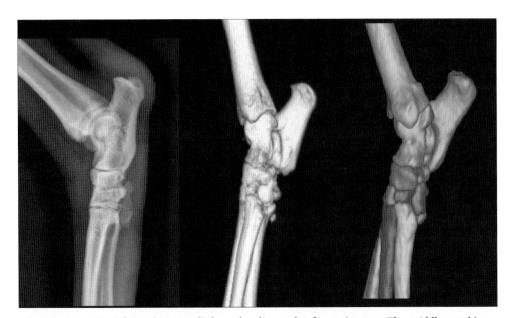

Figure 1-6. The *left panel* is a mediolateral radiograph of a canine pes. The *middle panel* is a three-dimensional rendering of a normal right canine pes as seen from the perspective of the x-ray beam when making a mediolateral radiograph. The *right panel* is also a three-dimensional rendering of a normal right canine pes, as seen from the perspective of the x-ray beam when making a mediolateral radiograph, but where each bone has been colorized (see Color Plate 2). The colorized version makes it easier to comprehend the extent of overlap. Note in the radiograph how the only aspects of the tarsal bones that are projected in an unobstructed fashion are the dorsal and plantar aspects of the tarsus and the cranial and caudal aspects of the tibia. The proximal surface of the calcaneus is also visible in this projection because it is not superimposed on any other structure.

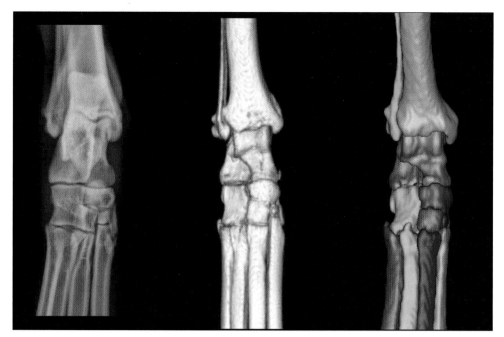

Color Plate 1. The *left panel* is a dorsoplantar radiograph of a canine pes. The *middle panel* is a three-dimensional rendering of a normal right canine pes as seen from the perspective of the x-ray beam when making a dorsoplantar radiograph. The *right panel* is also a three-dimensional rendering of a normal right canine pes, also as seen from the perspective of the x-ray beam when making a dorsoplantar radiograph, but where each bone has been colorized. The colorized version makes it easier to comprehend the extent of overlap. Note in the radiograph how the only aspects of the tarsal bones that are projected in an unobstructed fashion where the surface can be evaluated are the medial and lateral aspects of the tarsus. (See Figure 1-4.)

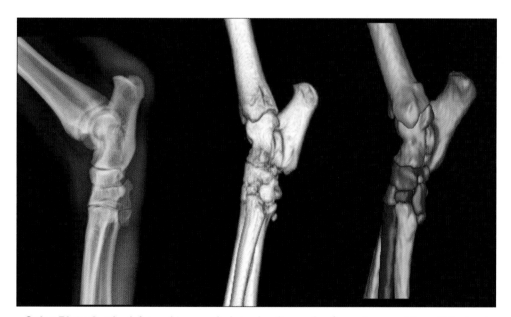

Color Plate 2. The *left panel* is a mediolateral radiograph of a canine pes. The *middle panel* is a three-dimensional rendering of a normal right canine pes as seen from the perspective of the x-ray beam when making a mediolateral radiograph. The *right panel* is also a three-dimensional rendering of a normal right canine pes, as seen from the perspective of the x-ray beam when making a mediolateral radiograph, but where each bone has been colorized. The colorized version makes it easier to comprehend the extent of overlap. Note in the radiograph how the only aspects of the tarsal bones that are projected in an unobstructed fashion are the dorsal and plantar aspects of the tarsus and the cranial and caudal aspects of the tibia. The proximal surface of the calcaneus is also visible in this projection because it is not superimposed on any other structure. (See Figure 1-6.)

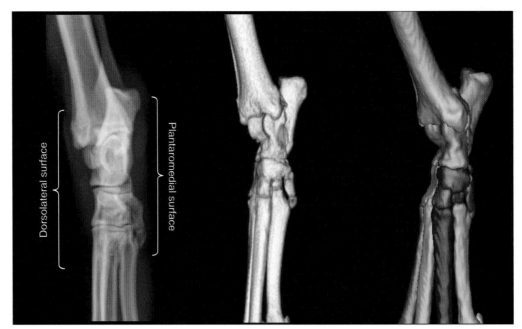

Color Plate 3. The *left panel* is a dorsal 45-degree medial-plantarolateral radiograph of a canine pes. The *middle panel* is a three-dimensional rendering of a normal right canine pes as seen from the perspective of the x-ray beam when making a dorsal 45-degree medial-plantarolateral radiograph. The *right panel* is also a three-dimensional rendering of a normal right canine pes, as seen from the perspective of the x-ray beam when making a dorsal 45-degree medial-plantarolateral radiograph, but where each bone has been colorized. The colorized version makes it easier to comprehend the extent of overlap. Note in the radiograph how the only aspects of the tarsal bones that are projected in an unobstructed fashion are the dorsolateral and plantaromedial aspects of the tarsus. Even though the proximal aspect of the calcaneus is plantarolateral, it can still be seen in this radiograph because it extends sufficiently proximal that it will not be superimposed on the tibia in either oblique view. (See Figure 1-8.)

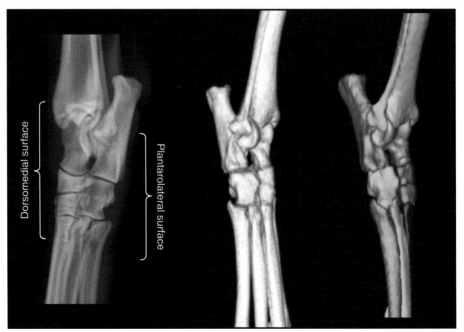

Color Plate 4. The *left panel* is a dorsal 45-degree lateral-plantaromedial radiograph of a canine pes. The *middle panel* is a three-dimensional rendering of a normal right canine pes as seen from the perspective of the x-ray beam when making a dorsal 45-degree lateral-plantaromedial radiograph. The *right panel* is also a three-dimensional rendering of a normal right canine pes, as seen from the perspective of the x-ray beam when making a dorsal 45-degree lateral-plantaromedial radiograph, but where each bone has been colorized. The colorized version makes it easier to comprehend the extent of overlap. It is important to note that the dorsal surface of the radiograph is oriented to the viewer's left, whereas the dorsal surfaces of the three-dimensional models are oriented to the viewer's right. As the three-dimensional models are anatomically correct models of a right tarsus, this is the orientation that the radiographer would see. However, when radiographs are viewed, the cranial or dorsal surface of the structure is always oriented to the viewer's left; this explains the difference in orientation of the radiograph versus the models in this figure. Note in the radiograph how the only aspects of the tarsal bones that are projected in an unobstructed fashion are the dorsomedial and plantarolateral aspects of the tarsus. (See Figure 1-10.)

Table 1-3 Correct Names for Oblique Radiographic Projections of a Limb Where the X-Ray Beam Strikes the Front Surface of the Limb Midway between the Front and Side and the Cassette or Imaging Plate Is behind the Limb and Perpendicular to the Primary X-Ray Beam

Correct name of View	Orientation
Dorsal 45-degree lateral-*palmaro*medial	Primary x-ray beam strikes front surface of *forelimb* midway between dorsal and lateral aspects, at antebrachiocarpal joint or distal. Cassette or imaging plate is perpendicular to primary x-ray beam. Results in projection of dorsomedial and palmarolateral aspects of region of interest.
Dorsal 45-degree lateral-*plantaro*medial	Primary x-ray beam strikes front surface of *hindlimb* midway between dorsal and lateral aspects, at tarsocrural joint or distal. Cassette or imaging plate is perpendicular to primary x-ray beam. Results in projection of dorsomedial and plantarolateral aspects of region of interest. (See Figures 1-9 and 1-10.)
Dorsal 45-degree medial-*palmaro*lateral	Primary x-ray beam strikes front surface of *forelimb* midway between dorsal and medial aspects, at antebrachiocarpal joint or distal. Cassette or imaging plate is perpendicular to primary x-ray beam. Results in projection of dorsolateral and palmaromedial aspects of region of interest.
Dorsal 45-degree medial-*plantaro*medial	Primary x-ray beam strikes front surface of *hindlimb* midway between dorsal and medial aspects, at tarsocrural joint or distal. Cassette or imaging plate is perpendicular to primary x-ray beam. Results in projection of dorsolateral and plantaromedial aspects of region of interest. (See Figures 1-7 and 1-8.)
Cranial 45-degree lateral-caudomedial	Primary x-ray beam strikes front surface of forelimb or hindlimb midway between cranial and lateral aspects, proximal to antebrachiocarpal or tarsocrural joint. Cassette or imaging plate is perpendicular to primary x-ray beam. Results in projection of craniomedial and caudolateral aspects of region of interest.
Cranial 45-degree medial-caudolateral	Primary x-ray beam strikes front surface of forelimb or hindlimb midway between cranial and medial aspects, proximal to antebrachiocarpal or tarsocrural joint. Cassette or imaging plate is perpendicular to primary x-ray beam. Results in projection of craniolateral and caudomedial aspects of region of interest.

and medial, but other angles can be used depending on the circumstances (Table 1-3).

The oblique view where the entrance point is shifted 45 degrees between dorsal and medial is termed a *dorsal 45-degree medial-plantarolateral oblique* (D45°M-PtLO, often abbreviated to DM-PtLO) view of a pes, for example (see Table 1-3). The x-ray beam strikes the dorsal surface of the pes midway between dorsal and medial with the image plate perpendicular to the primary x-ray beam (Figure 1-7). In this geometric arrangement, the dorsolateral and plantaromedial aspects of the pes are visualized in an unobstructed manner (see Figures 1-7 and 1-8).

The oblique view where the entrance point is shifted 45 degrees between dorsal and lateral is termed a *dorsal 45-degree lateral-plantaromedial oblique* (D45°L-PtMO, often abbreviated to DL-PtMO) view of a pes, for example (see Table 1-3). The x-ray beam strikes the dorsal surface of the pes midway between dorsal and lateral with the image plate perpendicular to the primary beam (Figure 1-9). In this geometric arrangement, the dorsomedial and plantarolateral aspects of the pes are visualized in an unobstructed manner (see Figures 1-9 and 1-10).

It is important to re-emphasize that the terminology used in this example of a pes applies to a pelvic limb and

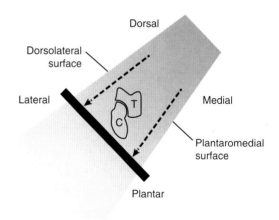

Figure 1-7. The level of the pes containing the calcaneus and talus was sliced transversely. The x-ray beam strikes the tarsal bones approximately midway between the dorsal and medial aspects; thus the correct name of this projection is a *dorsal 45-degree medial-plantarolateral view*. In this view, the dorsolateral surface of the talus *(T)* will be projected in an unobstructed fashion. The plantar aspect of the calcaneus *(C)* will also be seen due to it extending so far proximally; this is better comprehended in Figure 1-8. These are the only surfaces that can be evaluated for surface lesions, such as periosteal reaction or cortical lysis. Other surfaces will be superimposed on another structure.

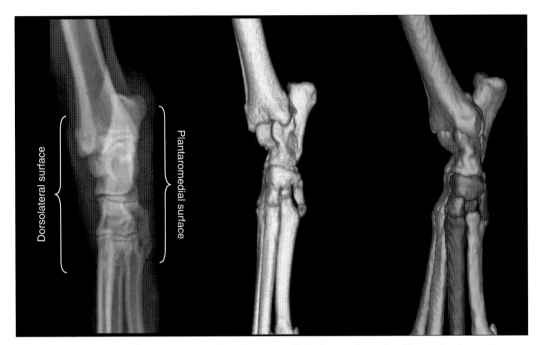

Figure 1-8. The *left panel* is a dorsal 45-degree medial-plantarolateral radiograph of a canine pes. The *middle panel* is a three-dimensional rendering of a normal right canine pes as seen from the perspective of the x-ray beam when making a dorsal 45-degree medial-plantarolateral radiograph. The *right panel* is also a three-dimensional rendering of a normal right canine pes, as seen from the perspective of the x-ray beam when making a dorsal 45-degree medial-plantarolateral radiograph, but where each bone has been colorized (see Color Plate 3). The colorized version makes it easier to comprehend the extent of overlap. Note in the radiograph how the only aspects of the tarsal bones that are projected in an unobstructed fashion are the dorsolateral and plantaromedial aspects of the tarsus. Even though the proximal aspect of the calcaneus is plantarolateral, it can still be seen in this radiograph because it extends sufficiently proximal that it will not be superimposed on the tibia in either oblique view.

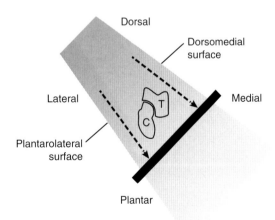

Figure 1-9. The level of the pes containing the calcaneus and talus was sliced transversely. The x-ray beam strikes the tarsal bones approximately midway between the dorsal and lateral aspects; this is a dorsal 45-degree lateral-plantaromedial view. As can be seen, the only surfaces that will be projected in an unobstructed fashion are the plantarolateral surface of the calcaneus *(C)* and the dorsomedial surface of the talus *(T); dotted arrows* indicate these surfaces. These are the only surfaces that can be evaluated for surface lesions, such as periosteal reaction or cortical lysis. Other surfaces will be superimposed on another structure.

if a thoracic limb were being radiographed, plantar would be replaced by palmar.

Not all oblique views involve the use of a primary x-ray beam angle between dorsal and lateral or dorsal and medial. For example, there are special oblique views of the bicipital groove of the humerus (cranioproximal-craniodistal flexed view of shoulder) and the proximal surface of the talus (dorsoplantar flexed tarsus) that are designed to make certain portions of these bones more conspicuous. Having a good understanding of how radiographs are named reduces confusion when assessing these more unconventional views and in understanding exactly why the images appear the way they do. These less frequently used oblique views are explained in more detail in the sections in which they are illustrated.

By using oblique radiographic views, all surfaces of a complex joint can be evaluated for periosteal reaction and cortical lysis, and small fragments can be localized accurately. It is important to understand the *anatomy* of oblique views to draw accurate conclusions regarding the location of any abnormality and to acquire the correct oblique view when interrogating specific anatomic regions.

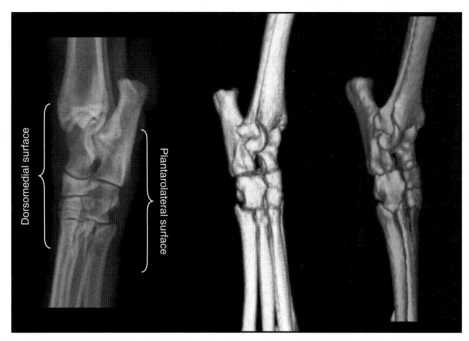

Figure 1-10. The *left panel* is a dorsal 45-degree lateral-plantaromedial radiograph of a canine pes. The *middle panel* is a three-dimensional rendering of a normal right canine pes as seen from the perspective of the x-ray beam when making a dorsal 45-degree lateral-plantaromedial radiograph. The *right panel* is also a three-dimensional rendering of a normal right canine pes, as seen from the perspective of the x-ray beam when making a dorsal 45-degree lateral-plantaromedial radiograph, but where each bone has been colorized (see Color Plate 4). The colorized version makes it easier to comprehend the extent of overlap. It is important to note that the dorsal surface of the radiograph is oriented to the viewer's left, whereas the dorsal surfaces of the three-dimensional models are oriented to the viewer's right. As the three-dimensional models are anatomically correct models of a right tarsus, this is the orientation that the radiographer would see. However, when radiographs are viewed, the cranial or dorsal surface of the structure is always oriented to the viewer's left; this explains the difference in orientation of the radiograph versus the models in this figure. Note in the radiograph how the only aspects of the tarsal bones that are projected in an unobstructed fashion are the dorsomedial and plantarolateral aspects of the tarsus.

PHYSEAL CLOSURE

Juvenile orthopedic disorders are common, particularly in dogs. Many arise from disruption to normal physeal development. Breed, genetics, nutrition, intercurrent disease, activity, and trauma can all affect skeletal development adversely. Some understanding of the radiographic appearance of normal physeal maturation and the age at which this occurs is a prerequisite to the identification and management of such disorders. Table 1-4 provides an overview of when the various ossification centers appear, and Table 1-5 shows when physes are typically radiographically closed. It should be noted that there is considerable variation in physeal closure, and these tables are designed to act only as guides. The tables reflect a compilation of data from multiple sources. Table 1-6 documents the approximate ages at which the fusion of skull bones occurs in both canines and felines. Figures 1-11 through 1-16 diagrammatically show the canine long bone and joint morphology from 1 month to 8.5 months. Figures 1-17 through 1-24 diagrammatically show the feline long bone and joint morphology from 3.5 weeks to 16.5 months.

Table 1-4 Approximate Ages at Which Ossification Centers Appear (Canine and Feline)

Site	Canine	Feline
Scapula		
Body	Birth	Birth
Supraglenoid tubercle	6-7 weeks	7-9 weeks
Humerus		
Proximal epiphysis (head and tubercles)	1-2 weeks	1-2 weeks
Diaphysis	Birth	Birth
Condyle	2-3 weeks	2-4 weeks
Medial epicondyle	6-8 weeks	6-8 weeks
Radius		
Proximal epiphysis	3-5 weeks	2-4 weeks
Diaphysis	Birth	Birth
Distal epiphysis	2-4 weeks	2-4 weeks
Ulna		
Olecranon tubercle	6-8 weeks	4-5 weeks
Diaphysis	Birth	Birth
Anconeal process	6-8 weeks	
Distal epiphysis	5-6 weeks	3-4 weeks
Carpus		
Radial carpal (3 centers)	3-6 weeks	3-8 weeks
Other carpal bones	2 weeks	3-8 weeks
Accessory carpal		
Diaphysis	2 weeks	3-8 weeks
Apophysis	6-7 weeks	3-8 weeks
Sesamoid bone in abductor pollicis longus	4 months	
Metacarpus/metatarsus		
Diaphysis of 1-5	Birth	Birth
Proximal epiphysis of MC1	5-7 weeks	
Distal epiphysis of MC2-5	3-4 weeks	3 weeks
Palmar sesamoid bones	2 months	2-2.5 months
Dorsal sesamoid bones	4 months	
Phalanges (fore and hind)		
P1		
Diaphysis of digits 1-5	Birth	Birth
Proximal epiphysis digit 1	5-7 weeks	3-4 weeks
Distal epiphysis digits 2-5	4-6 weeks	3-4 weeks
P2		
Diaphysis of digits 2-5	Birth	Birth
Proximal epiphysis of digits 2-5	4-6 weeks	4 weeks
P3 (one ossification center)	Birth	Birth
Pelvis		
Ilium/ischium/pubis	Birth	Birth
Acetabular bone	2-3 months	
Iliac crest	4-5 months	
Ischial tuberosity	3-4 months	
Ischial arch	6 months	

Table 1-4 Approximate Ages at Which Ossification Centers Appear (Canine and Feline)—cont'd

Site	Canine	Feline
Femur		
Greater trochanter	7-9 weeks	5-6 weeks
Lesser trochanter	7-9 weeks	6-7 weeks
Head	1-2 weeks	2 weeks
Diaphysis	Birth	Birth
Distal epiphysis	3-4 weeks	1-2 weeks
Stifle sesamoid bones		
Patella	6-9 weeks	8-9 weeks
Fabellae	3 months	10 weeks
Popliteal sesamoid	3-4 months	
Tibia		
Tibial tuberosity	7-8 weeks	6-7 weeks
Proximal epiphysis	2-4 weeks	2 weeks
Diaphysis	Birth	Birth
Distal epiphysis	2-4 weeks	2 weeks
Medial malleolus	3 months	
Fibula		
Proximal epiphysis	8-10 weeks	6-7 weeks
Diaphysis	Birth	Birth
Distal epiphysis	4-7 weeks	3-4 weeks
Tarsus		
Talus	Birth	Birth
Calcaneus		
Tuber calcanei	6 weeks	4 weeks
Diaphysis	Birth	Birth
Central tarsal bone	3 weeks	4-7 weeks
First and second tarsal bones	4 weeks	4-7 weeks
Third tarsal bone	3 weeks	4-7 weeks
Fourth tarsal bone	2 weeks	4-7 weeks
Spine		
Atlas, three centers of ossification		
Neural arch (bilateral)	Birth	
Intercentrum	Birth	
Axis, seven centers of ossification		
Centrum of proatlas	6 weeks	
Centrum 1	Birth	
Intercentrum 2	3 weeks	
Centrum 2	Birth	
Neural arch (bilateral)	Birth	
Caudal epiphysis	2 weeks	
Cervical, thoracic, lumbar, sacral vertebrae		
Paired neural arches and centrum	Birth	
Cranial and caudal epiphyses*	2 weeks	

*Epiphyses are often absent in the last 1-2 caudal vertebrae.

Table 1-5 Approximate Age When Physeal Closure Occurs (Canine and Feline)

Bone	Physis	Canine	Feline
Scapula	Supraglenoid tubercle	4-7 months	3.5-4 months
Humerus	Proximal	12-18 months	18-24 months
	Medial epicondyle	6-8 months	
	Condyle to shaft	6-8 months	3.5-4 months
	Condyle (lateral and medial parts)	*6-10 weeks*	3.5 months
Radius	Proximal	7-10 months	5-7 months
	Distal	*10-12 months*	14-22 months
Ulna	Anconeal process	<5 months	
	Olecranon tuberosity	7-10 months	9-13 months
	Distal	*9-12 months*	14-25 months
Metacarpus/metatarsus			
MC1	Proximal	6-7 months	4.5-5 months
MC2-5	Distal	6-7 months	4.5-5 months
Phalanges (fore and hind)			
P1 and P2	Proximal	6-7 months	
Pelvis	Acetabular	3-5 months	
	Ischiatic tuberosity	10-12 months	
	Ilial crest	24-36 months	
	Pubic symphysis	4-5 months	
Femur	*Head, capital physis*	*8-11 months*	7-11 months
	Greater trochanter	9-12 months	13-19 months
	Lesser trochanter	9-12 months	
	Distal physis	9-12 months	
Tibia	*Tibial tuberosity*	*10-12 months*	9-10 months
	Tibial plateau	9-10 months	12-19 months
	Distal physis	12-15 months	10-12 months
	Medial malleolus	3-5 months	
Fibula	Proximal	10-12 months	13-18 months
	Distal (lateral malleolus)	12-13 months	10-14 months
Tarsus			
Calcaneus	Tuberosity	6-7 months	
Spine ossification centers			
Axis	Arches fuse	3-4 months (106 days)	
	Intercentrum	3-4 months (115 days)	
Atlas	Centrum of proatlas + C1	100-110 days	
	Intercentrum 2, centrum 1, and centrum 2	3.3-5 months	
	Neural arches (bilateral)	30 days	
	Caudal physis	7-12 months	
Cervical, thoracic, lumbar	Cranial physis	7-10 months	7-10 months
	Caudal physis	8-12 months	8-11 months
Sacrum	Cranial and caudal physes	7-12 months	
Caudal	Cranial and caudal physes	7-12 months	

Physes most commonly associated with clinical disorders are shown in *italics*.

Table 1-6 Approximate Age When Fusion of Skull Bones Occurs (Canine and Feline)

Bone	Center of Ossification	Age
Occipital	Basilar part	2.5-5 months
	Squamous part	3-4 months
	Interparietal part	Before birth
Sphenoid	Body/wings of presphenoid	Before birth
	Body/wings of basisphenoid	3-4 years
	Basisphenoid and presphenoid	1-2 years
	Sphenobasilar suture	8-10 months
Parietal	Interparietal suture	2-3 years
Frontal	Interfrontal suture	3-4 years
Temporal	Petrosquamous suture	2-3 years
Mandible	Intermandibular symphysis	Never or very late

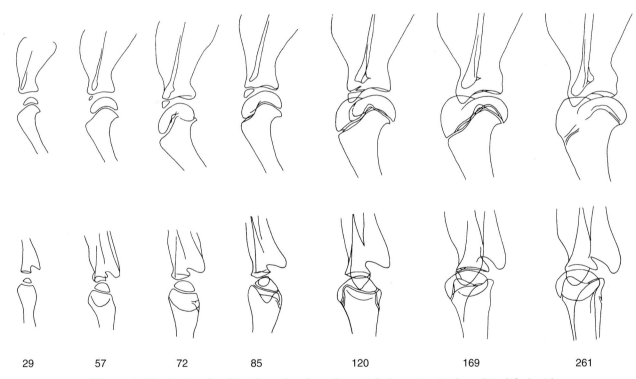

Figure 1-11. Canine shoulder, lateral and caudocranial views. Age in days. (Modified with permission from Schebitz H, Wilkens H: *Atlas of radiographic anatomy of the dog and cat*, ed 4, Philadelphia, 1986, Saunders.)

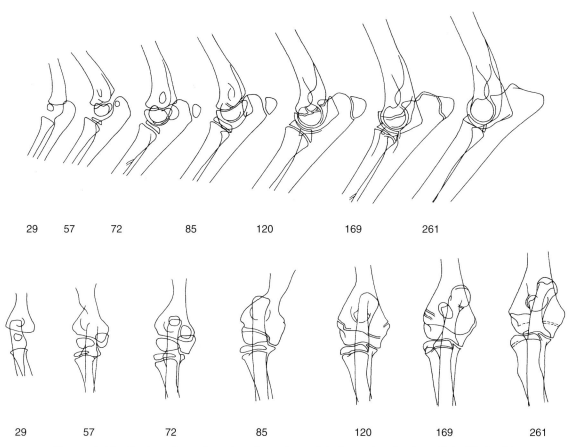

Figure 1-12. Canine elbow, lateral and craniocaudal views. Age in days. (Modified with permission from Schebitz H, Wilkens H: *Atlas of radiographic anatomy of the dog and cat*, ed 4, Philadelphia, 1986, Saunders.)

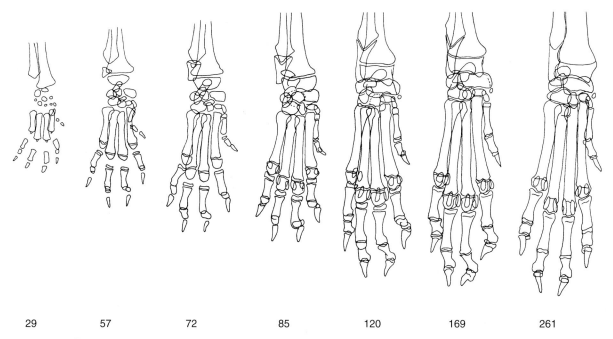

Figure 1-13. Canine manus, dorsopalmar view. Age in days. (Modified with permission from Schebitz H, Wilkens H: *Atlas of radiographic anatomy of the dog and cat*, ed 4, Philadelphia, 1986, Saunders.)

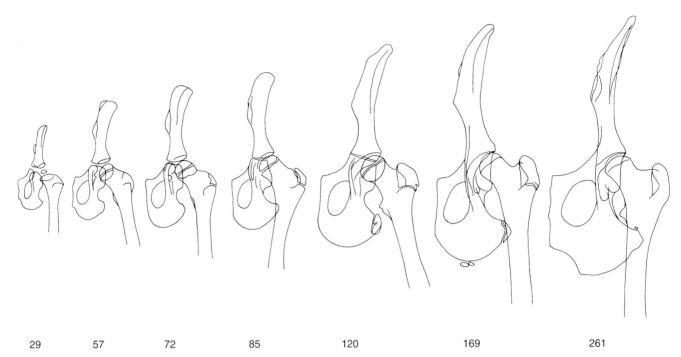

Figure 1-14. Canine left hemipelvis, ventrodorsal view. Age in days. (Modified with permission from Schebitz H, Wilkens H: *Atlas of radiographic anatomy of the dog and cat*, ed 4, Philadelphia, 1986, Saunders.)

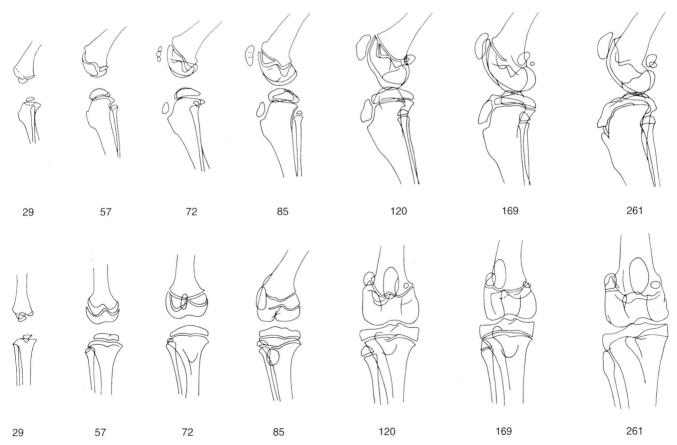

Figure 1-15. Canine stifle, lateral and craniocaudal views. Age in days. (Modified with permission from Schebitz H, Wilkens H: *Atlas of radiographic anatomy of the dog and cat*, ed 4, Philadelphia, 1986, Saunders.)

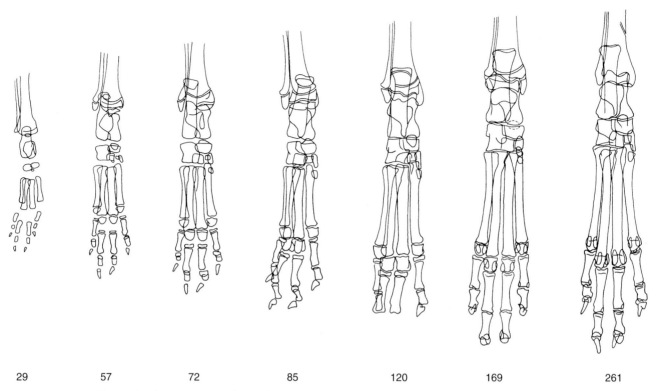

Figure 1-16. Canine pes, dorsoplantar view. Age in days. (Modified with permission from Schebitz H, Wilkens H: *Atlas of radiographic anatomy of the dog and cat,* ed 4, Philadelphia, 1986, Saunders.)

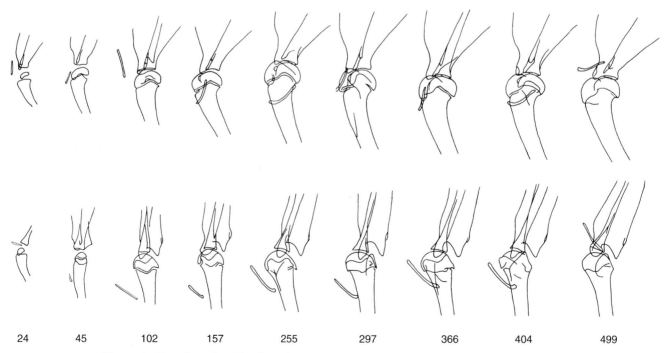

Figure 1-17. Feline shoulder, lateral and caudocranial views. Age in days. (Modified with permission from Schebitz H, Wilkens H: *Atlas of radiographic anatomy of the dog and cat,* ed 4, Philadelphia, 1986, Saunders.)

Chapter 1 ■ Basic Imaging Principles and Physeal Closure Time   17

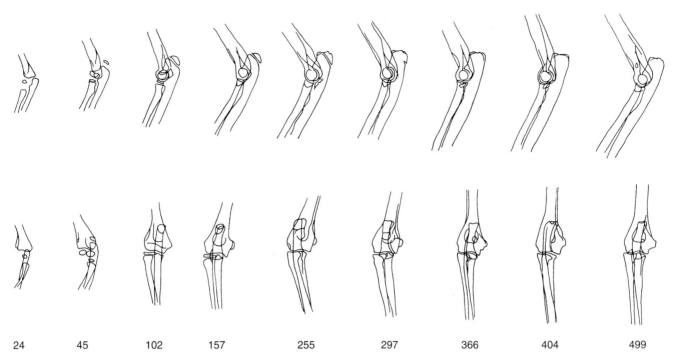

Figure 1-18. Feline elbow, lateral and craniocaudal views. Age in days. (Modified with permission from Schebitz H, Wilkens H: *Atlas of radiographic anatomy of the dog and cat*, ed 4, Philadelphia, 1986, Saunders.)

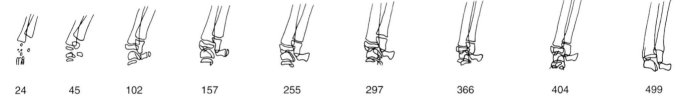

Figure 1-19. Feline distal antebrachium and carpus, lateral view. Age in days. (Modified with permission from Schebitz H, Wilkens H: *Atlas of radiographic anatomy of the dog and cat*, ed 4, Philadelphia, 1986, Saunders.)

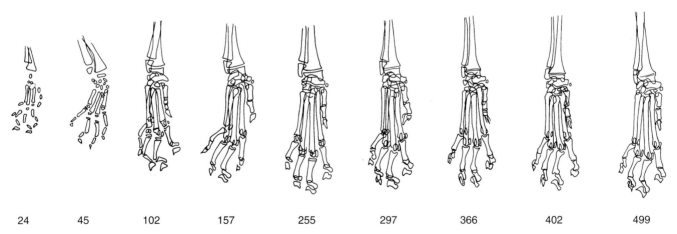

Figure 1-20. Feline manus, dorsomedial plantarolateral oblique view. Age in days. (Modified with permission from Schebitz H, Wilkens H: *Atlas of radiographic anatomy of the dog and cat*, ed 4, Philadelphia, 1986, Saunders.)

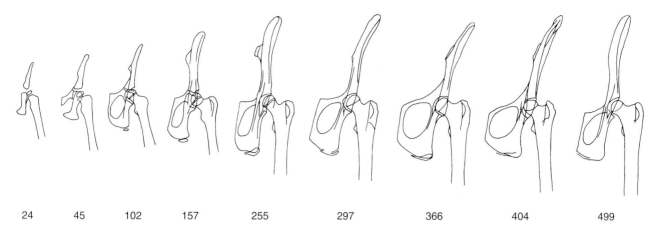

Figure 1-21. Feline left hemipelvis, ventrodorsal view. Age in days. (Modified with permission from Schebitz H, Wilkens H: *Atlas of radiographic anatomy of the dog and cat*, ed 4, Philadelphia, 1986, Saunders.)

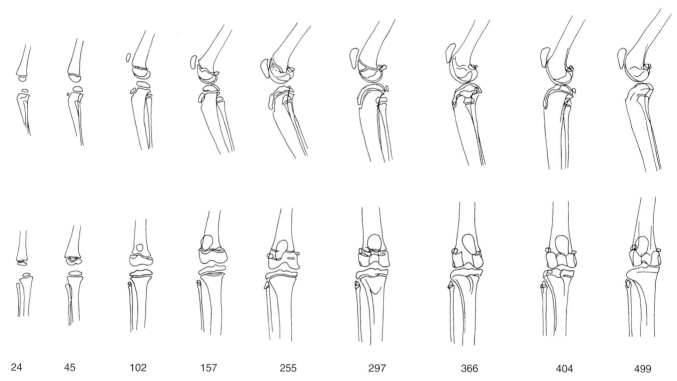

Figure 1-22. Feline stifle, lateral and craniocaudal views. Age in days. (Modified with permission from Schebitz H, Wilkens H: *Atlas of radiographic anatomy of the dog and cat*, ed 4, Philadelphia, 1986, Saunders.)

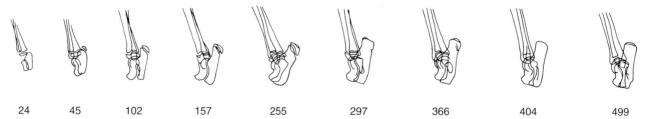

Figure 1-23. Feline distal crus, lateral view. Age in days. (Modified with permission from Schebitz H, Wilkens H: *Atlas of radiographic anatomy of the dog and cat*, ed 4, Philadelphia, 1986, Saunders.)

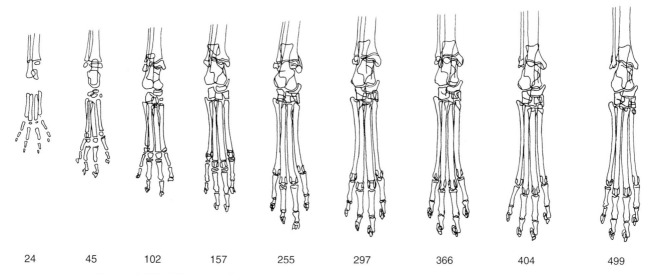

Figure 1-24. Feline pes, dorsoplantar view. Age in days. (Modified with permission from Schebitz H, Wilkens H: *Atlas of radiographic anatomy of the dog and cat*, ed 4, Philadelphia, 1986, Saunders.)

REFERENCE

1. Smallwood J, Shively M, Rendano V, et al: A standardized nomenclature for radiographic projections used in veterinary medicine. *Vet Radiol* 26:2-5, 1985.

SKELETAL MATURATION DATA COMPILED FROM THE FOLLOWING SOURCES:

1. Barone R: Caracteres Generaux Des Os. *Anatomie comparée des mammifères domestiques*, Paris, 1976, Vigot Frères.
2. Chapman W: Appearance of ossification centers and epiphyseal closures as determined by radiographic techniques. *J Am Vet Med Assoc* 147:138-141, 1965.
3. Constantinescu GM: The head. *Clinical anatomy for small animal practitioners*, Ames, IA, 2002, Iowa State Press.
4. Newton C, Nunamaker D: *Textbook of small animal orthopedics*, Philadelphia, 1985, JB Lippincott.
5. Owens J, Biery D: *Extremities in radiographic interpretation for the small animal clinician*, Baltimore, MD, 1999, Williams & Wilkins.
6. Smallwood J: *A guided tour of veterinary anatomy*, Raleigh, NC, 2010, Millenium Print Group.
7. Smith R: Appearance of ossification centers in the kitten. *J Small Anim Pract* 9:496-511, 1968.
8. Smith R: Fusion of ossification centers in the cat. *J Small Anim Pract* 10:523-530, 1969.
9. Smith R: Radiological observations on the limbs of young greyhounds. *J Small Anim Pract* 1:84-90, 1960.
10. Sumner-Smith G: Observations on epiphyseal fusion of the canine appendicular skeleton. *J Small Anim Pract* 7:303-312, 1966.
11. Watson A: *The phylogeny and development of the occipito-atlas-axis complex in the dog*, Ithaca, NY, 1981, Cornell University.
12. Watson A, Evans H: The development of the atlas-axis complex in the dog. *Anat Rec* 184:558, 1976.
13. Watson A, Stewart J: Postnatal ossification centers of the atlas and axis of miniature schnauzers. *Am J Vet Res* 51:264-268, 1990.
14. Frazho J, Graham J, Peck J, et al: Radiographic evaluation of the anconeal process in skeletally immature dogs. *Vet Surg* 39:829-832, 2010.

CHAPTER 2

The Skull

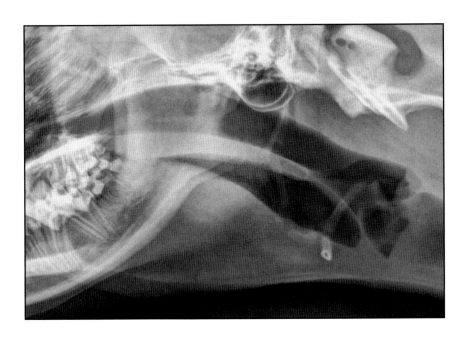

OVERVIEW

The skull is complex and challenging to evaluate with conventional radiography because of superimposition of the many complex osseous structures. This limits the value of radiography for assessing skull lesions. Imaging the skull is much better suited to the cross-sectional modalities of computed tomography (CT) and magnetic resonance imaging, which are tomographic and solve the issue of superimposition and also have superior contrast resolution. However, well-positioned radiographs with good detail and contrast can provide information that can help in directing appropriate patient management or selecting the best alternative imaging modality.

Although it is often more convenient to acquire radiographs of the skull with the patient conscious or minimally sedated, this rarely results in satisfactory skull radiographs. Poorly positioned skull radiographs are challenging to interpret and usually have no diagnostic value. Because the skull is so complex, standardization of views becomes more important than in any other location. Taking the time to ensure adequate patient restraint and optimal positioning significantly increases the chances of obtaining meaningful information.

The skull varies more in size and shape among domestic dogs than in any other mammalian species. Various measurement parameters are used to group dogs based on skull morphology. The term *dolichocephalic* describes long, narrow-headed breeds, like Collies and Wolfhounds. The term *mesaticephalic* describes breeds that have heads of medium proportions, such as German Shepherds, Beagles, and Setters. *Brachycephalic* describes breeds with short and wide heads, such as Boston Terriers and Pekingese.

A radiographic study of the skull should always include a lateral and dorsoventral (or ventrodorsal) view (Figures 2-1 and 2-2).

Additional views of the skull are often acquired depending on the region of interest. They should, however, always be acquired in concert with standard lateral and dorsoventral (or ventrodorsal) views. Oblique views should not be used to replace the standard views because standard views often provide information that allows more effective interpretation of oblique views. In addition, oblique views are impossible to interpret correctly unless a readily understandable external radiographic marking system is used. This allows correct identification of surfaces projected in profile. This is particularly important when there is no obvious unilateral disease. This is discussed in the following section on dentition.

Approximately 50 bones, many of which are paired, compose the skull. Many are not identifiable with survey radiographs, and most fuse with adjoining bones precluding specific delineation. The important larger bones of the skull that are visible radiographically are the incisive, nasal, maxillary, lacrimal, frontal, zygomatic, pterygoid, sphenoid, parietal, temporal, and occipital bones. Figure 2-3 identifies these bones diagrammatically.

DENTITION

The superior incisors (superior and inferior are acceptable terms with respect to the dental arcades) are located within the incisive bone, and the canine teeth, premolars, and molars are located within the maxilla bone (Figure 2-4). The maxilla is the large bone that, in addition to the much smaller incisive and nasal bones, encases the majority of the nasal cavity and the blind-ended maxillary recesses located ventrolaterally on each side of the nasal cavity. Ventrally, the maxilla, incisive, and palatine bones form the hard palate, which is the physical separation between the nasal and oral cavities. Caudally, the maxilla fuses with the frontal bone, which encompasses the frontal sinuses and the rostral aspect of the cranial vault.

In the dog, the anatomic dental formulas are as follows (*I*, Incisors; *C*, canine; *PM*, premolars; *M*, molars):
Deciduous: I3/3, C1/1, PM3/3
Permanent: I3/3, C1/1, PM4/4, M2/3.

The fourth maxillary premolar and first mandibular molar in the dog are sometimes called *carnassial teeth*, which refers to their shearing functions. Canine tooth eruption times are given in Table 2-1.

In the cat, the anatomic dental formulas are as follows:
Deciduous: I3/3, C1/1, PM3/2
Permanent: I3/3, C1/1, PM, 3/2, M1/1.

In the cat, the roots of the maxillary fourth premolar extend into the wall of the orbit, Figures 2-5 and 2-6. The carnassial teeth in the cat are the same as in the dog: the last upper premolar and the first lower molar. Feline tooth eruption times are given in Table 2-2. In dogs, the carnassial teeth are typically more rostral. The roots are separated from the nasal cavity by only a thin shelf of bone.

Table 2-1 Canine Dental Eruption Times

	Deciduous Teeth (Weeks)	Permanent Teeth (Months)
Incisors	3-4	3-5
Canines	3	4-6
Premolars	4-12	4-6
Molars	n/a	5-7

From Wiggs R, Lobprise H, editors: Oral Anatomy and Physiology. *Veterinary dentistry: Principles and practice*, Philadelphia, 1997, Lippincott-Raven.[1]

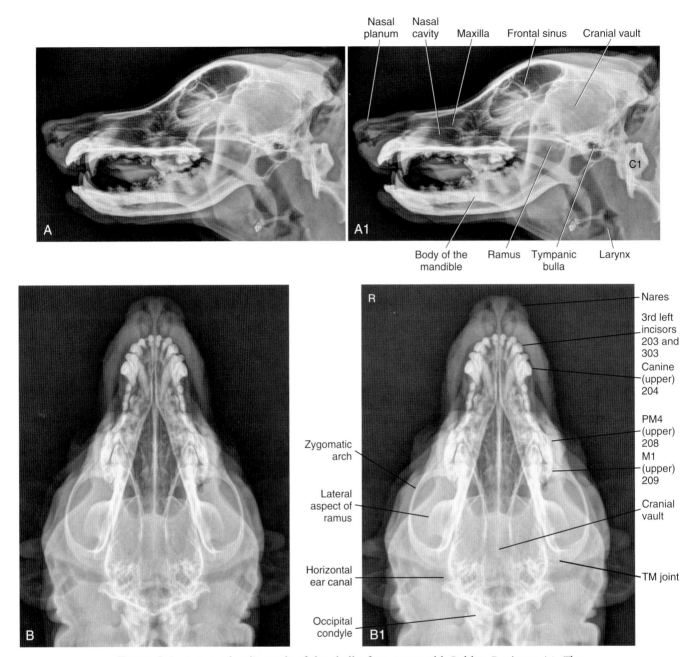

Figure 2-1. A, Lateral radiograph of the skull of an 8-year-old Golden Retriever. **A1,** The same image as in **A,** with labels identifying major structures. **B,** A dorsoventral radiograph of a 9-year-old mixed breed dog. **B1,** The same image as in **B,** with major structures identified. *TM,* Temporomandibular.

Table 2-2 Feline Dental Eruption Times

	Deciduous Teeth (Weeks)	Permanent Teeth (Months)
Incisors	2-3	3-4
Canines	3-4	4-5
Premolars	3-6	4-6
Molars	n/a	4-5

From Wiggs R, Lobprise H, editors: Oral anatomy and physiology. *Veterinary dentistry: Principles and practice,* Philadelphia, 1997, Lippincott-Raven.[1]

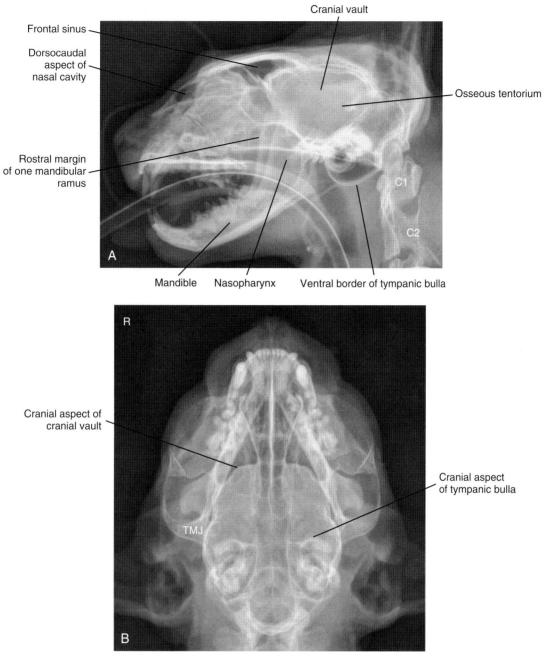

Figure 2-2. A, Lateral radiograph of a 3-year-old Domestic Shorthair cat. The osseous tentorium, which separates the caudal aspect of the cerebrum from the rostral aspect of the cerebellum, is particularly well-developed in cats compared to dogs. An endotracheal tube is present. **B,** Dorsoventral radiograph of an 18-year-old Domestic Shorthair cat. *TMJ,* Temporomandibular joint.

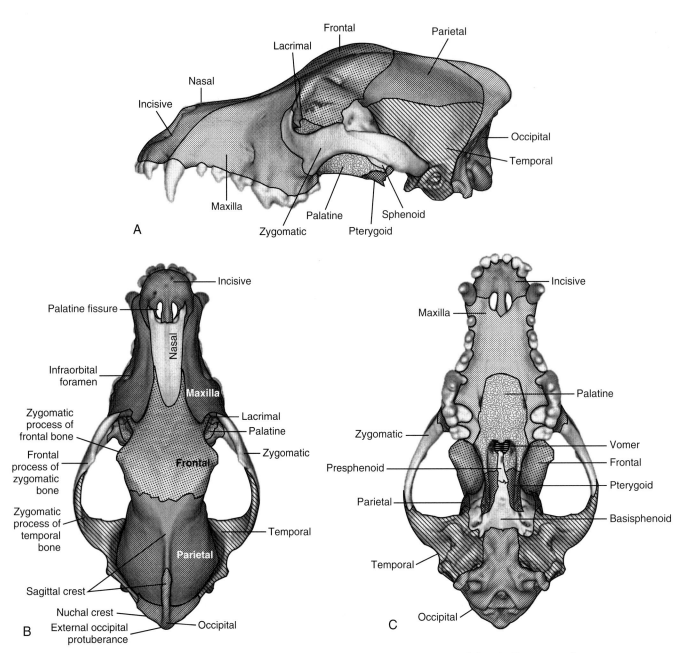

Figure 2-3. Three-dimensional (3-D), volume-rendered CT images of the skull. **A,** Lateral view; **B,** dorsal view; **C,** ventral view.

Figure 2-4. Oblique view of a maxillary dental arcade of an 8-month-old mixed breed dog. The four premolars and two molars are readily identifiable. The fourth upper premolar (08) and the first lower molar (09) are sometimes referred to as *carnassial teeth*.

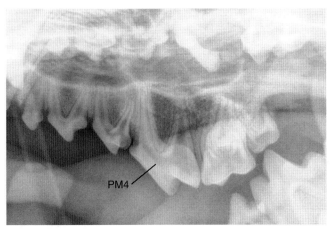

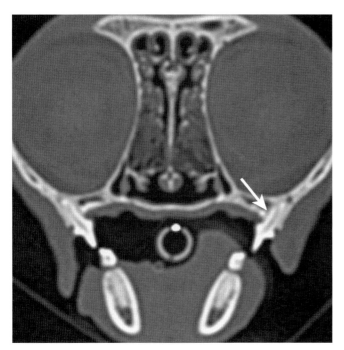

Figure 2-5. Transverse CT image at the level of the rostral aspect of the carnassial teeth of a cat. The slice is at right angles to the hard palate. The *white arrow* is a root of the left fourth upper premolar (tooth 208, using the modified Triadan system). Note the close association of the root with the orbit, the cavity in which the globe resides.

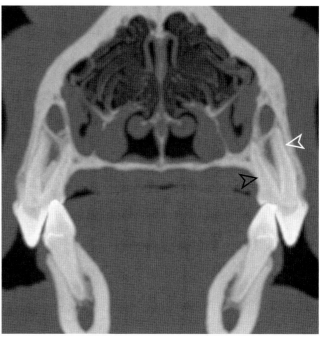

Figure 2-6. Transverse CT image at the level of the rostral aspect of the carnassial teeth of a dog. The *hollow black arrow* is the palatal surface; the *hollow white arrow* is the vestibular (buccal, labial) surface of the left fourth upper premolar (tooth 208, using the modified Triadan system). Note the close association of the mesial root with the nasal cavity. Dental abscessation is an important differential diagnosis in canine nasal disease.

The modified *Triadan system* of dental nomenclature is an alternative system for tooth identification[2] (Figure 2-7). The first digit denotes the quadrant. The right upper quadrant is designated *1*, the left upper quadrant is designated *2*, the left lower quadrant is designated *3*, and the right lower quadrant is designated *4*. The second and third digits denote positions within the quadrant, and numbering always begins at the midline; thus the central incisor is always *01*, the canine is always *04*, and the first molar is always *09*. For example, the right upper fourth premolar would be tooth number 108 in the Triadan system.

Radiography of the dental arcade is commonly performed using dedicated dental x-ray machines, which generally have relatively low mA and kVp capabilities, used with intraoral film or a small dedicated intraoral dental imaging plate. In the absence of a dental radiographic system, lateral oblique views can be used to assess each dental arcade.

As previously noted, it is critical that labeling of oblique radiographs be accurate and that the radiographer understands what will be projected when the skull is angled. This has been described elsewhere,[3] but a brief example follows. Assume the dog is in right lateral recumbency on the x-ray table and the goal is to evaluate the right maxillary arcade. By elevating the mandible with a nonradiopaque triangular sponge and using a vertical x-ray beam, the right maxillary arcade will be projected ventral to the rest of the maxilla without superimposition (Figure 2-8). To evaluate the right mandibular arcade, use the triangular sponge to elevate the maxilla, not the mandible. To evaluate the left arcades, reposition the dog in left recumbency and repeat the procedure.

When a tooth is evaluated, the structures that can be visualized are the roots and crown, the pulp cavity, and the lamina dura. Although the periodontal membrane cannot be visualized directly, its location can be inferred (Figure 2-9).

Visualization of both deciduous teeth and permanent teeth buds is normal in adolescent patients (Figure 2-10). Molars have no associated deciduous tooth. Normally, the deciduous teeth are shed as permanent teeth erupt. A retained deciduous tooth can impede eruption of permanent incisors, canines, or premolars. Alternatively, retained deciduous teeth may not impede eruption of the subjacent permanent tooth but become displaced as the permanent tooth erupts (Figure 2-11). Retained deciduous teeth are readily apparent upon visual inspection of the oral cavity. Radiographs are not needed for their detection but will be helpful in determining whether the

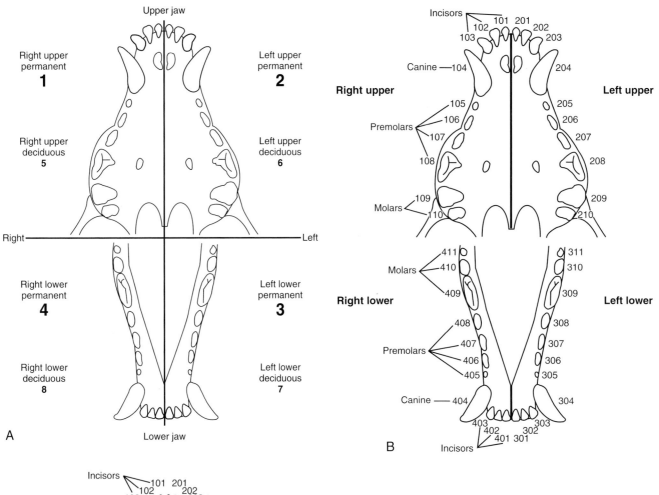
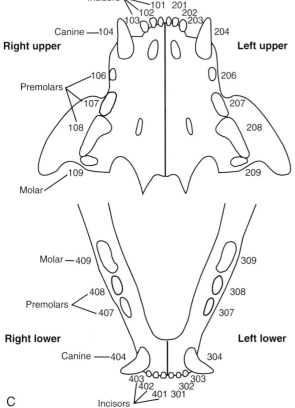

Figure 2-7. The modified Triadan system of dental identification. In **A**, each quadrant of the mouth is designated by a number; this is the first digit in identification. With respect to permanent teeth, the right maxilla is designated *1*, the left maxilla *2*, the left mandible *3*, and the right mandible *4*. When describing deciduous teeth, the right maxilla is designated *5*, the left maxilla *6*, the left mandible *7*, and the right mandible *8*. **B**, Schematic model of a canine jaw with each permanent tooth identified. The numbering always begins on midline; the central incisor is designated *01*, the canine *04*, and the first molar *09*. **C**, Schematic model of a feline jaw, showing tooth identification.

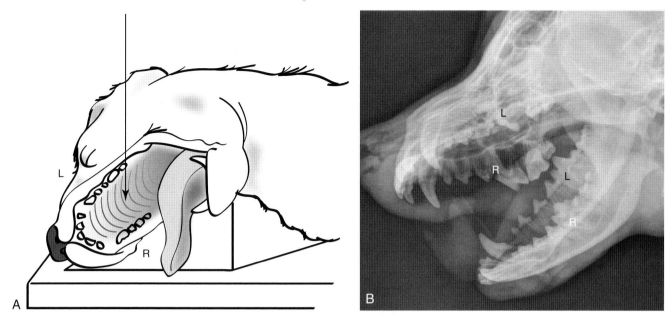

Figure 2-8. **A,** Schematic showing the oblique positioning for radiography of the right maxillary dental arcade. Padding has been placed under the mandible to elevate the mandible and to rotate the left side of the patient's head dorsally. The left side of the patient is subsequently projected dorsally, and the right maxillary and mandibular arcades are projected ventrally. Both a left and a right marker should be used to avoid confusion as to which arcade is highlighted. **B,** Radiograph of an 8-month-old mixed breed dog, positioned as in **A.** A small plastic sleeve has been inserted between the left canines to hold the mouth open. This reduces mandibular and maxillary superimposition and maximizes the chances of obtaining an unobstructed view of the dental arcade of interest. (**A,** From Owens JM, Biery DN: *Radiographic interpretation for the small animal clinician,* Baltimore, 1999, Williams & Wilkins.)

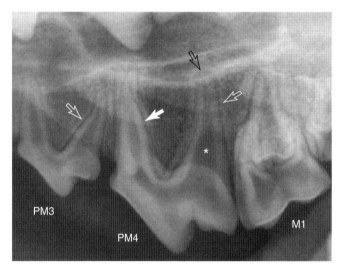

Figure 2-9. Oblique view of the maxilla of an 8-month-old mixed breed dog. The thin radiopaque lines delineated by the *hollow white arrows* are the laminae dura, which consist of solid compact bone that surrounds the dental alveolus and provides an attachment surface for the periodontal ligament. The periodontal ligament in health is not visible radiographically but is in the radiolucent space between the tooth root and lamina dura *(solid white arrow).* The *hollow black arrow* is the apex of the tooth and the base of the alveolus or socket. An *asterisk* is superimposed over the lucent pulp cavity within the fourth premolar.

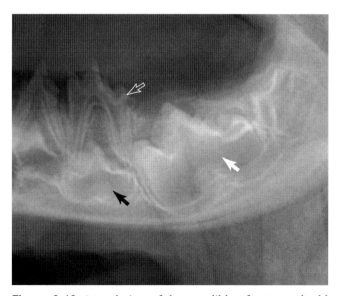

Figure 2-10. Lateral view of the mandible of a 3-month-old Basset Hound. The *hollow white arrow* is the deciduous fourth lower premolar. Subjacent to the deciduous fourth lower premolar is the tooth bud for the permanent fourth lower premolar *(solid black arrow).* The *solid white arrow* is the tooth bud for the permanent first lower molar.

28 Atlas of Normal Radiographic Anatomy and Anatomic Variants in the Dog and Cat

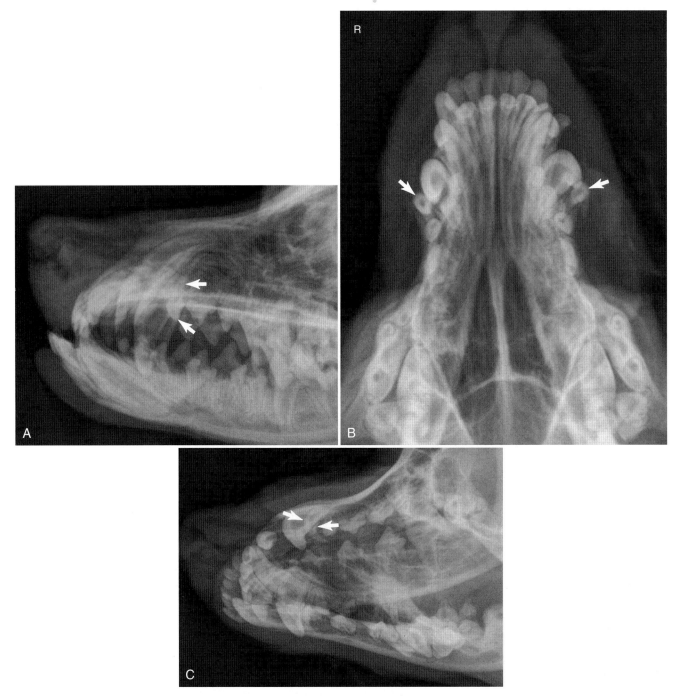

Figure 2-11. Lateral (**A**), dorsoventral (**B**), and oblique (**C**) radiographs of a 10-month-old Bichon Frise. There is bilateral retention of the deciduous maxillary canine tooth *(solid white arrow)*.

eruption disorder is altering the underlying tooth roots or bone structure.

NASAL CAVITY AND SINUSES

The nasal cavity is normally bilaterally symmetric, divided in the midsagittal plane by the vomer bone and cartilaginous nasal septum. Each nasal cavity contains a myriad of fine turbinate bones, collectively known as the *dorsal* and *ventral nasal conchae* rostrally and the *ectoturbinates* and *endoturbinates* caudally (Figure 2-12). These fine turbinates support the large mucosal surface area that is responsible for filtering, warming, and moisturizing inspired air. Standard views for the nasal cavity are lateral and dorsoventral views (see Figures 2-1 and 2-13), intraoral views with the film or imaging plate placed in the mouth, and oblique open mouth views. With *analog* (film-based) imaging systems, the best way to evaluate

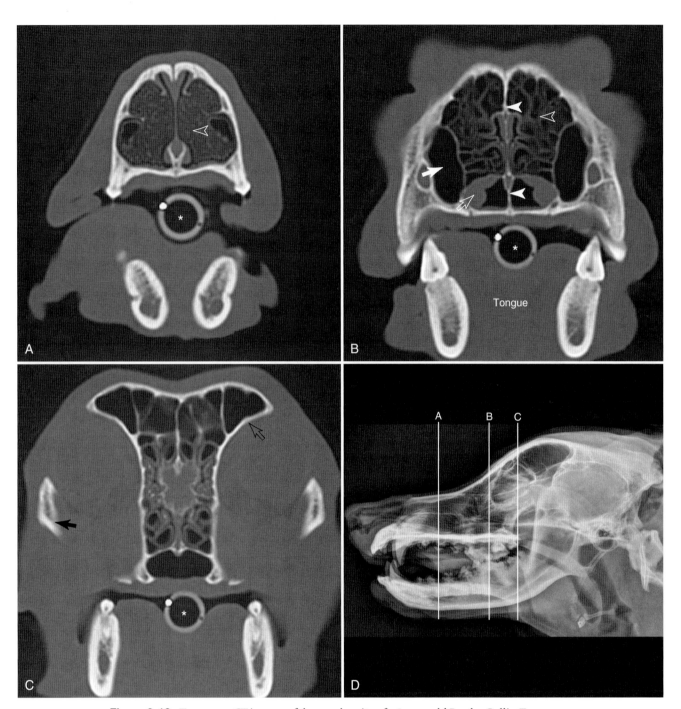

Figure 2-12. Transverse CT images of the nasal cavity of a 2-year-old Border Collie. Transverse images **A**, **B**, and **C** are at the levels indicated by the *vertical lines* in the lateral radiograph **D**. In A, B, and C, an *asterisk* is in the lumen of an endotracheal tube. The *hollow white arrowheads* in A and B are fine turbinate structures that support the nasal mucosa. In B, the *solid white arrow* is the right maxillary recess and the *hollow white arrow* is the nasal organ, a region of highly vascular nasal mucosa. The *solid white arrowheads* are the vomer bone and nasal septum. In C, the *solid black arrow* is the zygomatic arch and the *hollow black arrow* is the ventral border of the left frontal sinus.

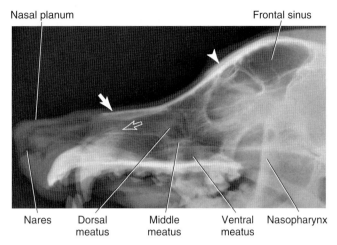

Figure 2-13. Lateral view of the nasal cavity of an 8-year-old Golden Retriever. Unilateral nasal disease is often difficult to detect in lateral views of the nasal cavity due to superimposition of the diseased side with the contralateral normal air-containing side. The *solid white arrow* is the small nasal bone. The *hollow white arrow* is the root of a canine tooth. The *solid white arrowhead* is the frontal bone over the dorsal aspect of the frontal sinuses.

the nasal cavity is by using an intraoral film encased in a vinyl cassette system (3M Company, St Paul, Minn.) (Figure 2-14). By placing the film in the mouth and directing the x-ray beam onto the dorsal aspect of the maxilla, superimposition of the mandible is avoided. With standard digital systems, because the plate cannot be placed in the oral cavity, the best view to evaluate the nasal cavity is an open-mouth ventrodorsal oblique view (V20°R-DCdO). To acquire this view, the dog is anesthetized and placed in dorsal recumbency; the mouth is pulled open by placing tape or a small rope around the maxillary and mandibular incisors while maintaining a parallel relationship between the hard palate and x-ray table; the tongue is held against the mandible using mandibular gauze; and the x-ray beam is angled 20 degrees rostrally and directed into the mouth. This eliminates superimposition due to the mandible but does result in some anatomic distortion (Figure 2-15).

The hard palate is the osseous demarcation between the nasal cavity and the oral cavity. Caudal to the last molar, the hard palate transitions into the soft palate. The region dorsal to the soft palate is the nasopharynx, and the region ventral to the soft palate is the oropharynx. The soft palate extends caudally to the level just cranial to the epiglottis and hyoid apparatus. When a lateral radiograph is assessed, it is extremely important to closely evaluate the size and overall opacity of the nasopharynx, particularly in patients with signs of upper respiratory disease (Figures 2-16 and 2-17). Suboptimally positioned lateral views can give the false impression of soft palate thickening. The soft palate lying in the dorsal

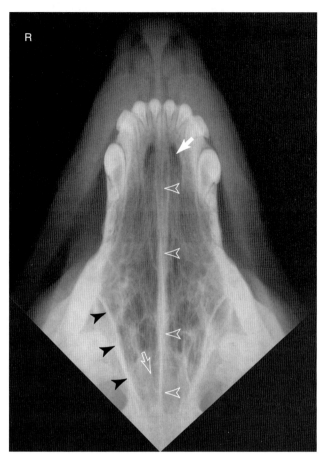

Figure 2-14. Intraoral radiograph of the maxilla of an 8-month-old mixed breed dog. Radiographic film in a slim vinyl cassette containing standard intensifying screens has been placed diagonally in the mouth and a dorsoventral image acquired. The mandible will not be included in a radiograph made in this fashion, allowing an unobstructed view of the nasal cavity. Most nasal cavity disorders are unilateral, and side-to-side comparison is readily possible in this view. The *hollow white arrowheads* designate the midline between the nasal cavities that comprise the vomer bone and osseous and cartilaginous nasal septum. The *solid black arrowheads* are the medial wall of the orbit. The *hollow white arrow* is the region of the cribriform plate. The *solid white arrow* is the paired palatine fissure within the incisive bone, the most rostral aspect of the hard palate. The fine turbinate structures surrounded by air within the nasal cavity are readily apparent in this view.

plane appears wider in oblique views, giving the false impression of thickening. In addition, the soft palate can appear thickened in some brachycephalic breeds, which have a combination of soft palate redundancy and a large tongue base (Figure 2-18).

The frontal sinuses are paired and located dorsal and caudal to the nasal cavity. They extend caudolaterally over the rostrodorsal aspect of the cranial vault. They should be evaluated in both the lateral and dorsoventral views (or ventrodorsal) (Figure 2-19). A rostrocaudal view of the frontal sinuses allows interrogation of the sinuses

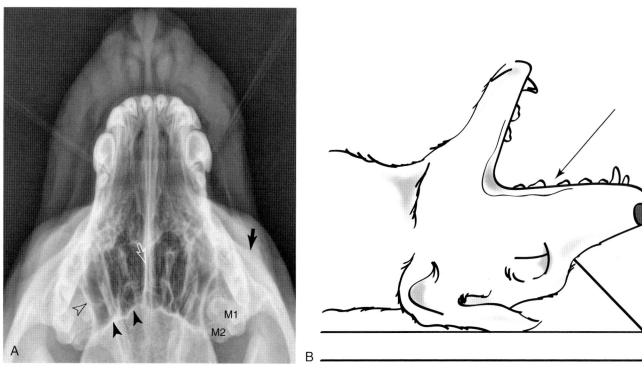

Figure 2-15. A, V20°R-DCdO open-mouth view of the nasal cavity of an 8-month-old mixed breed dog. This is an alternative technique to intraoral radiography that is well-suited for digital systems in which it is impossible to insert the imaging plate into the patient's mouth. Some distortion of the nasal cavity occurs as a result of beam angulation. Tape or a small rope is usually used to hold the maxilla in place and can be seen crossing caudal to the canines. The *solid black arrow* is the frontal process of the zygomatic bone. The *solid black arrowhead* is the rostral aspect of the cranial vault. The *hollow black arrowhead* is the lateral margin of the frontal sinus. The *hollow white arrow* is the nasal septum. **B,** Diagram of patient positioning for image **A**. (**B,** From Owens JM, Biery DN: *Radiographic interpretation for the small animal clinician,* Baltimore, 1999, Williams & Wilkins.)

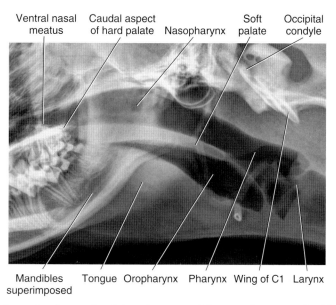

Figure 2-16. Lateral radiograph of the pharynx and larynx of a 6-month-old mixed breed dog. The major structures are identified.

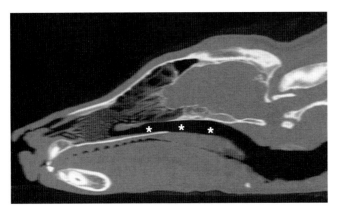

Figure 2-17. True sagittal slice CT image of a dog with a dolichocephalic conformation, optimized for bone. The *white asterisks* are the nasopharynx, dorsal to the hard palate rostrally and soft palate caudally. An air-filled nasopharynx will be readily visible in a lateral skull radiograph, despite patient conformation.

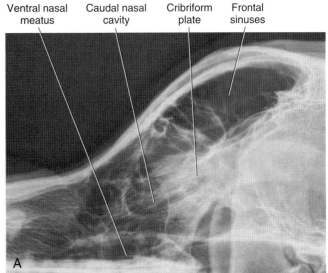

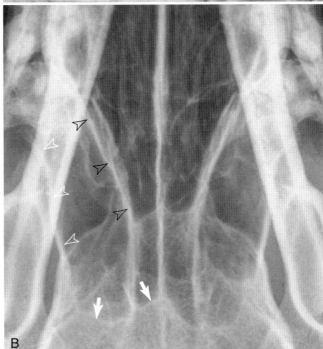

Figure 2-19. Lateral view (A) and dorsoventral view (B) of a 9-year-old Golden Retriever. In B, the lateral aspect of the frontal sinus is delineated by the *hollow white arrowheads*. The *hollow black arrowheads* are the medial wall of the pterygopalatine fossa, the orbit. The frontal sinus is dorsal and medial to the pterygopalatine fossa and extends to midline over the rostral aspect of the calvarium. It is superimposed over the cribriform plate and rostral aspect of the frontal lobe. The *solid white arrows* delineate the caudomedial aspect of the right frontal sinus.

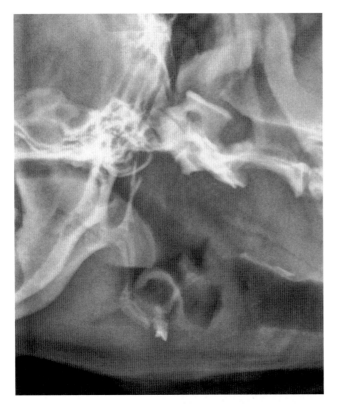

Figure 2-18. Lateral view of an 8-year-old Cavalier King Charles Spaniel. The soft palate appears thickened. This is due to a combination of soft tissue redundancy of the soft palate and because the base of the tongue is contacting the majority of the palate. This appearance is, for the most part, breed related and common in mesaticephalic and brachycephalic breeds.

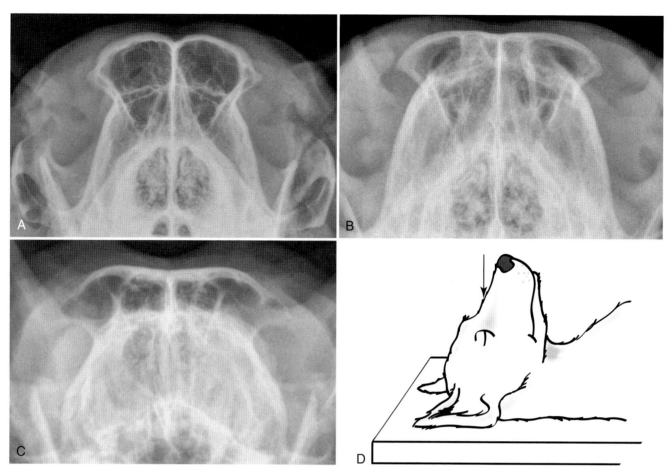

Figure 2-20. Rostrocaudal view (**A**) of the frontal sinuses of an 8-year-old Golden Retriever. The frontal sinuses are large, symmetric, and air filled. The air-filled nasal cavity is also readily visible. **B,** Rostrocaudal view of the frontal sinuses of an 8-month-old mixed breed dog. The frontal sinuses appear different when compared with **A** because of both beam angulation and patient conformation. There is considerable variability in the appearance of the frontal sinuses with this view due to a combination of patient morphologic variation and differences in beam angulation during radiography. Comparing side-to-side symmetry is the primary objective with this view. **C,** Rostrocaudal view of the frontal sinuses of a 3-year-old domestic shorthair cat. Optimal positioning is paramount to allow side-to-side comparisons. **D,** Schematic of method to obtain the rostrocaudal view of the frontal sinuses. (**D,** From Owens JM, Biery DN: *Radiographic interpretation for the small animal clinician,* Baltimore, 1999, Williams & Wilkins.)

for symmetry, opacification, and osseous remodeling (Figure 2-20). In many brachycephalic breeds, the frontal sinuses are rudimentary or absent (Figure 2-21).

The parietal, temporal, and occipital bones combined with the smaller sphenoid and pterygoid bones encompass the middle and caudal aspect of the cranial vault. In the juvenile patient, the associated sutures are conspicuous radiographically and are easily confused with traumatic fractures. At the ventral aspect of the temporal bone is the bilateral retroarticular process, which forms the temporal component of each temporomandibular joint. In addition, the ventral aspect of the temporal bone provides osseous support for the middle and inner ear structures and the base of the cranial vault. The zygomatic process of the temporal bone extends rostrally to fuse with the temporal process of the zygomatic bone to form the zygomatic arch (see Figures 2-1 and 2-22). The coronoid process of the mandible extends between the medial aspect of the zygomatic arch and the lateral aspect of the calvarium and is surrounded by extensive masticatory musculature. The zygomatic process of the frontal bone is larger in the cat than in the dog (see Figure 2-2). A

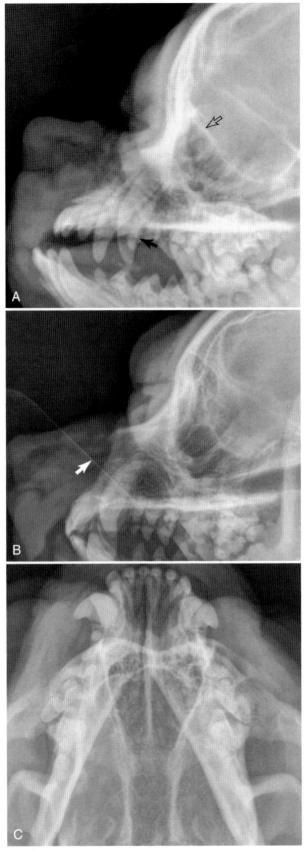

Figure 2-21. **A,** Lateral view of an 8-month-old Shih Tzu. A gas-filled frontal sinus is not apparent. This is common in brachycephalic breeds. The *hollow black arrow* is the caudodorsal margin of the nasal cavity. Incidentally, a retained deciduous maxillary canine is present *(solid black arrow)*. Lateral view **(B)** and dorsoventral view **(C)** of a 3-year-old Boston Terrier. No frontal sinuses are apparent on either the lateral or dorsoventral view. The sharp oblique radiopaque line superimposed over the rostral aspect of the nasal cavity *(solid white arrow)* is the edge of padding used to support the rostral aspect of the nasal cavity during radiography.

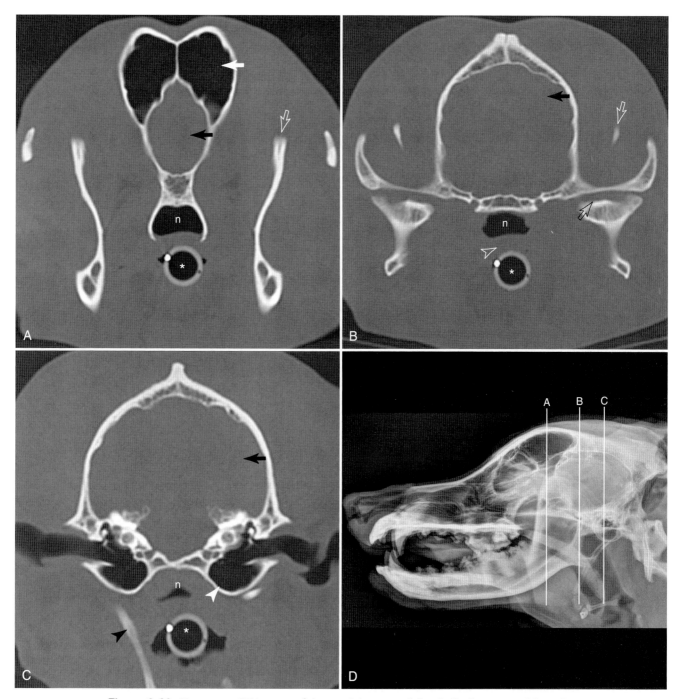

Figure 2-22. Transverse CT images of the caudal aspect of the skull of a 2-year-old Border Collie, the same patient as in Figure 2-12. Transverse images **A**, **B**, and **C** are at the levels indicated by the vertical lines in the lateral radiograph **D**. In **A**, **B**, and **C**, an *asterisk* is in the lumen of an endotracheal tube, *n* is the nasopharynx, and the *solid black arrow* is the brain. The *solid white arrow* is the left frontal sinus (**A**), and the *hollow white arrow* (**A-B**) is the dorsal aspect of the coronoid process of the mandible. In **B**, the *hollow black arrow* is the temporomandibular joint, and the *hollow white arrowhead* is the soft palate. In **C**, the *solid white arrowhead* is the ventral border of the left tympanic bulla, and the *solid black arrowhead* is the right stylohyoid bone, which is positioned slightly more rostral than the left stylohyoid bone.

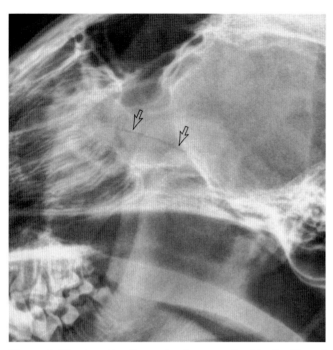

Figure 2-23. Lateral radiograph of a 6-month-old mixed breed dog. *Hollow black arrows* outline the symphysis between the temporal process of the zygomatic bone and the zygomatic process of the temporal bone. In many patients, this rarely fuses completely. This should not be confused with a fracture.

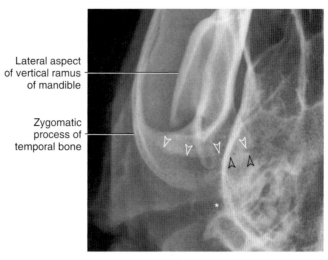

Figure 2-25. Dorsoventral view of the right temporomandibular joint of a 9-year-old Golden Retriever. The *hollow white arrowheads* are the articular margin of the condylar process *(articular process)* of the mandible. The *hollow black arrowheads* are the mandibular fossa of the zygomatic process of the temporal bone formed by the retroarticular process. An *asterisk* overlies the lumen of the horizontal ear canal.

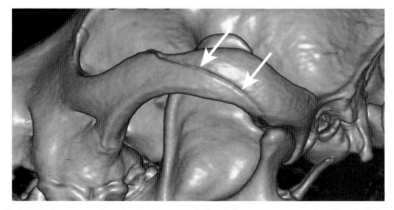

Figure 2-24. Volume rendered 3-D image of the left zygomatic arch of a dolichocephalic dog. The suture between the temporal process of the zygomatic bone and zygomatic process of the temporal bone, due to incomplete fusion, is readily evident *(white arrows)*.

suture between the zygomatic process of the temporal bone and the temporal process of the zygomatic bone is often confused with a fracture in young dogs (Figures 2-23 and 2-24). The suture becomes less radiographically apparent with age but often never completely closes in both dogs and cats.

TEMPOROMANDIBULAR JOINTS AND TYMPANIC BULLAE

The temporomandibular joints are best evaluated with a dorsoventral (or ventrodorsal) view (Figure 2-25) and specifically positioned lateral oblique views. Oblique

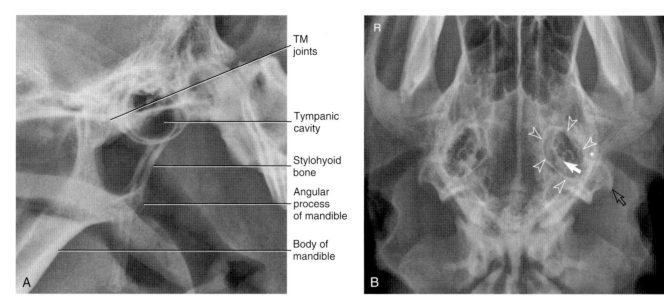

Figure 2-26. A, Lateral radiograph of an 8-year-old Golden Retriever. The thin-walled tympanic bullae are nearly perfectly superimposed, and gas is readily visible within the tympanic bullae. Although the tympanic bullae may appear thin-walled and of the expected radiolucency on this view, fluid within one tympanic bulla may go undetected due to the overlap. Additional views are required to rule out unilateral disease. **B,** Ventrodorsal view of a 1-year-old Labrador Retriever. The wall of one tympanic bulla is outlined by the *hollow white arrowheads*. The *solid white arrow* is the dense petrous temporal bone that contains the bony labyrinth of the inner ear. The *hollow black arrow* is the mastoid process of the temporal bone. An *asterisk* is at the level of the tympanic membrane and the most medial aspect of the horizontal canal. This view allows evaluation of each tympanic bulla and associated ear canal independently. *TM,* Temporomandibular.

views are required to minimize unwanted superimposition. The caveat is that the degree of obliquity should lead to minimal distortion of joint morphology. The most consistently reliable method is the *nose up* technique. With this method, one should begin with true lateral positioning and then elevate the rostral aspect of the skull (the nose) approximately 30 degrees while avoiding any rotational change. The degree of inclination is related to the shape of the skull. Elevation of the nose 10 to 30 degrees in mesaticephalic and dolichocephalic breeds and 20 to 30 degrees in brachycephalic breeds usually gives the most consistent results[4] (Figure 2-26, *A*). The dependent temporomandibular joint that is closest to the x-ray table is displaced rostrally, which minimizes superimposition and, importantly, minimizes distortion (Figure 2-27). In some patients, slight rotation along the sagittal plane (10-degree rotation of the nondependent surface of the head dorsally) results in the rostral aspect of the joint being displayed more conspicuously (Figure 2-28). However, excessive rotation in the sagittal plane can result in excessive distortion and superimposition of the opposite angular process over the temporomandibular joint.

An alternative method to image the temporomandibular joint is simply to lay the patient's head on the x-ray table in a natural position. This results in rotation in both the sagittal plane and the transverse plane with sufficient obliquity to profile one temporomandibular joint without superimposition. The dependent tympanic bulla closest to the x-ray table is usually positioned more dorsal relative to the upper nondependent tympanic bulla, and it is the nondependent tympanic bulla that is projected without superimposition (Figure 2-29). The success of this positioning method depends primarily on the shape of the skull and is more unreliable than the technique of elevating the rostral aspect of the skull, as described previously. In addition, it can become confusing as to which temporomandibular joint is being profiled; for this reason, this alternate technique is not recommended.

In both the cat and the dog, as in many species, a thin articular disc lies between the articular surface of the condylar process of the mandible and the mandibular fossa of the temporal bone. This is not apparent radiographically unless mineralized pathologically, which is rare.

The tympanic bullae are best evaluated using the following views (Figure 2-30):
- dorsoventral (or ventrodorsal) view
- lateral view
- minimally oblique lateral views
- open-mouth rostrocaudal view with the beam centered at the level of the tympanic bullae

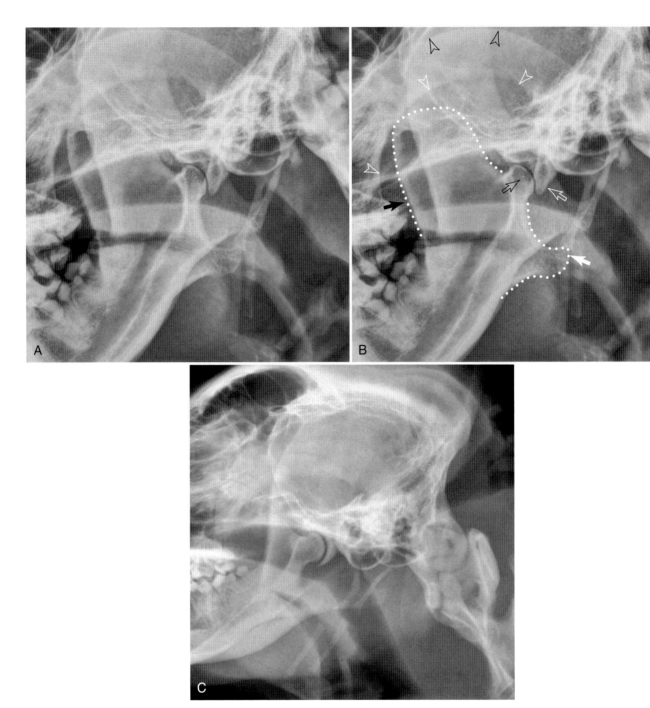

Figure 2-27. **A,** Lateral oblique view of the right temporomandibular joint. The rostral aspect of the skull has been elevated approximately 30 degrees, without concurrent rotation (the nose up technique). The objective of this oblique view is to image the dependent temporomandibular joint with minimal distortion and minimal superimposition of adjacent structures. **B,** The same image as in **A;** the *hollow black arrow* is the condylar process (articular process) of the mandible. The *hollow white arrow* is the retroarticular process of the temporal bone; the temporomandibular joint is formed between the condyloid process of the mandible and the mandibular fossa on the ventral surface of the zygomatic process of the temporal bone. The retroarticular process serves to enlarge this joint on its medial aspect. The *solid white arrow* is the angular process of the mandible. The *dotted line* outlines the ramus; the *solid black arrow* at the rostral border of the coronoid process is the coronoid crest. The *hollow white arrowheads* delineate the ventral border and the *hollow black arrowheads* show the dorsal border of the zygomatic arch. Each tympanic bulla is also readily visible in this view. **C,** Nose up view in a 10-month-old Labrador Retriever mix, as in **A,** characterized by a slightly different appearance to the temporomandibular joint.

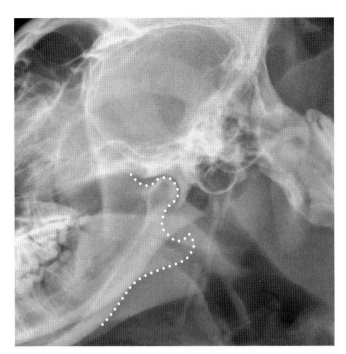 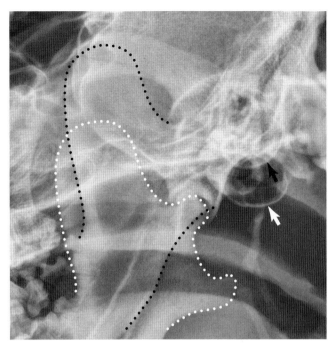

Figure 2-28. Thirty-degree nose up view of a 10-month-old Labrador Retriever mix. The nondependent surface of the head has also been rotated dorsally 10 degrees. In some patients, slight rotation is required to enable visualization of the rostral aspect of the temporomandibular joint.

Figure 2-29. Lateral oblique radiograph of a 1-year-old Labrador Retriever. The patient's head is lying naturally on the table, resulting in rotation of the skull, in addition to the nares being closer to the table (nose down). With the patient lying naturally on the table, there is often sufficient obliquity to separate both the tympanic bullae and the temporomandibular joints. However, this positioning technique is less reliable and depends heavily on the skull conformation. In addition, it is usually the nondependent temporomandibular joint that is projected more rostrally and ventrally. The *dotted white line* is the nondependent vertical ramus. The *dotted black line* is the contralateral dependent ramus. The *solid white arrow* is the nondependent tympanic bulla. The *solid black arrow* is the contralateral dependent tympanic bulla. This method of obtaining oblique images of the bulla and temporomandibular joints is not recommended.

With respect to the lateral oblique views, they are very similar to the views used for evaluation of the temporomandibular joints. Usually 10 to 20 degrees of rotation in the long axis provides sufficient obliquity to eliminate superimposition of the tympanic bullae. The patient should be absolutely lateral and then the nondependent mandible and maxilla should be rotated dorsally. This results in the lower tympanic bulla being positioned more ventrally. One should ensure each tympanic bulla is labeled appropriately to eliminate the possibility of misidentification (Figure 2-31). Both insufficient and excessive rotation can result in distortion of the bullae that can confound radiographic interpretation (Figure 2-32).

When the open-mouth view is performed, the tongue is best restrained ventrally toward the mandible to reduce superimposition with the tympanic bullae (Figure 2-33). The open-mouth view is particularly important when evaluating the feline tympanic bullae, due to the bulla compartmentalization present in this species. In the cat, each tympanic bulla is divided into a small dorsolateral compartment and a large ventromedial compartment (Figure 2-34). It is important that both compartments be identified radiographically. Ideally, the endotracheal tube should be removed immediately before radiography to reduce potential superimposition over the tympanic bullae (Figure 2-35). In the cat, an alternative view is often used to evaluate the tympanic bullae. In this view, the patient is positioned similarly for the open-mouth view, but the mouth is closed and the head extended such that the plane of the ventral aspect of the horizontal ramus of the mandible is approximately 10 degrees rostral to the vertical plane. The x-ray beam is directed vertically, immediately caudal to the temporomandibular joints[5] (Figure 2-36).

The occipital bone encases the caudal aspect of the cranial vault, and the spinal cord exits the calvarium via the foramen magnum, which is located at the caudoventral aspect of the skull. This is an important region morphologically but difficult to assess radiographically (Figure 2-37). An open-mouth rostrocaudal view centered at the level of the foramen magnum can sometimes help detect morphologic abnormalities associated with the occipital bone (Figure 2-38). Caution should be used

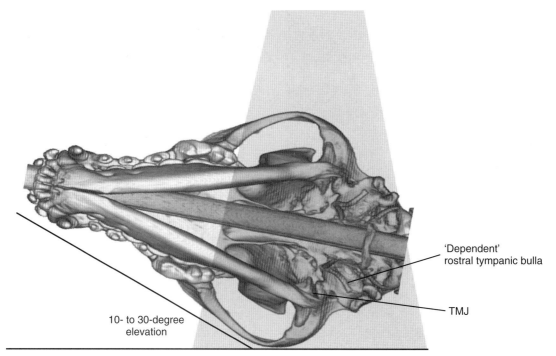

Figure 2-30. Schematic of nose up positioning for evaluation of the temporomandibular joints (TMJs). Elevation of the nose approximately 30 degrees with no rotation around the sagittal plane usually results in sufficient obliquity to eliminate unwanted superimposition of the nondependent TMJ as it is displaced caudally relative to the dependent TMJ of interest.

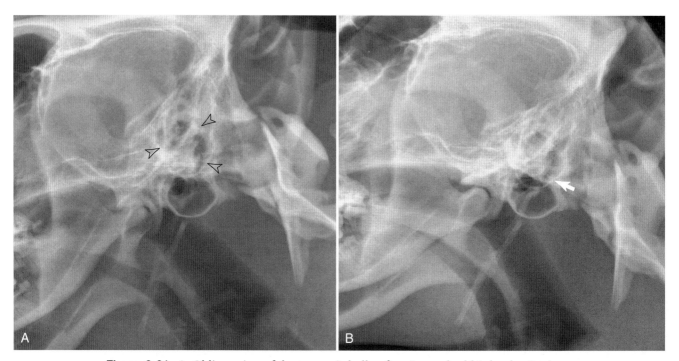

Figure 2-31. **A,** Oblique view of the tympanic bulla of an 8-month-old Labrador Retriever mix. The nondependent surface of the head has been rotated dorsally 20 to 30 degrees. The dependent tympanic bulla closest to the x-ray table is now ventral and imaged without superimposition. The tympanic cavity and horizontal ear canal are readily visible. The nondependent tympanic bulla is dorsal and superimposed over adjacent structures *(hollow black arrowheads)*. **B,** The same patient as in **A**; there has been insufficient rotation to remove superimposition of the nondependent tympanic bulla. The ventral border of the nondependent tympanic bulla *(solid white arrow)* is superimposed over the dependent tympanic bulla of interest.

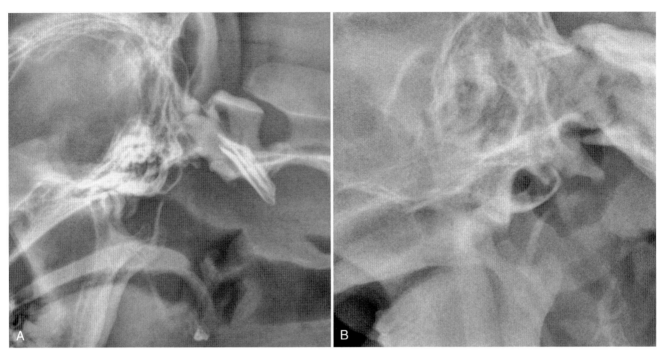

Figure 2-32. Lateral radiograph (**A**) of an 8-year-old Scottish Terrier. The slight obliquity in this view results in the false impression of thickening of the more ventrally positioned tympanic bulla. **B,** Oblique view of the tympanic bulla of an 8-month-old mixed breed dog. There is excessive obliquity of the skull resulting in distortion of the bulla.

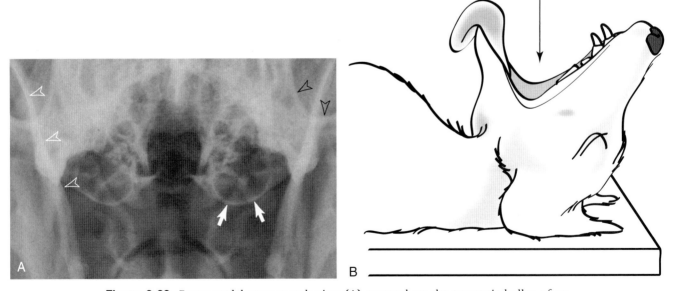

Figure 2-33. Rostrocaudal open-mouth view (**A**) centered on the tympanic bullae of an 8-month-old mixed breed dog. The *solid white arrows* are the ventral margin of the left tympanic bulla. Gas within the tympanic bulla is readily visible. The *hollow white arrowheads* are the ramus of the right mandible, and the *hollow black arrowheads* are the dorsal aspect of the zygomatic process of the temporal bone. **B,** Positioning schematic for open-mouth rostrocaudal view of the tympanic bullae. (**B,** From Owens JM, Biery DN: *Radiographic interpretation for the small animal clinician,* Baltimore, 1999, Williams & Wilkins.)

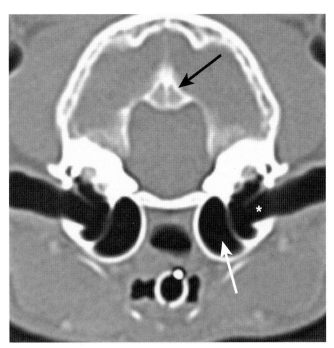

Figure 2-34. Transverse CT image at the level of the tympanic bullae of a cat optimized for bone. The *solid white arrow* is the ventromedial compartment and the *white asterisk* is the dorsolateral compartment of the left tympanic bulla. The tympanic membrane is not evident because the image is not optimized for soft use structures. The *black arrow* is the osseous tentorium. This is well developed in cats and is an osseous barrier between the caudal aspect of the cerebrum and the rostral aspect of the cerebellum over which are draped the leptmeninges.

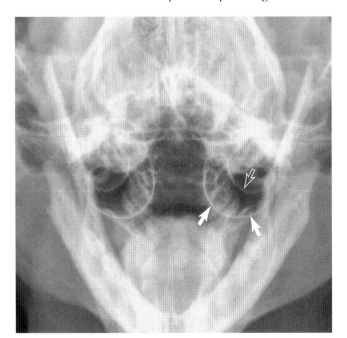

Figure 2-35. Rostrocaudal open-mouth view of a 3-year-old Domestic Shorthair cat, centered on the tympanic bullae. The endotracheal tube was removed immediately before radiography. The *solid white arrows* are the wall of the larger ventromedial compartment of the tympanic bulla, and the *hollow white arrow* is the border of the smaller dorsolateral compartment.

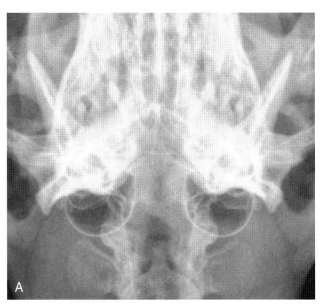

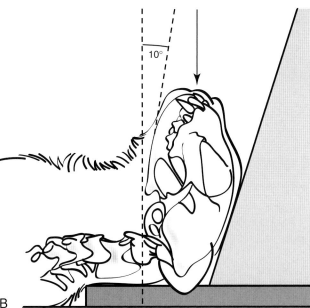

Figure 2-36. A, R10°V-CdO view of a 6-year-old Domestic Shorthair cat. Patient positioning for this projection is easier than for the open-mouth view in Figure 2-35. This is considered a satisfactory alternative to assess the larger ventromedial compartment of the tympanic bulla, but the dorsolateral compartment of the tympanic bulla can be harder to see in this view. **B,** Schematic of patient positioning for the R10°V-CdO view. (**B,** From Owens JM, Biery DN: *Radiographic interpretation for the small animal clinician,* Baltimore, 1999, Williams & Wilkins.)

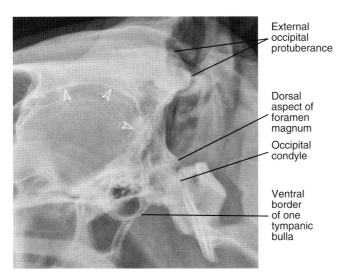

Figure 2-37. Lateral image of the caudal aspect of the cranial vault and cranial cervical region in an 8-year-old Golden Retriever. The *hollow white arrowheads* are the caudodorsal border of the cranial vault. The ears are extended dorsally, and the irregular gas opacity superimposed over the caudodorsal aspect of the calvarium is air within the vertical ear canals.

when acquiring this view: the very condition one may be attempting to rule out, such as traumatic instability or congenital atlantoaxial instability, is potentially exacerbated by this positioning.

THE MANDIBLES AND LARYNX

There are two mandibles, left and right, united rostrally at the mandibular symphysis (Figure 2-39). Each mandible is divided into a body and, more caudally, the ramus. All teeth are located within the body. The ramus comprises primarily the *coronoid process,* which is a relatively flat plate in the sagittal plane and is the insertion site for muscles that close the mouth. The condylar process is the most caudal aspect of the mandible and is the transversely elongated centrally convex condylar process, which forms the temporomandibular joint by articulating with the zygomatic process of the temporal bone (see Figure 2-25). The *angle of the jaw* is the caudal aspect of the junction between the ramus and the body. The angular process is a small extension caudally and an

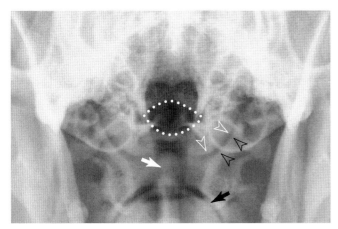

Figure 2-38. Rostrocaudal open-mouth view of an 8-month-old mixed breed dog. In addition to the tympanic bullae being readily apparent in this view, the occipital condyles and foramen magnum are visible. The *hollow white arrowheads* are the left occipital condyle. The *hollow black arrowheads* are cranial articular fovea of the atlas; this is the atlanto-occipital joint. The *dotted white line* is the foramen magnum; the *solid white arrow* is the odontoid process (dens) of C2. The *solid black arrow* is the atlantoaxial joint. Although this view allows optimal visualization of these structures, it requires at least 90-degree head flexion and should not be used when occipito-atlantoaxial instability is suspected.

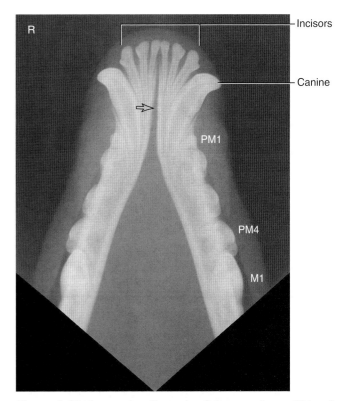

Figure 2-39. Intraoral radiograph of the rostral mandible of an 8-month-old mixed breed dog. Radiographic film in a thin vinyl cassette containing standard intensifying screens was placed diagonally in the mouth and a ventrodorsal image acquired. This eliminates superimposition of the maxilla on the mandible. The *hollow black arrow* is the mandibular symphysis. This persists as a fibrous union throughout life, and a radiolucent zone at the symphysis is normal.

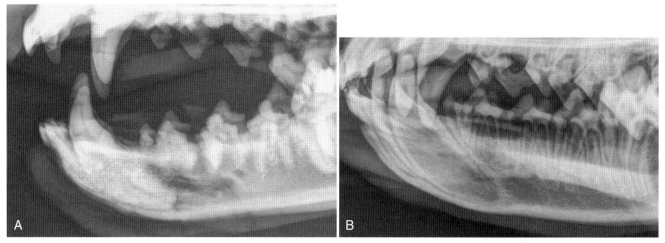

Figure 2-40. **A,** Lateral radiograph of the rostral aspect of the mandible of a 9-year-old Golden Retriever. There is an ill-defined lucency within the mandible caudal to the canine teeth and ventral to the first and second premolars. This is due to overlapping of the two (sometimes more) mental foramina present in each mandible. The superimposition results in an irregular region of decreased opacity and should not be confused with an aggressive lesion. If there is doubt, an intraoral or open-mouth oblique view should be acquired. **B,** Lateral radiograph of the rostral mandible of a 6-month-old mixed breed dog. An ill-defined lucency caudal to the mandibular canine due to overlapping mental foramina is readily apparent.

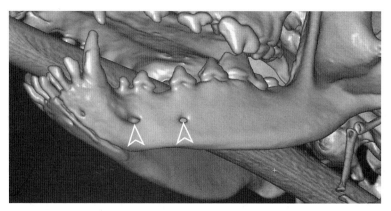

Figure 2-41. Volume rendered 3-D image of the rostral aspect of the left mandible of a dolichocephalic dog. The two focal defects in the rostral aspect of the mandible, delineated by the *hollow white arrows*, are the mental foramina. As these foramina extend obliquely through the lateral cortex, a poorly defined region of radiolucency is created.

important site for muscle attachment (see Figure 2-30). At the rostral aspect of each body, immediately ventral to the first and second premolars, are the mental foramina, usually two associated with each mandible. These foramina, primarily because of the obliquity with which they exit the cortex, create a poorly defined region of lucency immediately caudal to the mandibular canines bilaterally. This should not be confused with an aggressive lesion (Figures 2-40 to 2-42).

Dental and maxillomandibular malocclusion is common in dogs, particularly in the brachycephalic breeds. Gross malocclusion is readily identified radiographically on standard views. Subtle malocclusion is better assessed by direct visualization. Prognathism of the mandible is common in brachycephalic breeds, particularly the bulldog (Figure 2-43). Most other breeds have *brachygnathic mandibles,* that is, receding lower jaws (Figure 2-44).

The larynx is the musculocartilaginous organ at the entrance of the trachea. It is composed of the epiglottic, thyroid, cricoid, arytenoid, sesamoid, and paired interarytenoid cartilages (Figure 2-45). The sesamoid cartilage

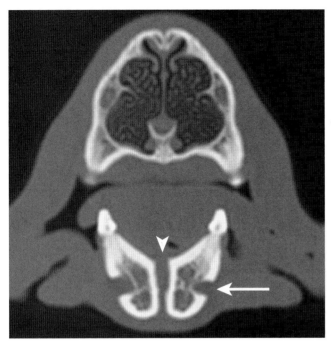

Figure 2-42. Transverse CT image at the level of the second mandibular premolars (306 and 406 by the modified Triadan system) optimized for bone. The two focal defects in the lateral aspect of the rostral aspect of the mandible, delineated by the *solid white arrow*, are the mental foramina. As these foramina extend obliquely through the lateral cortex, a poorly defined region of radiolucency is created. The *solid white arrowhead* is the mandibular symphysis.

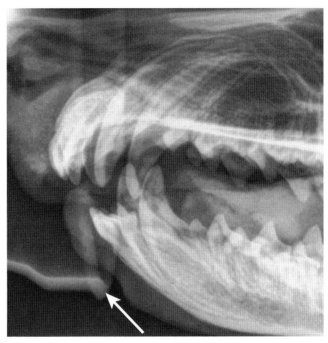

Figure 2-44. Lateral radiograph of a Cocker Spaniel with brachygnathism. The *white arrow* is an oxygen mask. Its presence is unrelated to the malocclusion.

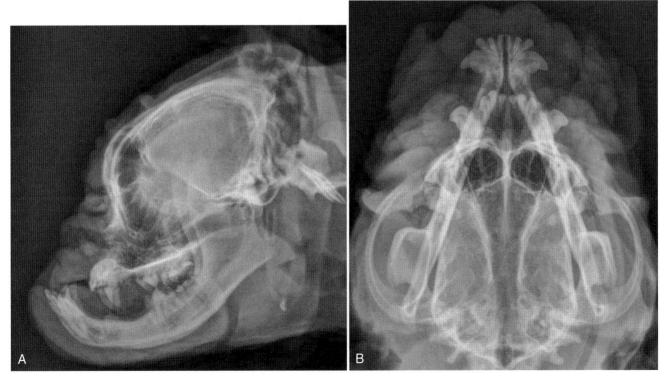

Figure 2-43. Lateral (**A**) and dorsoventral (**B**) images of a 2-year-old Boxer. There is significant prognathism. This is common in brachycephalic breeds. In this patient, multiple mandibular premolars are absent.

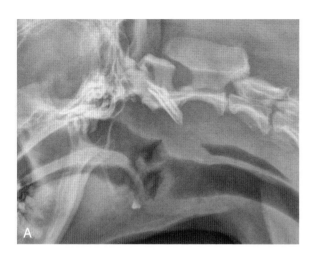
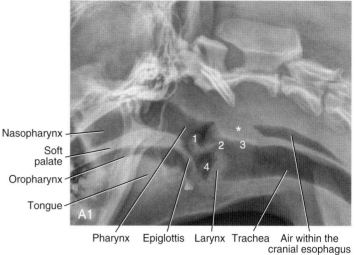
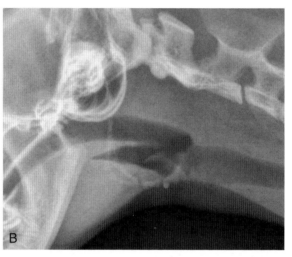

Figure 2-45. **A,** Lateral radiograph of the pharynx and larynx of an 8-year-old Scottish Terrier. **A1,** The same image as in **A,** with labels: *1,* the cuneiform process of the arytenoid cartilage; *2,* the thyroid cartilage; *3,* the cricoid cartilage; *4,* the laryngeal saccules. An *asterisk* is superimposed over the upper esophageal sphincter. **B,** Lateral radiograph of an 8-month-old Tonkinese cat. The soft palate is positioned dorsal to the epiglottis; this is of no clinical significance in most patients.

and the interarytenoid cartilages are not radiographically visible, and only part of the arytenoid cartilage is visible. The hyoid apparatus comprises the paired stylohyoid, epihyoid, ceratohyoid, and thyrohyoid bones and the unpaired basihyoid bone (Figure 2-46). The basihyoid bone lies in the transverse plane, giving it depth in the lateral view. This results in increased opacity relative to the remaining bones of the larynx and hyoid apparatus. The epiglottis, with its wide base caudally and thin apex cranially, extends craniodorsally from the cranioventral aspect of the larynx. Its association with the soft palate is variable, influenced primarily by sedation and the presence or absence of an endotracheal tube. It is usually completely surrounded by air and is a constant landmark. However, occasionally, the oral surface will silhouette with the tongue, particularly if the head is flexed, which reduces conspicuity. In the cat, the corniculate and cuneiform processes of arytenoid cartilage are absent.

The morphology of the hyoid apparatus and associated larynx is best evaluated on a well-positioned lateral view with the head moderately extended. Excessive head flexion results in unwanted superimposition of the mandible over the larynx. When evaluating the larynx, take care to ensure that the pinnae are positioned dorsally and are not superimposed over the region of interest, which can result in additional soft tissue and gas opacities and complicate radiographic interpretation (Figure 2-47).

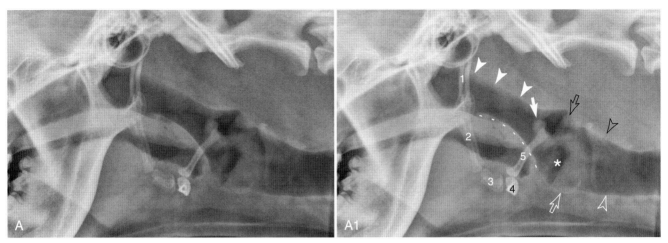

Figure 2-46. **A,** Lateral view of the larynx of an 8-year-old Golden Retriever. **A1,** The same image as **A,** with labels: *1,* the paired stylohyoid bone; *2,* the paired epihyoid bone; *3,* the paired ceratohyoid bone; *4,* the single transverse basihyoid bone; *5,* the paired thyrohyoid bone. The *dotted white line* is the epiglottis. The *solid white arrow* is the cuneiform process of the arytenoid cartilage. The *hollow black arrow* is the dorsocranial aspect of the thyroid cartilage. The *hollow white arrow* is the ventral aspect of the thyroid cartilage. The *solid white arrowheads* are the dorsal aspect of the pharynx. The *hollow black arrowhead* is the dorsal aspect of the cricoid cartilage, and the *hollow white arrowhead* is the ventral aspect of the cricoid cartilage. The *asterisk* is gas within the lateral saccules.

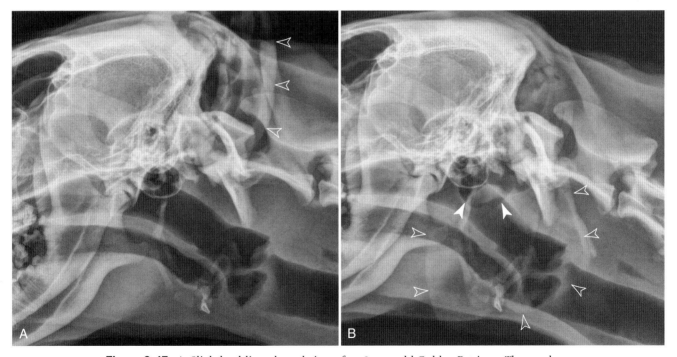

Figure 2-47. **A,** Slightly oblique lateral view of an 8-year-old Golden Retriever. The ears have been pulled dorsally. Alternating soft tissue and gas opacities superimposed over the caudal aspect of the calvarium is trapped air within the vertical canals. The *hollow white arrowheads* are the caudal border of the pinnae. The ears should be positioned dorsally, particularly when the larynx is imaged, because superimposition can complicate radiographic interpretation. **B,** The pinnae are falling ventrally. There are multiple gas and soft tissue opacities superimposed over the larynx, thus complicating radiographic assessment. The *solid white arrowheads* are the ventral border of an artifact created by the superimposed pinnae. The *hollow white arrowheads* are the edges of the pinnae.

REFERENCES

1. Wiggs R, Lobprise H, editors: *Oral anatomy and physiology. Veterinary dentistry: principles and practice*, Philadelphia, 1997, Lippincott-Raven.
2. Floyd MR: The modified Triadan system: nomenclature for veterinary dentistry. *J Vet Dent* 8:18-19, 1991.
3. Thrall D, editor: Technical issues and interpretation principles relating to the axial skeleton. *Textbook of veterinary diagnostic radiology*, ed 5, Philadelphia, 2007, Saunders.
4. Dickie A, Sullivan M: The effect of obliquity on the radiographic appearance of the temporomandibular joint in dogs. *Vet Radiol Ultrasound* 42:205-217, 2001.
5. Hofer P, Bartholdi B, Kaser-Hotz B: Radiology corner: a new radiographic view of the feline tympanic bulla. *Vet Radiol Ultrasound* 36:14-15, 1995.

CHAPTER 3

The Spine

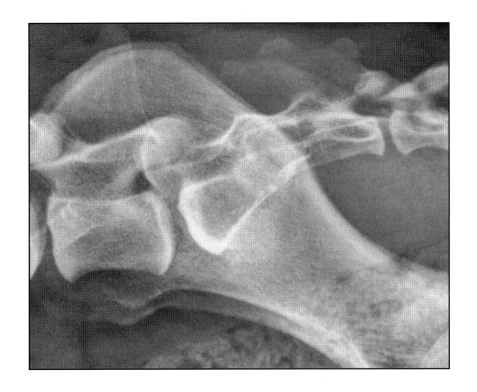

Survey radiography of the spine typically involves acquisition of lateral and ventrodorsal projections, and those are the views emphasized in this chapter. To interpret spinal radiographs accurately, the patient must be sedated or anesthetized to optimize positioning. Optimal positioning is critical for spinal radiographs, and this cannot be achieved in patients who are not sedated or anesthetized. The anatomic complexity of the spine can easily lead to an incorrect interpretation if positioning is not optimal.

For lateral views, it is critical that the midsagittal plane of the patient, and thus of the spine, be parallel to the x-ray table top. This may require elevating the sternum slightly with a foam positioning sponge so that the spine and sternum are in the same plane. Even if the sternum is elevated, the patient cannot be allowed to lie unsupported on the x-ray table for the lateral projection or sagging will occur. If spinal regions that *sag* are not elevated with nonradiopaque sponges, the natural undulation of the vertebrae will lead to distortion (Figure 3-1).

Another important fact with regard to positioning for lateral spinal radiographs is that the divergent nature of the primary x-ray beam will lead to only a few intervertebral disc spaces being projected at a size that is representative of their true size. Beam divergence results in angulation of the x-ray beam with respect to the disc spaces toward the periphery of the field, and this creates a false appearance of disc space narrowing (Figure 3-2). To compensate for this, multiple lateral projections of any spinal region are needed, with the center of the x-ray beam located at multiple locations: that is, the center of the x-ray beam at C3 and C6 for lateral cervical spinal series, and the center of the x-ray beam at T5, T9, T13, L3, and L7 for thoracolumbar series.

Positioning for a ventrodorsal radiograph of the spine is less tedious, with superimposition of the sternum and spine being the most important consideration. As with lateral views, the use of multiple centering points for each spinal region is also important for the ventrodorsal view.

Although size and shape of vertebrae along the spine vary considerably, each vertebra has a set of core components. The *vertebral body* forms the bulk of each vertebra. *Transverse processes* extend laterally from the body. The *pedicles*, dorsal extensions from each side of the body, form the lateral boundaries of the *vertebral canal*. The pedicles are joined dorsally by the *lamina*, a bony shelf that forms the dorsal limit of the vertebral canal. A *spinous process* extends dorsally from the lamina. The pedicles, lamina, and spinous process are collectively termed the *vertebral arch* (Figure 3-3). Embryologically, the vertebral arch comprises two *neural arches*, one on each side, that subsequently fuse.

Vertebrae articulate at two points, the *intervertebral disc*, which joins adjacent vertebral bodies, and the *articular process joints*, which join the lamina. The intervertebral disc is a cartilaginous and fibrous structure that allows some motion and provides cushioning. The normal intervertebral disc is nonmineralized and therefore of soft tissue opacity, creating a region of soft tissue opacity between vertebrae.

There are four articular processes on most vertebrae: left and right cranial articular processes and left and right

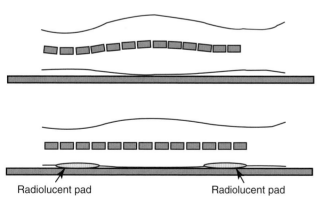

Figure 3-1. Elevation of sagging portions of the vertebral column with nonradiopaque pads leads to improved vertebral alignment. The perspective of this image is looking at the dorsal aspect of a dog while it is lying on the x-ray table. In the *top panel*, the dog is allowed to lie unrestrained on the table. The natural curve of the body results in the vertebrae not being aligned in one plane. In this instance, the varying orientation of vertebrae with respect to the primary x-ray beam will lead to distortion. In the *bottom panel*, sagging portions of the vertebrae are elevated with nonradiopaque pads, which result in all vertebrae being aligned more parallel with the top of the x-ray table. This leads to a less distorted lateral projection of the vertebrae.

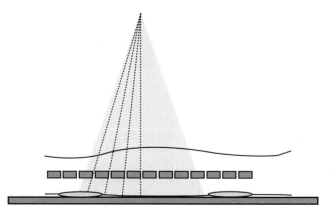

Figure 3-2. Diagram illustrating the effect of divergence of the x-ray beam on radiographic disc space width. The *gray shaded area* represents the diverging x-ray beam. The *dotted lines* represent x-ray photons that will strike four adjacent intervertebral disc spaces. The vertical photon will pass through the central disc space, and the image of the disc space will be representative of the actual size of the disc space. Farther peripherally, to the reader's left, the photon becomes more angled with reference to the disc space, which leads to an image of the disc space that is narrower than its actual width and not representative of the true size of the disc space.

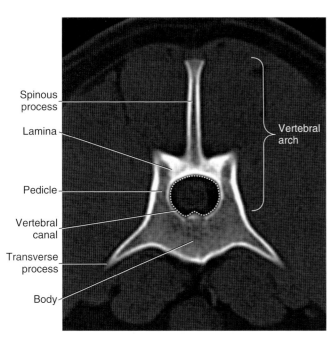

Figure 3-3. Transverse CT image of a lumbar vertebra. The core components common to all vertebrae are identified.

caudal articular processes. The articular processes are located on the vertebral arch at the junction of the pedicles and lamina. A cranial articular process joins with a caudal articular process from the vertebra located immediately cranially to form an articular process joint (Figure 3-4). The cartilage-covered surface of the articular process is the *facet*, or face; the term *facet* should not be used to encompass the entire articular process or to describe the joint. The articular process joints are *synovial joints*, with articular cartilage, a joint capsule, and synovial fluid.

From a lateral perspective, notches in the pedicles of adjoining vertebrae create a foramen, the *intervertebral foramen*, through which spinal nerves, arteries, and veins pass (Figure 3-4).

Except for C1 and C2, cervical, thoracic, and lumbar vertebrae have three primary centers of ossification: one for the *centrum*, or body, and one for each neural arch.[1] The centrum has a growth plate, or *physis*, and adjacent secondary centers of ossification, the *epiphyses*, on the cranial and caudal ends. Vertebral physes, being cartilaginous, appear radiographically as a more radiolucent region of soft tissue opacity between the body and the

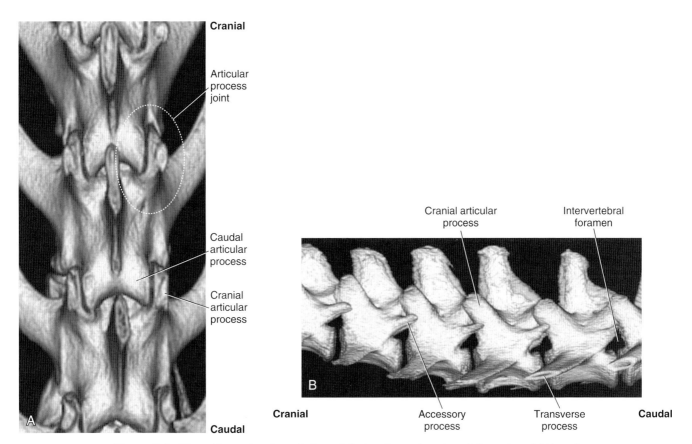

Figure 3-4. Three-dimensional rendering of the lumbar spine viewed from dorsal (**A**) and left (**B**) aspects. The intervertebral foramen (**B**) and the anatomic relationship of the articular processes and the joint formed by their articulation are noted. The cranial and caudal aspects of the spine are indicated in each part of this figure.

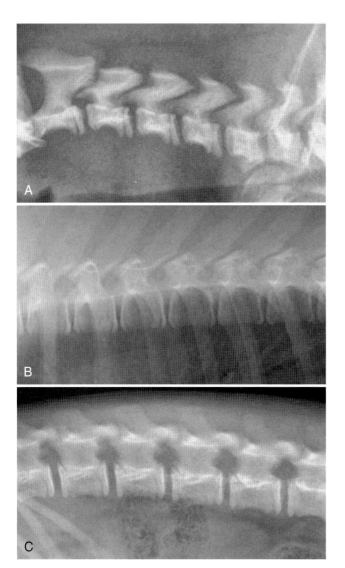

Figure 3-5. Lateral radiographs of the midportion of the cervical spine from a 4-month-old Maltese (**A**), the midportion of the thoracic spine from a 5-month-old Labrador Retriever (**B**), and the midportion of the lumbar spine from a 4-month-old Labrador Retriever (**C**). The vertebral physes are visible in each of these images. The junction of the epiphysis, physis, and vertebral body creates an irregular margin to the ventral aspect of the vertebral body. This ventral margin becomes more smooth following physeal closure.

epiphysis. When vertebral physes are open, the ventral aspect of the vertebra has an irregular, jagged margin (Figure 3-5).

CERVICAL SPINE

There are seven cervical vertebrae, and the variation in size and shape of individual cervical vertebrae is greater than in any other spinal region. The first cervical vertebra, the *atlas*, articulates with the occipital condyles at the atlanto-occipital articulation. There is no intervertebral disc between the occipital condyles and the atlas. The atlanto-occipital articulation provides for flexion and extension but not lateral or rotary movement; thus, this joint has been referred to as the *yes* joint. There are no typical cranial and caudal articular processes on the atlas; instead, there are articular fovea cranially and caudally to accommodate articulation with the occipital condyles and centrum 1 of the second cervical vertebra, respectively. The atlas also does not possess a spinous process, and the body is abbreviated compared with other cervical vertebrae.[2] The transverse processes of the atlas are large and are termed *wings*[3] (Figure 3-6). The atlas has two foramina on the left and right sides, the lateral vertebral foramina, which are located in the craniodorsal part of the vertebral arch and the transverse foramina, located in the wing (Figures 3-6 and 3-7). The vertebral artery and vein pass through the lateral and transverse vertebral foramina, and the first cervical spinal nerve exits through the lateral vertebral foramen.

The atlas arises from three centers of ossification, a pair of neural arches and intercentrum 1 that forms the body. The neural arches fuse dorsally shortly after 100 days, and the body fuses[4,5] (Figures 3-8 and 3-9) later, in the range of 110 to 120 days.[2] Intercentrum 1 is rarely seen radiographically as a separate structure due to its overlap with the ventral aspect of the atlas, but, occasionally, intercentrum 1 may appear as an isolated fragment, especially when the patient is positioned slightly obliquely (see Figure 3-9).

The second cervical vertebra, the *axis*, is the largest cervical vertebra. Its most conspicuous feature is the large spinous process, the cranial portion of which overlaps the lamina of the atlas (see Figure 3-6, *A*). Just as at the atlanto-occipital junction, there is no intervertebral disc between C1 and C2. Given the unique articulation between C1 and C2, there are no cranial articular processes on C2; however, caudal articular processes are present and they constitute part of the left and right articular process joints at C2-C3. The lack of cranial articular processes on C2 allows an unobstructed view of the intervertebral foramen at C1-C2 in lateral radiographs (see Figure 3-6, *A* and *A1*). As discussed later in this section, this is contrary to sites more caudal in the cervical spine where the articular processes are superimposed on intervertebral foramina in lateral radiographs.

Spinal nerves exiting intervertebral foramina are typically designated corresponding to the vertebra forming the cranial aspect of the foramen; for example, the third lumbar spinal nerve exits the foramen formed by notches in the adjacent pedicles of L3 and L4. However, the first cervical spinal nerve exits the lateral vertebral foramen, cranial to the intervertebral foramen between C1 and C2, and the second cervical spinal nerve exits through the intervertebral foramen between C1 and C2, and so on. Thus, although there are only seven cervical vertebrae,

Chapter 3 ■ The Spine   53

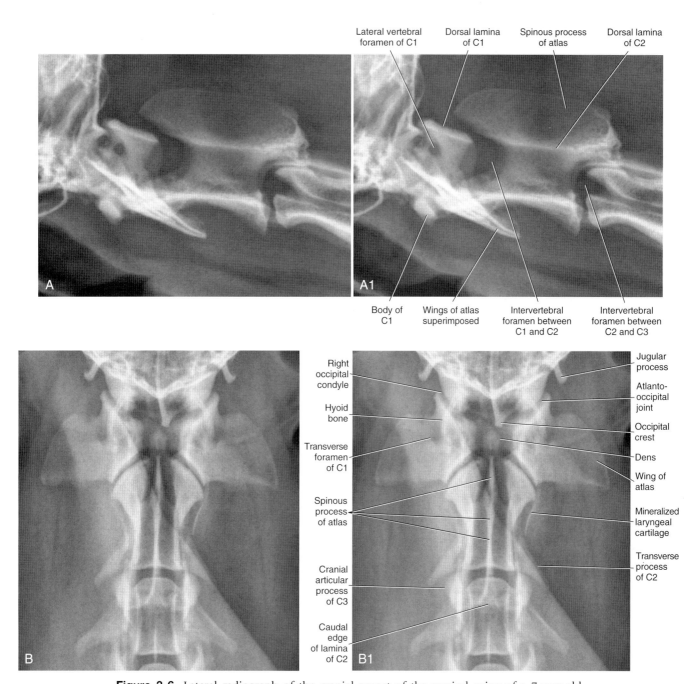

Figure 3-6. Lateral radiograph of the cranial aspect of the cervical spine of a 7-year-old Shetland Sheepdog (**A**), a ventrodorsal radiograph of a 6-year-old beagle (**B**), and corresponding labeled radiographs (**A1, B1**).

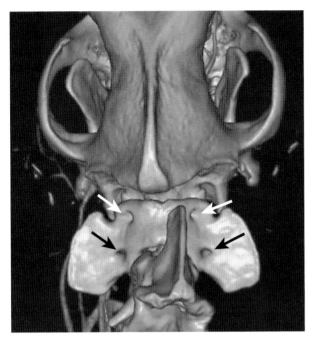

Figure 3-7. Three-dimensional rendering of the dorsal aspect of the atlantoaxial region of a 9-year-old Rottweiler. The lateral vertebral foramina *(white arrows)* and the transverse foramina *(black arrows)* are visible in C1. The dorsal spine of C2 is rotated to the right because the spine was not perfectly aligned for imaging. Note the large transverse processes, or wings, on C1.

can also usually be identified in ventrodorsal radiographs (see Figure 3-6, B and B1). If visualization of the dens from a lateral perspective is a priority, it can usually be seen in a lateral view with the skull intentionally rotated slightly dorsally or ventrally (Figure 3-10). The dens will usually be very conspicuous in this slightly rotated view as it is no longer superimposed on the wings of the atlas.

Normally, there is essentially no flexion possible at the atlantoaxial joint due to stabilization provided by the apical ligament of the dens, the transverse atlantal ligament, and the dorsal atlantoaxial ligament. The apical ligament of the dens has three pillars that extend cranially: the middle pillar runs to the basihyoid bone at the ventral aspect of the foramen magnum (see Figure 3-11), and the two lateral, more robust pillars attach to the occipital bone. The transverse atlantal ligament is strong and connects one side of the ventral arch of the atlas to the other by crossing dorsally over the dens, securing it against the body of the atlas.[6] Lateral or rotary movement is possible at the atlantoaxial joint. Lateral movement but no flexion or extension at the atlantoaxial joint has led to it being referred to as the *no* joint. Occasionally, due to either a congenital malformation or trauma, or both, there is instability at the atlantoaxial joint, allowing an abnormal range of flexion. A good method to evaluate whether the atlantoaxial joint is malaligned is to compare the orientation of the dorsal lamina of the atlas with respect to the dorsal lamina of the axis (Figure 3-12). Under normal circumstances, these laminae should align in a nearly linear fashion (see Figure 3-12). The amount of overlap of the spinous process of C2 with the dorsal lamina of C1 is not a good indicator of malalignment because this overlap can vary considerably between subjects. The space between the lamina of C1 and the spinous process of C2 also varies. However, the parallel, or linear, orientation of the lamina of C1 and the lamina of C2 is relatively constant and is a good landmark for assessing the normality of the C1-C2 alignment (Figure 3-13).

there are eight cervical spinal nerves. This is in contrast to the thoracic and lumbar spinal regions where the number of spinal nerves is equal to the number of vertebrae.

Another unique feature of the axis is the *dens,* an oblong protuberance that extends into the ventral aspect of the vertebral canal of C1 (Figure 3-10). The dens is difficult to identify in lateral radiographs due to its superimposition with portions of C1, but the dens can easily be identified in computed tomography (CT) images and

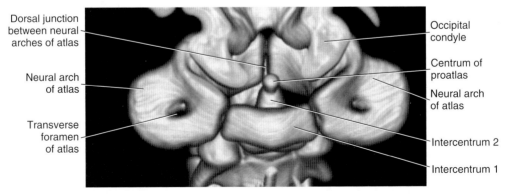

Figure 3-8. Three-dimensional rendering of the ventral aspect of the atlantoaxial region of an 11-week-old Akita. The osseous components of the immature atlantoaxial region are labeled.

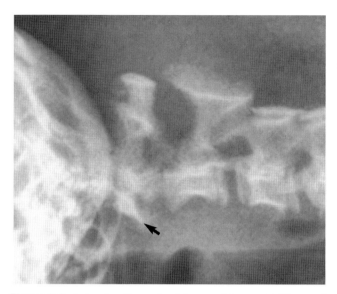

Figure 3-9. Lateral radiograph of the cranial aspect of the cervical spine of an 8-week-old Chihuahua. The body of C1, termed *intercentrum 1*, has not yet fused with the vertebral arch and appears as a bone opacity at the ventral aspect of the vertebra (*black arrow*). The conspicuity of intercentrum 1 of the atlas in this dog is enhanced by the slight obliquity present; note the tympanic bullae are not superimposed.

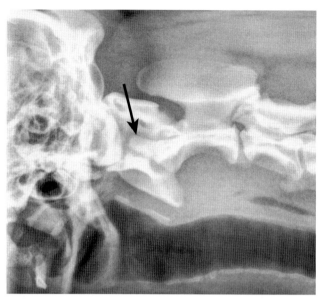

Figure 3-10. Slightly rotated lateral view of the cranial aspect of the cervical spine in a 2-year-old Dachshund. Note the tympanic bullae are not superimposed due to the obliquity, but the dens (*black arrow*) is conspicuous because it is no longer superimposed on the wings of the atlas.

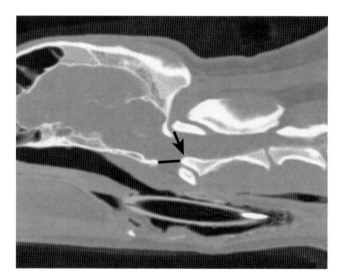

Figure 3-11. Sagittally reformatted CT image of the atlantoaxial region of a 7-year-old German Shepherd. The dens (*black arrow*) extends from the cranioventral aspect of C2 into the ventral aspect of the vertebral canal over C1. The structure ventral to the dens is the body of C1. The *black line* indicates the position of the apical ligament of the dens, connecting the dens to the basihyoid bone and contributing to ligamentous stabilization of the atlantoaxial joint.

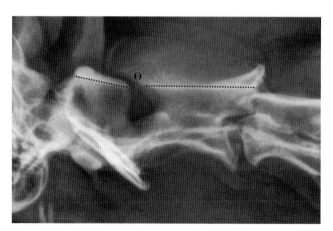

Figure 3-12. Lateral radiograph of the cranial aspect of the cervical spine of a 7-year-old Beagle. The dorsal laminae of C1 and C2 have been outlined with a *dotted line*. The relationship of the laminae of these vertebrae should be parallel or only slightly angled, as seen here. The normal slight angular relationship, designated here as *angle theta* (θ), may be either slightly acute or slightly obtuse at the intersection; in this dog, *angle* θ is slightly acute.

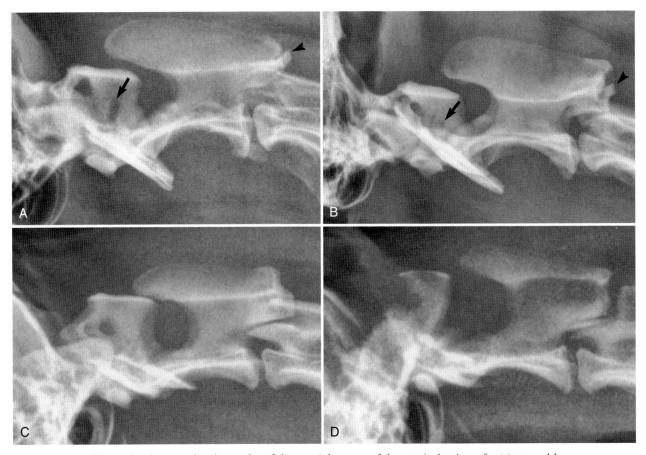

Figure 3-13. Lateral radiographs of the cranial aspect of the cervical spine of a 12-year-old Dachshund (**A**), a 7-year-old mixed breed dog (**B**), a 7-year-old Chihuahua (**C**), and a 3-year-old Toy Poodle (**D**). The amount of overlap of the spinous process of the axis with the dorsal lamina of the atlas and the space between these structures varies between dogs, but the linear relationship of the dorsal aspect of the lamina of the atlas to the dorsal lamina of the axis is constant. This linear relationship is a good landmark to use to assess the alignment of the atlantoaxial joint. The lateral vertebral foramen is visible in each dog. In **D**, the cranial aspect of this foramen is incomplete. The transverse foramen can be seen in **A** and **B** *(black arrows)* but is less conspicuous in **C** and **D**. In **A** and **B**, note the irregular margin present on the most dorsocaudal aspect of the vertebral arch of C2 *(black arrowhead)*. This normal appearance can be confused with new bone formation resulting from trauma or degenerative disease.

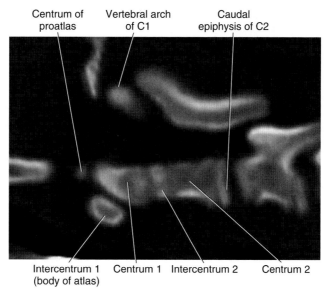

Figure 3-14. Sagittally reconstructed CT image of the cranial aspect of the cervical spine of a 4-month-old Akita. Some of the bony elements of the axis are labeled in this image.

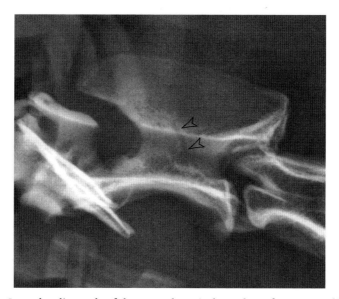

Figure 3-15. Lateral radiograph of the second cervical vertebra of a 10-year-old Dalmatian. There is a large vascular channel, appearing as a curving radiolucent region, in the vertebral arch of C2 *(hollow black arrowheads)*. This should not be confused with a fracture or a destructive process.

The axis, or C2, arises from seven centers of ossification: two vertebral arches, centrum 2, the caudal epiphysis, intercentrum 2, centrum 1, and the centrum of the proatlas (Figure 3-14). Centrum 1 and the centrum of the proatlas form the dens; the centrum of the proatlas is embryologically part of the atlas[4,5] (see Figure 3-8). The centrum of the proatlas fuses with centrum 1 at approximately 100 to 110 days, intercentrum 2 fuses with centrum 1 and centrum 2 in the range of 115 to 150 days, and the caudal epiphysis fuses with the body of C2 in the range of 220 to 400 days.[2]

Occasionally, a conspicuous linear radiolucent region will be visible in the vertebral arch of C2 in mature dogs; this is due to a vascular channel and should not be confused with a fracture or a destructive process (Figure 3-15). In addition, the trabecular pattern in parts of the

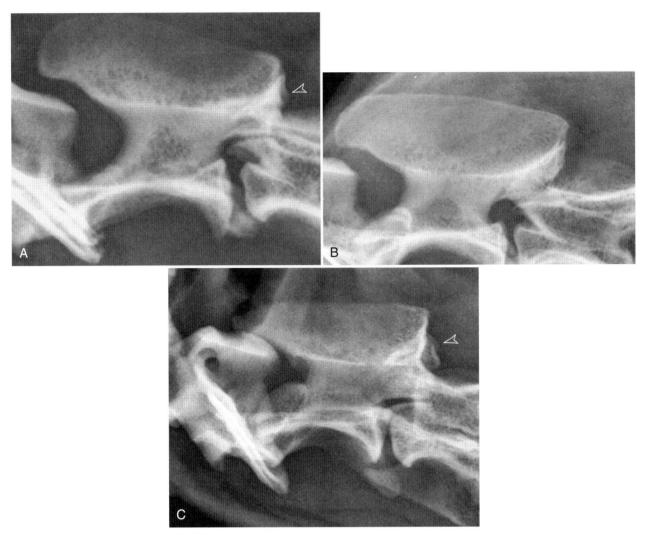

Figure 3-16. Lateral radiographs of the second cervical vertebra of a 15-year-old mixed breed dog (**A**), an 8-year-old Basset Hound (**B**), and a 10-year-old Boston Terrier (**C**). In each dog, portions of the vertebral arch of the axis are characterized by a coarse trabecular pattern giving the appearance of multiple regions of bone lysis. This is a normal appearance that should not be confused with disease causing bone effacement. The exact association between patient age and the appearance of the trabecular bone pattern in C2 has not been characterized. The caudal aspect of the vertebral arch of C2 that forms the dorsal aspect of the articular process joint at C2-C3 is often irregular in appearance; this normal appearance could be confused with hyperostosis from degenerative disease or trauma (*arrowheads* in **A** and **C**).

vertebral arch of C2 is often very heterogeneous, with coarse round radiolucent regions; this appearance should also not be confused with a destructive process, such as lysis due to a reticuloendothelial tumor (Figure 3-16). The dogs in Figure 3-16 are all elderly and, although not documented, this coarse trabecular pattern could be due to a degree of geriatric osteopenia.

The caudal aspect of the vertebral arch that forms the dorsal aspect of the articular process joint between C2 and C3 is often irregular. This normal appearance could be misinterpreted as new bone formation secondary to trauma or degenerative processes (see Figures 3-13, *A-B*, and 3-16, *A* and *C*).

The third, fourth, and fifth cervical vertebrae are similar to each other and generally devoid of distinguishing features (Figure 3-17). As with the atlas and the axis, each of the third, fourth, and fifth cervical vertebrae has a transverse foramen through each pedicle, through which the

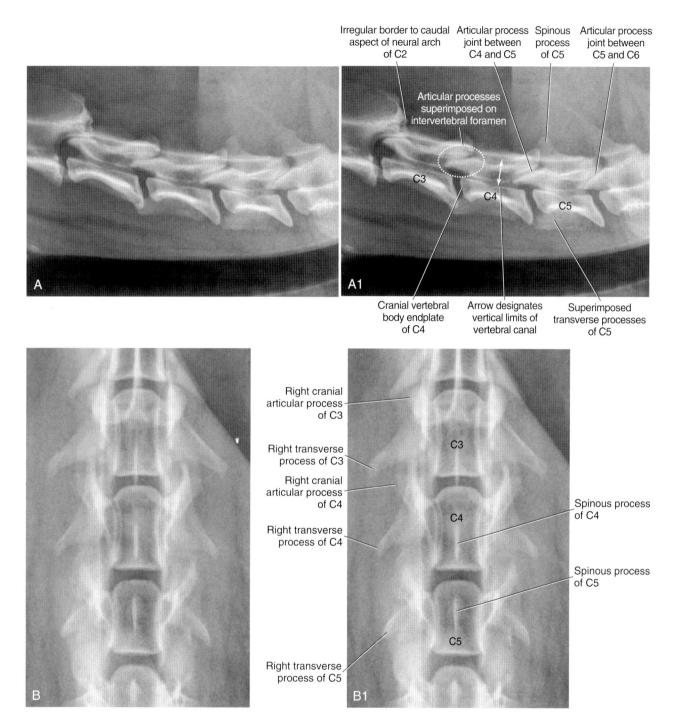

Figure 3-17. Lateral radiograph of the midportion of the cervical spine of a 7-year-old Shetland Sheepdog (A), ventrodorsal radiograph of the midportion of the cervical spine of a 6-year-old Beagle (B), and corresponding labeled radiographs (A1, B1).

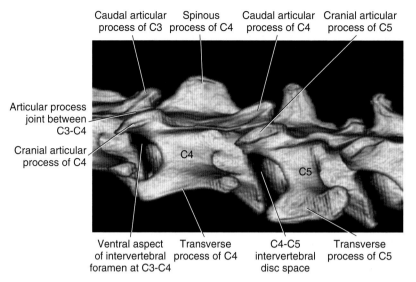

Figure 3-18. Three-dimensional rendering of the mid-cervical region of an 8-year-old Greyhound. Note the relatively large cranial articular process that obscures radiographic visualization of the intervertebral foramina in lateral radiographs of the cervical spine compared with thoracic and lumbar regions, where the articular processes are smaller.

vertebral artery passes. The cranial articular processes of C3, C4, and C5 are relatively large compared with articular processes in the thoracic and lumbar region. Thus, in the cervical spine, these large cranial articular processes result in the articular process joint being superimposed on the intervertebral foramen in lateral radiographs (see Figures 3-17 and 3-18), and an unobstructed view of the intervertebral foramen at C3-C4 through C6-C7 is not possible. The cranial articular processes of C3 are smaller than at C4 through C7, allowing for less superimposition of the articular process joint on the intervertebral foramen at C2-C3 (see Figure 3-6, *A* and *A1*).

Embryologic failure of separation of vertebrae, or *block vertebra*, is a common developmental anomaly in the cranial and mid-cervical regions. A block vertebra result from disturbance in segmentation of vertebral somites. The intervertebral disc between the fused vertebrae is incompletely developed, although a residual vertical radiolucent disc space is often present. Therefore the width of the disc space encompassed by the block vertebra is decreased[7] (Figure 3-19). The block vertebra usually affect two adjacent vertebrae only, but involvement of three adjacent vertebrae can also occur (see Figure 3-19, *E-F*). The altered biomechanics associated with a block vertebra can predispose to degenerative vertebral and disc disease at immediately adjacent disc interspaces (see Figure 3-19), but the block vertebra itself is not associated with abnormal signs.

Figure 3-19. Lateral (**A**) and ventrodorsal (**B**) radiographs of a 5-year-old Dachshund with a block vertebra at C2-C3. In situ, disc mineralization is present at C3-C4. Lateral (**C**) radiograph of a 10-year-old Labrador Retriever with a block vertebra at C3-C4. There is spondylosis at C2-C3 and C4-C5. Lateral (**D**) radiograph of a 5-year-old Australian Shepherd with a block vertebra at C4-C5. Lateral (**E**) and ventrodorsal (**F**) radiographs of a 12-year-old Bichon Frise with a block vertebra at C2-C3-C4. In **E**, the focal radiolucency in the disc at C4-C5 is gas from the vacuum phenomenon created by abnormal traction associated with the block vertebra.

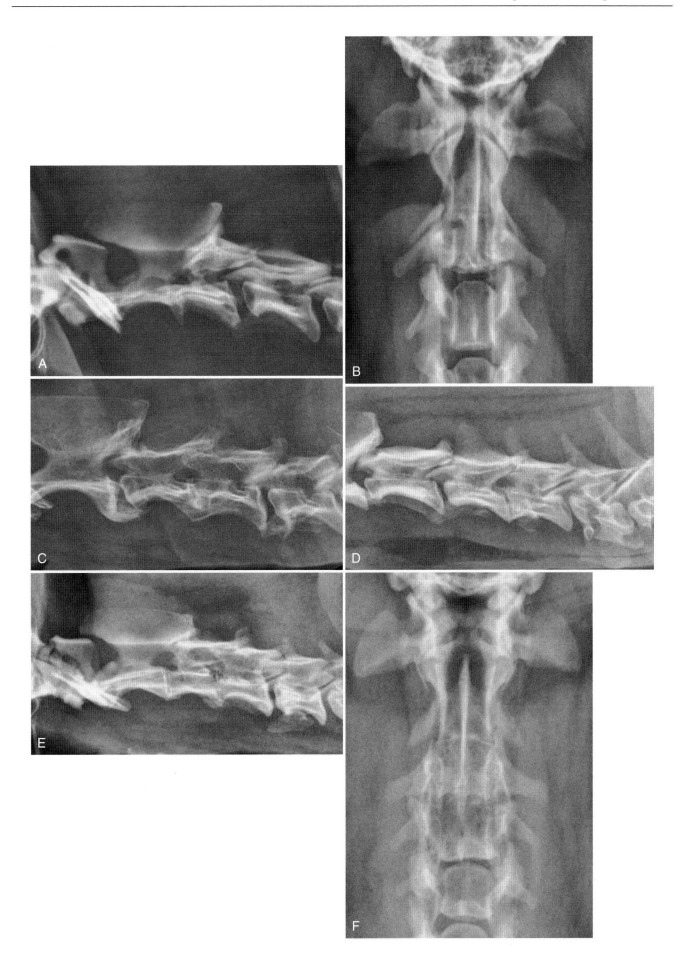

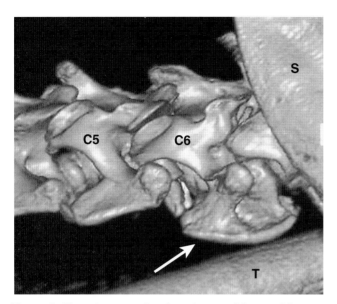

Figure 3-20. Volume rendered CT image of the caudal aspect of the cervical spine of a 6-year-old Doberman Pinscher. Note the relatively large transverse processes on C6 *(white arrow)* compared to C5. S, spine of scapula superimposed on caudal aspect of cervical spine; T, trachea.

Compared to C3, C4, and C5, the sixth cervical vertebra has a slightly larger spinous process and significantly larger transverse processes (Figures 3-20 and 3-21). The large transverse processes are excellent landmarks for identification of C6 if a limited field of view is present in the radiographic image (Figures 3-22 and 3-23). C6 also has a transverse foramen through each pedicle, through which the vertebral artery passes.

The seventh cervical vertebra is similar in appearance to the third through the fifth cervical vertebrae, except that the spinous process is larger (see Figure 3-22). There is no transverse foramen in the seventh cervical vertebra because the vertebral artery is ventral to the vertebral body at this level. The intervertebral foramina at C5-C6 and C6-C7 are obscured in lateral radiographs by their large articular processes (see Figure 3-23).

Although the transverse foramina in the caudal aspect of the cervical spine are relatively large, they cannot be seen as distinct entities radiographically because they are surrounded by the dense bone of the body/pedicle region (Figure 3-24). However, the reduced amount of bone created by these large foramina creates focal, sometimes

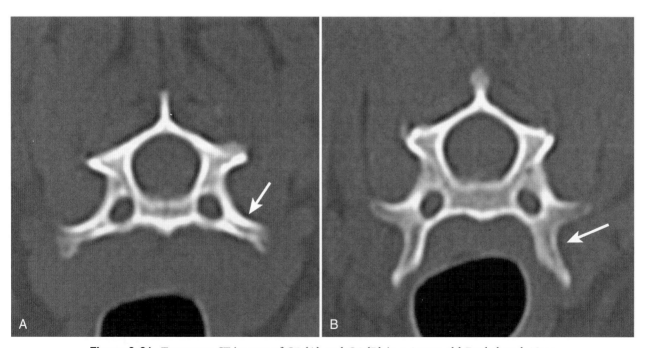

Figure 3-21. Transverse CT images of C5 **(A)** and C6 **(B)** in a 9-year-old Dachshund. Note the larger, and ventrally projecting, transverse processes on C6 *(arrow* in **B)** compared to those on C5 *(arrow* in **A)**.

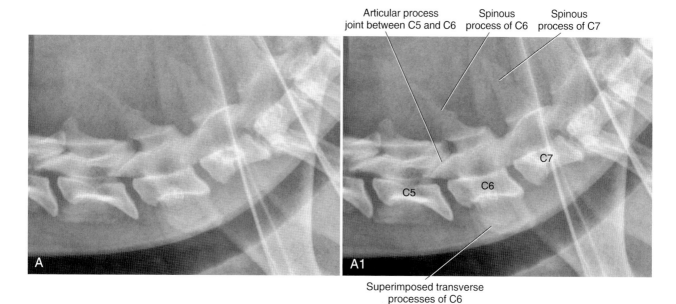

Figure 3-22. Lateral radiograph of the caudal aspect of the cervical spine of a 7-year-old Shetland Sheepdog **(A)**, and corresponding labeled radiograph **(A1)**.

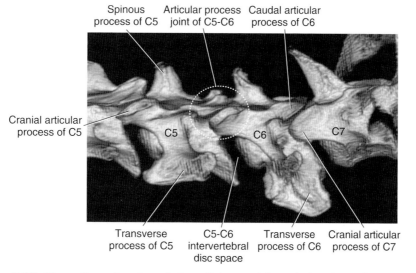

Figure 3-23. Three-dimensional rendering of the caudal cervical region of an 8-year-old Greyhound. Note the large transverse processes on C6; these constitute a good landmark for identification of C6 in cervical radiographs.

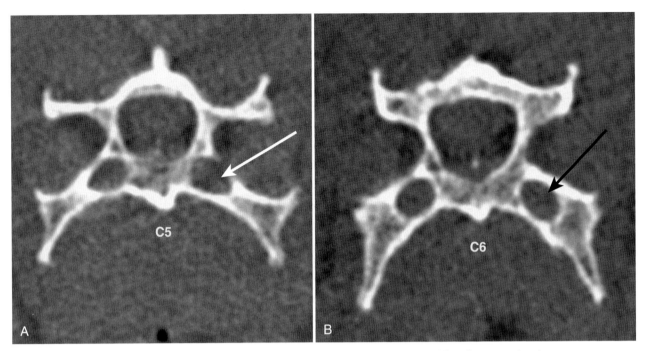

Figure 3-24. CT images through C5 **(A)** and C6 **(B)** in an 8-year-old Doberman Pinscher. Note the large transverse foramen *(arrows)* that is present in these vertebrae. In a lateral radiograph of this region, the reduced amount of bone due to the presence of the foramen will create a radiolucency that can be confused with effacement from an aggressive bone lesion.

relatively well-defined, radiolucent regions that can be confused with bone loss due to an aggressive process (Figure 3-25).

Congenital *transitional anomalies* are common in the cervicothoracic, thoracolumbar, and lumbosacral junctional regions. These transitional anomalies are due to the vertebra at the junction having features of each adjoining region. The most common transitional anomaly at the cervicothoracic junction is the presence of ribs on C7. Cervical ribs are likely due to anomalous transverse processes and may be small isolated structures (Figure 3-26) or more developed structures that articulate with the first thoracic ribs, which are also usually malformed in this instance (Figure 3-27). Cervical ribs have no clinical significance.

The cervical spine in the cat has no unique anatomic features that differ from the dog. In general, in comparison with the dog, feline cervical vertebrae are more rectangular, the trabecular bone pattern is coarser, and the transverse processes of C6 are angled more laterally rather than projecting ventrally as in the dog (Figures 3-28 and 3-29). In addition, the transverse processes of C4, C5, and C6 in the cat are typically less distinct than in the dog and in lateral radiographs may be misinterpreted as new bone formation from the ventral aspect of these vertebrae (see Figure 3-29, *A*).

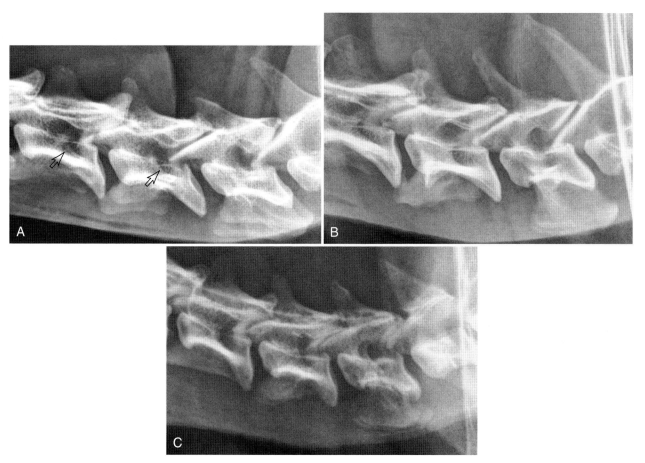

Figure 3-25. Lateral radiographs of the caudal aspect of the cervical spine of an 8-year-old mixed breed dog **(A)**, a 14-year-old mixed breed dog **(B)**, and a 15-year-old mixed breed dog **(C)**. In each radiograph, there are well-defined radiolucent regions in the dorsal aspect of the vertebral bodies of C4, C5, and C6 that extend dorsally into the pedicles *(hollow black arrows in A)*. These are created by the transverse foramina and the irregular shape of the transverse processes of these vertebrae and should not be confused with lysis due to an aggressive process.

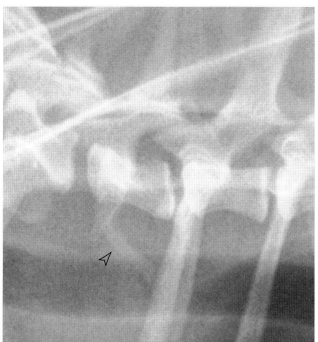

Figure 3-26. Lateral radiograph of the cervicothoracic junction of a 9-year-old Shih Tzu. There are anomalous ribs on the seventh cervical vertebra *(hollow black arrowhead)*.

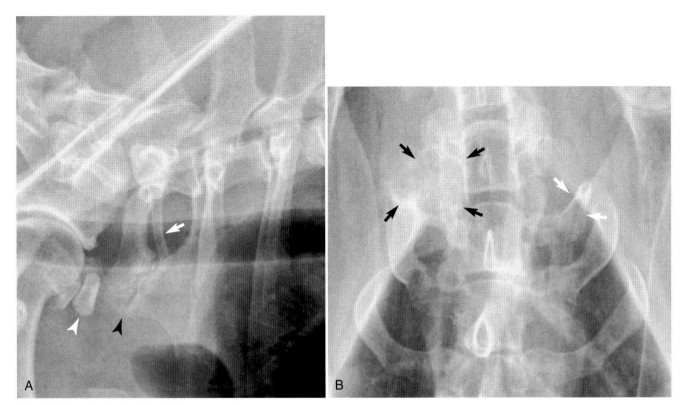

Figure 3-27. Lateral (**A**) and ventrodorsal (**B**) radiographs of the cervicothoracic junction of a 14-year-old Golden Retriever. There is a small cervical rib on the left aspect of C7 *(white arrows)* and a large cervical rib on the right aspect of C7. The large cervical rib on the right is articulating with an anomalous right first thoracic rib *(black arrowhead* in **A**), resulting in a large amorphous osseous structure *(black arrows* in **B**). There are multiple mineralized fragments in the shoulder joint *(white arrowhead* in **A**); these are not related to the cervical rib anomaly.

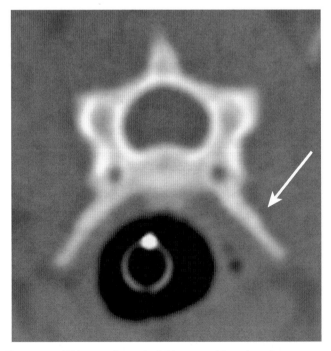

Figure 3-28. Transverse CT image through C6 in a cat. Note that the transverse processes of C6 in the cat are angled more laterally and do not project as far ventrally compared to the dog. This renders them slightly less conspicuous on the cat versus the dog. Compare to Figure 3-21, *B*.

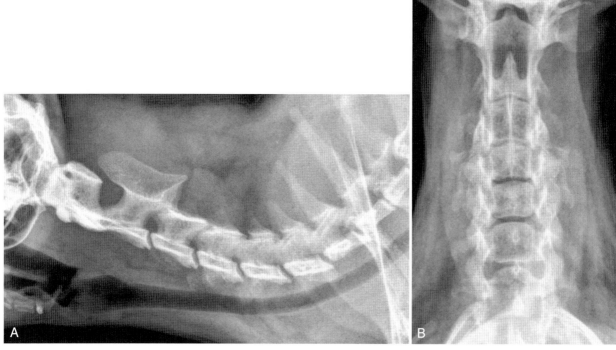

Figure 3-29. A, Lateral radiograph of the cervical spine of a 1-year-old domestic cat. B, Ventrodorsal radiograph of a 7-year-old domestic cat. In general, feline vertebral bodies are more rectangular than in the dog, and the trabecular pattern is more coarse. The transverse processes of C6 are less conspicuous in the cat than in the dog. Transverse processes of C4, C5, and C6 are also less distinct than in the dog and, in the lateral view, can be misinterpreted as new bone formation from the ventral aspect of these vertebrae.

THORACIC SPINE

There are 13 vertebrae in the normal thoracic spine. The features of the prototypical vertebra described previously are found in the thoracic spine (Figure 3-30). Paired ribs articulate with the cranial aspect of each thoracic vertebra. The heads of the first pair of ribs are more lobular than found on other ribs (Figure 3-31). These lobulated rib heads are very conspicuous radiographically and can be misinterpreted as new bone formation as the result of a pathologic process (Figure 3-32).

The height of the spinous processes in the thoracic spine decreases gradually from T1 caudally. Proceeding caudally, the spinous processes also angle caudally until they reach a point where the spinous process is essentially vertical, or perpendicular, to the vertebral body. The vertebra on which the spinous process is vertical is termed the *anticlinal vertebra* (Figure 3-33). Other definitions for the anticlinal vertebra have been proposed,[8] but this is the most accepted definition. The importance of the anticlinal vertebra is that it serves as an anatomic landmark.

T11 is the anticlinal vertebra in most dogs.[8] Breed can influence which vertebra has the features of the anticlinal vertebra, with T11 being the anticlinal vertebra most often in large dogs and T10 being the anticlinal vertebra most often in small dogs[8] (see Figure 3-33).

From the midthoracic region caudally, accessory processes arise from the caudal border of the pedicles (see Figure 3-4, *B*). These are typically not visible in the thoracic spine due to superimposition from the ribs.

Transitional anomalies are common at the thoracolumbar junction. These typically involve anomalous ribs on T13. Ribs can be missing, hypoplastic, or malformed, appearing as anomalous transverse processes (Figure 3-34). The clinically important feature of thoracolumbar transitional anomalies relates to the use of the most caudal pair of ribs as anatomic landmarks to plan the site of an invasive procedure, such as a needle aspiration or spinal decompressive surgery. Failure to recognize a unilateral asymmetry in the development of the last pair of ribs could result in the procedure being performed at the incorrect site.

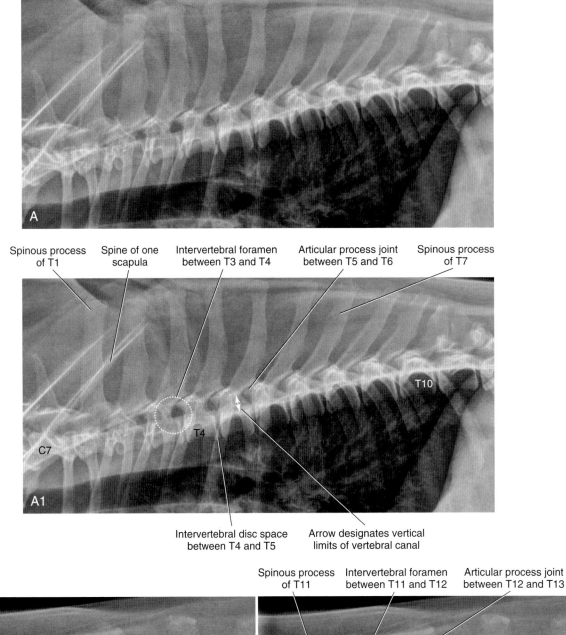
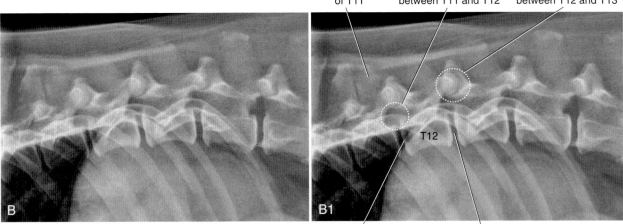

Figure 3-30. Lateral (**A, B**) and ventrodorsal (**C, D**) radiographs, and corresponding labeled radiographs (**A1-D1**), of the thoracic spine of an 11-year-old Golden Retriever.

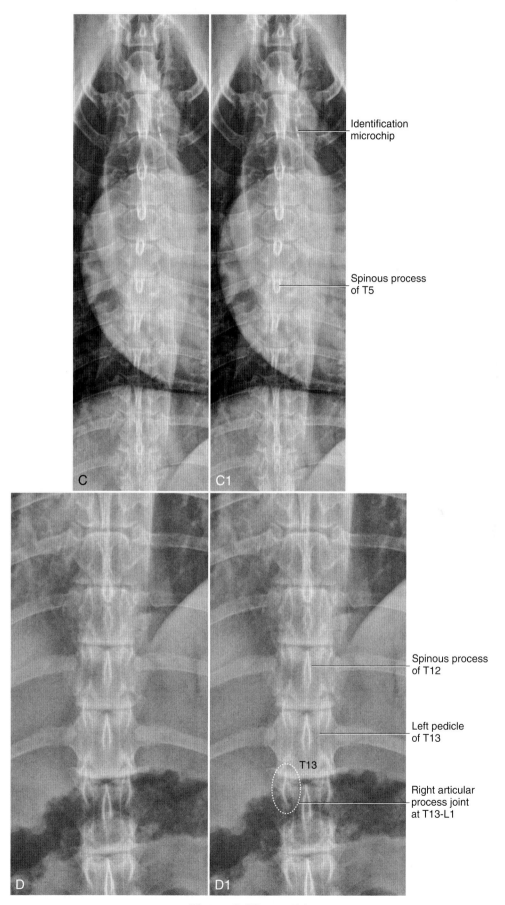

Figure 3-30, cont'd.

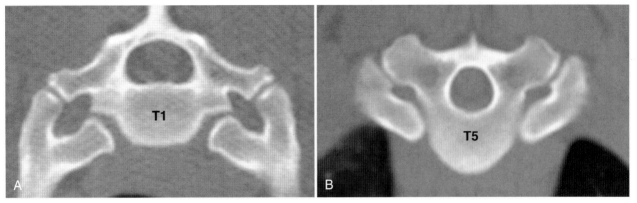

Figure 3-31. Transverse CT images through T1 (A) and T5 (B) of a 7-year-old German Shepherd. Note the difference in morphology of the head of rib 1 versus rib 5. The lobular nature of the head of rib 1 can be very conspicuous radiographically and misinterpreted as an expansile bone lesion.

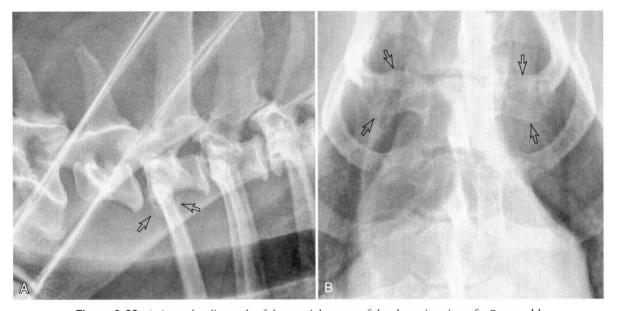

Figure 3-32. A, Lateral radiograph of the cranial aspect of the thoracic spine of a 7-year-old Rottweiler. B, Ventrodorsal radiograph of the cranial aspect of the thoracic spine of a 10-year-old mixed breed dog. In A, the large, lobular heads on the first rib extend ventral to the spine (*hollow black arrows*) creating an opacity that could be misinterpreted as abnormal new bone formation. Notwithstanding the bilateral symmetry, the lobular heads in B on the first rib (*hollow black arrows*) could be misinterpreted as an expansile lesion, such as from a low-grade infection or a benign tumor.

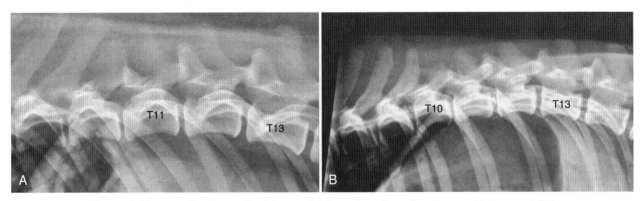

Figure 3-33. Lateral radiograph of the caudal thoracic region of an 8-year-old Samoyed (A) and a 7-year-old Pekingese (B). In the larger dog (A), the anticlinal vertebra is T11; in the smaller dog (B), it is T10.

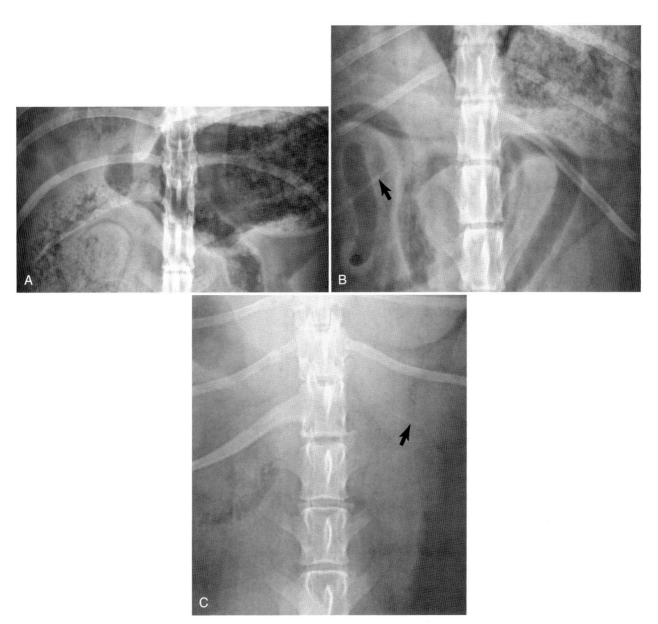

Figure 3-34. **A,** Ventrodorsal radiograph of the thoracolumbar junction of an 8-year-old Pug. The left thirteenth rib is absent and the right thirteenth rib is hypoplastic. **B,** Ventrodorsal radiograph of the thoracolumbar junction of a 3-year-old American Cocker Spaniel. The right thirteenth rib is hypoplastic *(solid black arrow)* and the left thirteenth rib has developed as an anomalous transverse process. **C,** Ventrodorsal radiograph of the thoracolumbar junction of a 7-year-old Maltese. The left thirteenth rib is hypoplastic *(solid black arrow)*, and the right thirteenth rib has developed as an anomalous transverse process.

Hemivertebra is another common congenital vertebral anomaly. A hemivertebra is a wedge-shaped vertebra that results from a developmental error. When the hemivertebra malformation is pronounced, there may be clinical signs of thoracic spinal cord compression. In many dogs, however, hemivertebrae are an incidental finding. Hemivertebrae are commonly encountered in the thoracic spine in the Bulldog, French Bulldog, Boston Terrier, and Pug. Hemivertebrae are also commonly found in the caudal spine in these breeds. A hemivertebra is typically narrower ventrally than dorsally, as seen in lateral radiographs, and can be butterfly shaped, as seen in ventrodorsal views (Figure 3-35).

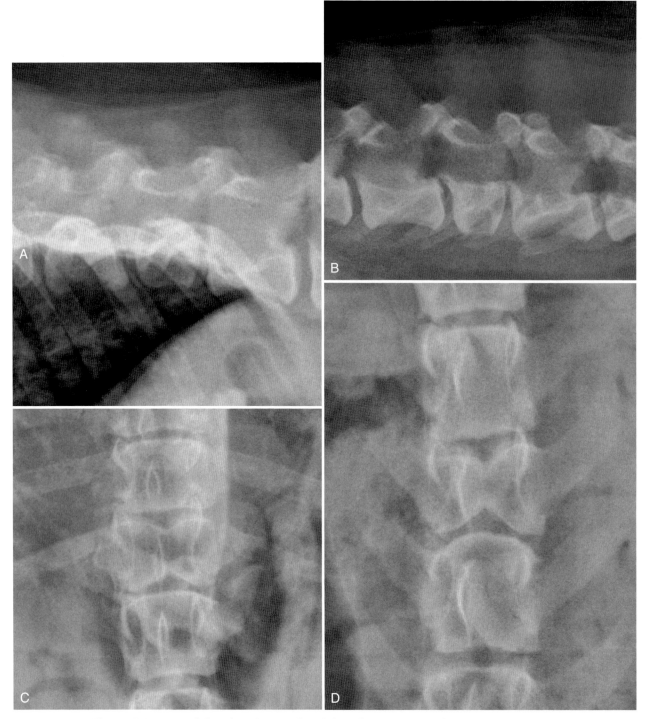

Figure 3-35. Lateral (**A, B**) and ventrodorsal (**C, D**) radiographs of a 3-year-old Bulldog. There are hemivertebrae at T13 (**A, C**) and L4 (**B, D**). The shape of a hemivertebra in lateral radiographs is more apparent at L4 due to the lack of superimposed ribs. Hemivertebrae are typically more narrow ventrally than dorsally and, in ventrodorsal views, can assume a butterfly appearance, as seen here (**C**).

Chapter 3 ■ The Spine 73

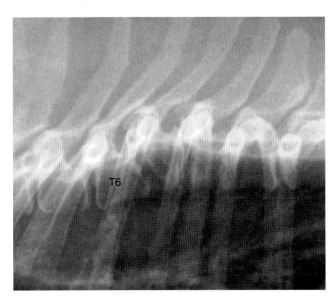

Figure 3-36. Lateral view of a 3-year-old Boston Terrier. There is a hemivertebra at T7. The malformed vertebra has resulted in a dorsal deviation of the thoracic spine at this location, and there is secondary remodeling of the caudal aspect of T6.

The most common locations for hemivertebrae in the thoracic spine are T7, T8, and T12[9] (Figure 3-36). In this location, there is often a dorsal deviation of the spine, termed *kyphosis*, due to the altered vertebral morphology. Remodeling of normally shaped adjacent vertebrae can also occur due to altered biomechanics created by the hemivertebrae (see Figure 3-36). When multiple adjoining hemivertebrae are present in the thoracic spine, the thoracic ribs are usually positioned more closely than normal, which creates a crowded appearance (Figure 3-37).

The thoracic spine in the cat has no unique anatomic features that were not described for the dog. In general, in comparison with the dog, the ribs do not extend as far dorsally in the cat, which allows for a less obstructed view

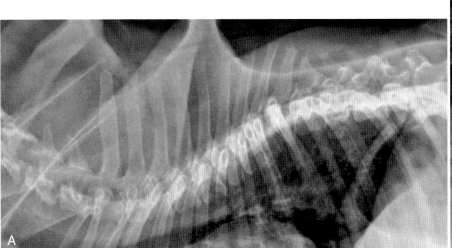
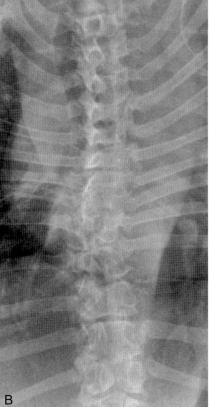

Figure 3-37. Lateral (**A**) and ventrodorsal (**B**) radiographs of the thoracic spine of a 9-year-old Bulldog. There are multiple hemivertebrae in the thoracic spine. As a result of the decreased length of the affected vertebrae, the proximal aspects of the adjoining ribs are more closely positioned than normal.

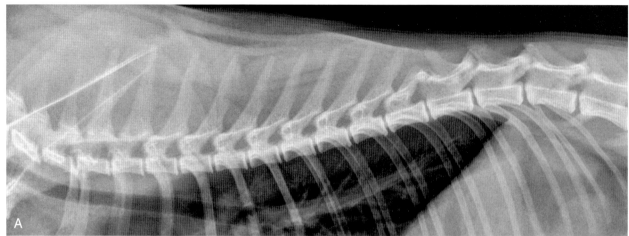

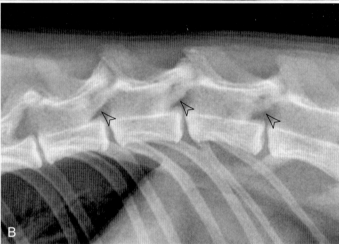

Figure 3-38. Lateral radiograph of the thoracic spine of a 1-year-old domestic cat (**A**), and a close-up view of the caudal thoracic region (**B**). The ribs do not extend as far dorsally in the cat as in the dog, allowing for a less obstructed view of the intervertebral foramina in the cat. This also provides for visualization of the accessory process of the thoracic vertebrae (*hollow black arrowheads* in **B**), which are rarely seen in the dog.

of the intervertebral foramina (Figure 3-38). This also provides for easier visualization of the accessory processes of the caudal thoracic vertebrae (see *open black arrowheads* in Figure 3-38, *B*).

LUMBAR SPINE

There are seven vertebrae in the lumbar spine. The features of the prototypical vertebra described previously are also found in the lumbar spine (Figure 3-39). Accessory processes of the cranial few lumbar vertebrae are well developed and can often be identified in lateral radiographs (see Figures 3-4 and 3-39, *B1*). The ventral cortex of L3 and L4 can appear less distinct than the ventral cortex of other lumbar vertebrae, particularly in large dogs, and be misinterpreted as an aggressive vertebral lesion (see Figure 3-39, *A-B*). Intervertebral foramina in the lumbar spine are clearly seen in lateral radiographs (see Figure 3-39, *A*).

Occasionally, there will appear to be eight lumbar vertebrae, but this may be due to the absence of ribs on T13. To actually have more than 20 vertebrae composing the thoracolumbar region, typically 13 thoracic and 7 lumbar, is very uncommon[3] (Figure 3-40). To ascertain whether there is a supernumerary lumbar vertebra, it is necessary to be able to view the entire thoracic spine to eliminate the possibility that both the left and right thirteenth ribs are absent. There is no clinical significance to a supernumerary lumbar vertebra.

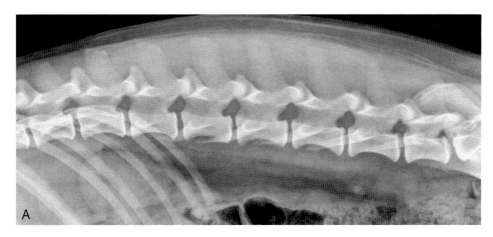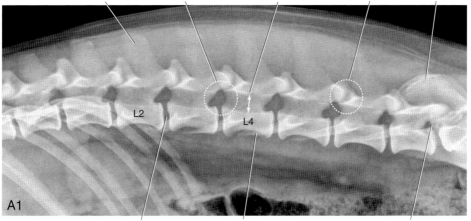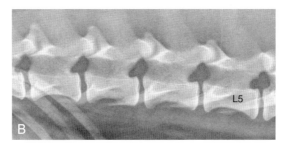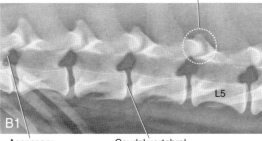

Figure 3-39. Lateral (**A, B**) and ventrodorsal (**C**) radiographs of the lumbar spine of a 10-year-old Rottweiler and corresponding labeled radiographs (**A1-C2**). In **A** and **B**, note the indistinct ventral cortex on the vertebral body of L3 and L4, compared with L2 and L5. This normal appearance should not be confused with an aggressive destructive process.

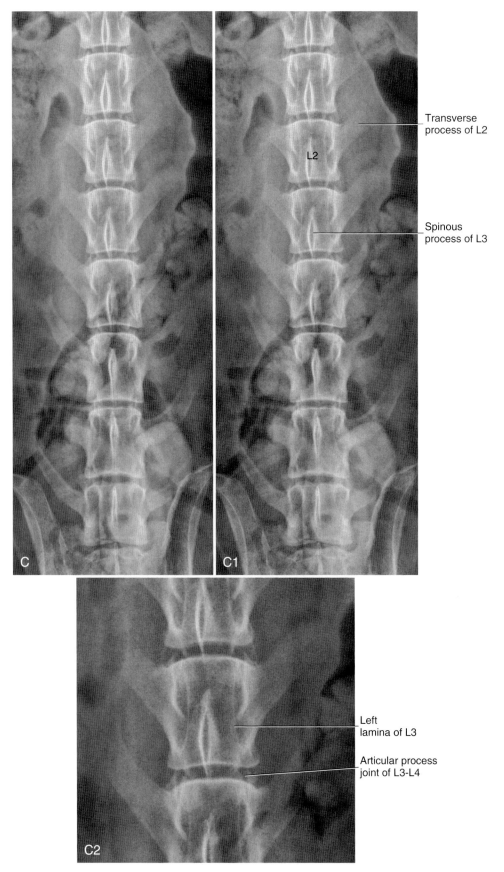

Figure 3-39, cont'd.

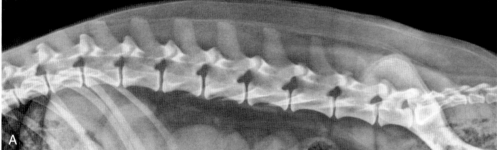
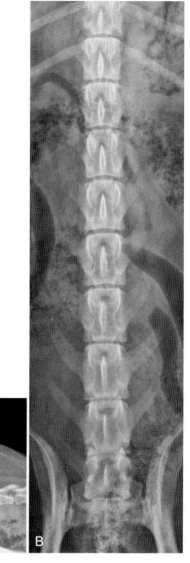

Figure 3-40. Lateral (**A**) and ventrodorsal (**B**) radiographs of the lumbar spine of a 3-year-old Nova Scotia Duck Tolling Retriever. There appear to be eight lumbar vertebrae.

A small amount of flexion and extension is normal at the lumbosacral junction. Rarely is this assessed radiographically; however, occasionally views made with the pelvic limbs pulled cranially or caudally are acquired to attempt to assess the range of motion present at the lumbosacral junction in dogs suspected of having lumbosacral instability. The range of motion shown in Figure 3-41 is normal. Strict guidelines for the amount of normal range to be expected have not been defined.

The lumbosacral junction is a common location for transitional anomalies. Most of these involve the last lumbar vertebra assuming some characteristics of the sacrum, termed *sacralization*. With sacralization of L7, the most common malformation is for one side of L7 to form a sacroiliac junction and the other side to have a transverse process (Figures 3-42 and 3-43). With sacralization, the angle between L7 and the sacrum in the lateral view is typically straighter than normal; compare the relatively straight lumbosacral angle in a dog with a transitional anomaly (see Figure 3-43, *A*) with the lumbosacral angle in a normal dog (see Figure 3-41, *A*).

Another manifestation of sacralization of L7 is a more caudal position of the apparent lumbosacral articulation in lateral radiographs (Figure 3-44). Because of L7 articulating with the ilium, the sacral articulation with the ilium is displaced to a more caudal position. As noted previously, the asymmetry of the articulation of the spine with the pelvis can also create asymmetry in the development of the pelvic bones (see Figure 3-44).

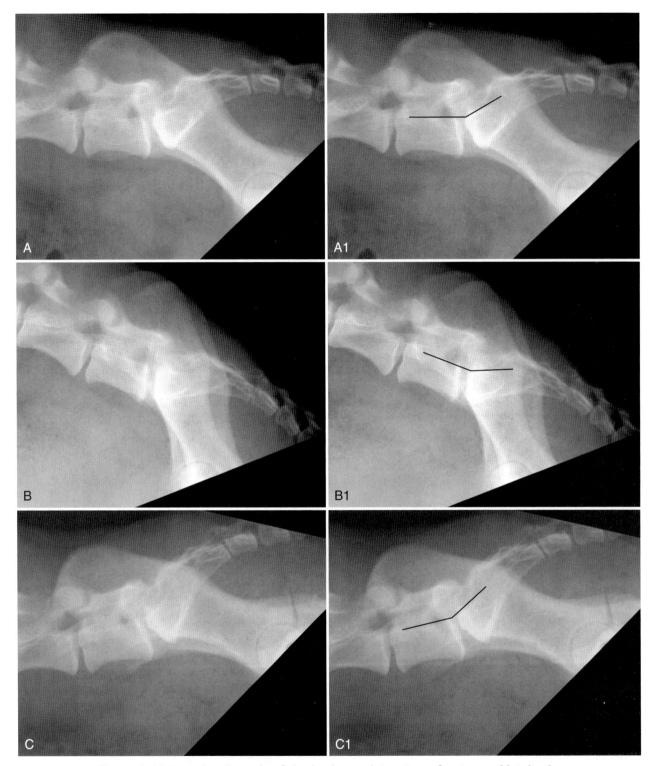

Figure 3-41. Lateral radiograph of the lumbosacral junction of a 2-year-old Labrador Retriever with the pelvic limbs in a neutral position (**A**), with the pelvic limbs pulled cranially to flex the lumbosacral junction (**B**), and with the pelvic limbs pulled caudally to extend the lumbosacral junction (**C**). Parts **A1**, **B1**, and **C1** are corresponding images to **A**, **B**, and **C**, respectively, where a *black line* has been added along the floor of the lumbosacral vertebral canal to enhance visualization of the angle formed between L7 and S1. Some range of motion at the lumbosacral joint is normal, but the exact limits for normal have not been quantified.

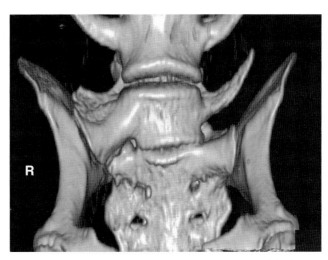

Figure 3-42. Three-dimensional volume rendered ventral view of the lumbosacral junction of a dog with a transitional anomaly of L7. On the *right (R)*, L7 is fused to the ilium but on the *left* there is a normal transverse process. This asymmetry results in altered biomechanical loading of the lumbosacral junction and can lead to clinically significant degenerative changes at L7-S1.

The pelvic asymmetry that is associated with sacralization of L7 can also lead to difficulty in positioning the pelvis for an extended hip ventrodorsal view. The altered alignment of the left versus the right side of the sacroiliac articulation can make it impossible to obtain a symmetric ventrodorsal radiograph of the pelvis (Figure 3-45). In addition, sacralization of L7 has been shown to increase the likelihood of developing cauda equina syndrome, likely resulting from altered biomechanical forces leading to degenerative disc disease and spondylosis with subsequent neural compression.[10]

In some dogs, the length of L7 will be notably shorter than other lumbar vertebrae. This is a normal variant and not clinically significant (Figure 3-46).

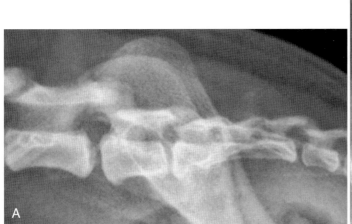
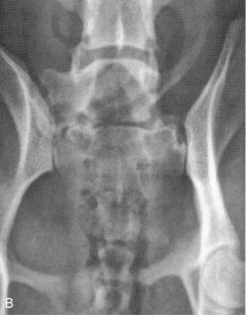

Figure 3-43. Lateral (**A**) and ventrodorsal (**B**) radiograph of the lumbosacral region of a 4-year-old mixed breed dog. There is sacralization of the right aspect of L7. In **B**, on the *right side*, L7 is enlarged laterally and articulates with the sacrum forming a sacroiliac joint, and, on the *left side* of L7, there is a normal transverse process. The abnormal sacroiliac joint on the *right* is also associated with an abnormal shape of the right ilium compared with the *left*.

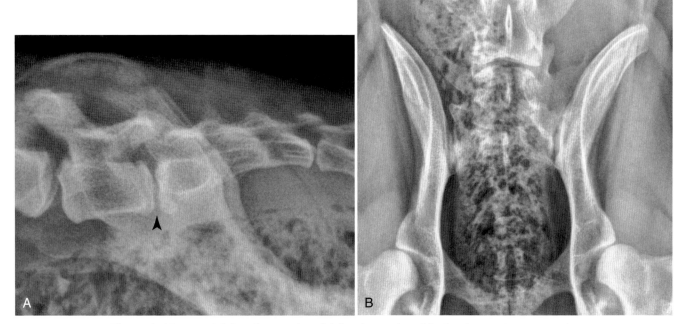

Figure 3-44. Lateral (**A**) and ventrodorsal (**B**) radiographs of the lumbosacral junction of a 1-year-old German Shepherd. There is sacralization of L7. On the *right side*, L7 has a transverse process, but, on the *left side*, L7 has a broad articulation with the sacrum. In the lateral view, the angle between L7 and the apparent sacrum is relatively straight, as described in the text, and the apparent lumbosacral articulation is more caudally positioned than normal (*black arrowhead* in **A**). Compare the position of this apparent lumbosacral articulation with that of a normal dog in Figure 3-41, *A*. Also note the broader appearance of the left iliac crest in **B** in spite of symmetric positioning of the caudal aspects of the pelvis. The broad attachment of L7 to the left ilium has led to remodeling creating this asymmetry.

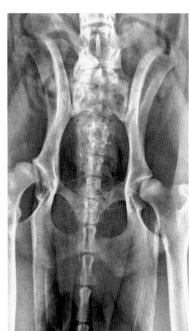

Figure 3-45. Ventrodorsal radiograph of the pelvis of a 1-year-old mixed breed dog. There is a transitional lumbosacral anomaly. The transverse processes on L7 are asymmetric, and the sacrum has a longer sacroiliac articulation on the right side. These changes have resulted in angulation of the pelvis in relation to the midsagittal axis of the spine; the ischial region is angled toward the right. Note the bilateral coxofemoral subluxation, greater on the right side, and more advanced degenerative joint disease on the right. The chronic angulation and altered biomechanics caused by the transitional anomaly likely contributed to this asymmetric coxofemoral disease. Vertical white streaks in the caudal part of the image are from the radiographic positioning pad.

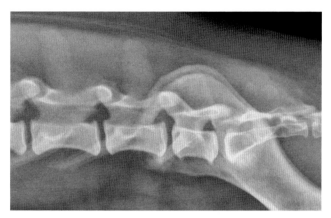

Figure 3-46. Lateral radiograph of the caudal lumbar spine of a 3-year-old Labrador Retriever. Note the short L7 vertebral body compared with other visible lumbar vertebrae. This is a normal variant and not clinically significant. There is mild spondylosis at L7-S1, but this is not related to the shorter L7 compared with other lumbar vertebrae.

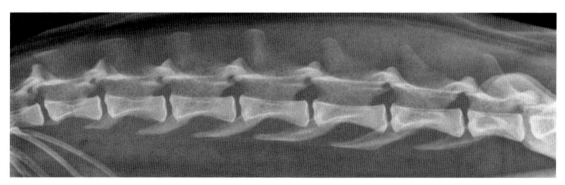

Figure 3-47. Lateral radiograph of the lumbar spine of a 3-year-old domestic cat. Note the relatively long vertebral bodies in the lumbar spine compared with their height; the relationship of vertebral length to height is larger in cats than in most dogs.

The lumbar spine in the cat has no unique anatomic features that were not described for the dog. In general, the vertebral body length-to-height ratio is larger in the cat than in the dog, and the transverse processes are relatively longer (Figure 3-47).

SACRAL SPINE

The sacrum is composed of three fused segments that become smaller progressively from cranial to caudal. The sacrum is fused to the ilium, thus connecting the spine to the pelvis and the pelvic limbs. The normal fusion of the three sacral segments defines the sacrum as a block vertebra (Figure 3-48). Congenital anomalies of the sacrum occur, as with failure of complete fusion of individual sacral segments (Figure 3-49), sacrocaudal transitional vertebrae (Figure 3-50), and lumbosacral transitional vertebrae (as described previously). To accurately characterize a sacral transitional anomaly, it is necessary to have a radiograph of the entire lumbosacral spine.

Occasionally, in lateral radiographs of the pelvic region, the articulation between the auricular surface of the wing of the sacrum and the ilium appears as an ill-defined region of opacity superimposed on the ilium. Care should be taken not to misdiagnose this opacity as a hyperostotic sacral lesion (Figure 3-51). In addition, the

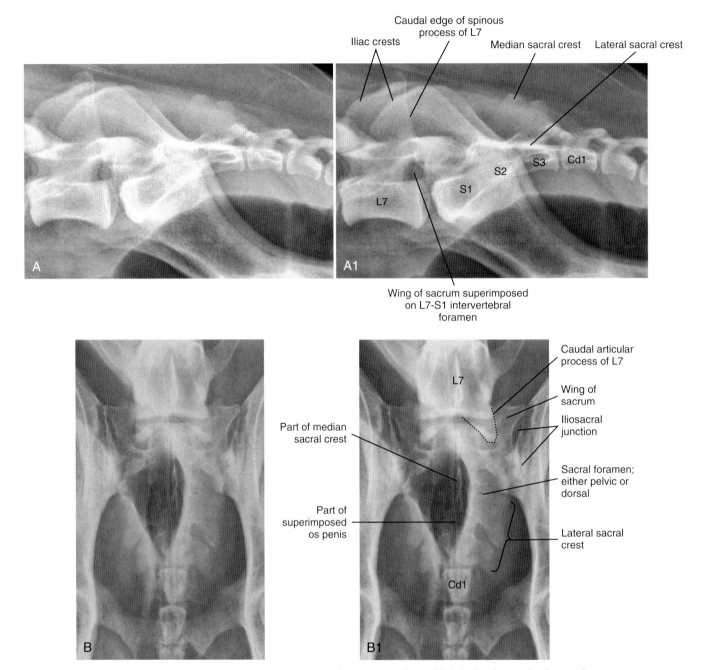

Figure 3-48. Lateral (A) and ventrodorsal (B) radiographs and labeled radiographs (A1, B1) of the sacral region of a 4-year-old German Shepherd.

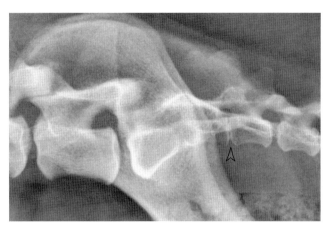

Figure 3-49. Lateral radiograph of the sacral region of a 7-year-old Golden Retriever. Failure of fusion of the second and third sacral segment is visible *(hollow black arrowhead)*.

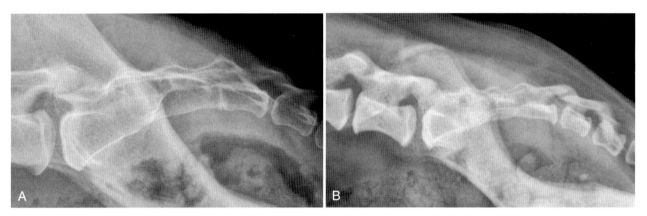

Figure 3-50. Lateral radiograph of the sacral region of a 7-year-old Greyhound **(A)** and a 7-year-old Miniature Dachshund. There is incomplete separation of the third sacral and first caudal segments in each dog. The sacrocaudal junction is more completely fused in **B**. This transitional anomaly is not clinically significant.

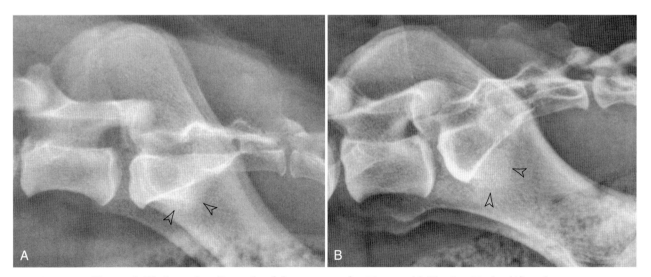

Figure 3-51. Lateral radiograph of the sacrum of a 13-year-old Siberian Husky **(A)** and a 9-year-old mixed breed dog **(B)**. In each dog, the auricular surface of the wing of the sacrum that articulates with the ilium creates an ill-defined opacity *(hollow black arrowheads)* that could be confused with a hyperostotic sacral lesion.

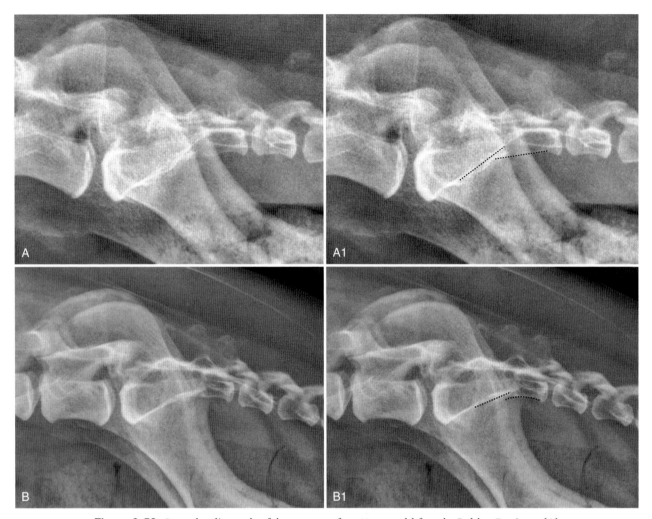

Figure 3-52. Lateral radiograph of the sacrum of an 11-year-old female Golden Retriever (**A**) and an 11-year-old male Golden Retriever (**B**). In each radiograph, the pelvic surface of the sacrum is very conspicuous and the nonlinear relationship between this surface and the caudal aspect of the sacrum can be misdiagnosed as a fracture. In **A1** and **B1**, corresponding images show where the pelvic surface of the sacrum and the caudal portion of the sacrum have been outlined to emphasize the nonlinear relationship between these structures.

pelvic surface of the sacrum can sometimes create a very conspicuous linear opacity that does not line up with the caudoventral aspect of the sacrum; this nonlinear arrangement creates an optical illusion of sacral discontinuity that can be misdiagnosed as a sacral fracture (Figure 3-52). Finally, most dogs will have fecal material in the descending colon and rectum when radiographs of the lumbar and sacral regions are made, and superimposition of fecal material in ventrodorsal radiographs can create lucent lines that are easily confused with fractures (Figure 3-53).

Manx cats have a mutation that results in a tailless phenotype in some subjects. Radiographically, these cats may also have an absence of a portion of the sacrum or other coexisting vertebral anomalies (Figures 3-54 and 3-55). The osseous anomalies in this anatomic region can also lead to neural dysfunction causing fecal retention and fecal impaction.

In some cats, the caudoventral aspect of the sacrum has a smooth protuberance that can be confused with callus; this is a normal variant and is not evidence of prior trauma (Figure 3-56).

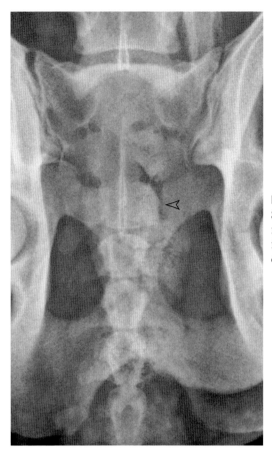

Figure 3-53. Ventrodorsal radiograph of the sacrum of a 10-year-old German Shepherd. Superimposition of fecal material with bony aspects of the pelvis can result in lucencies due to entrapped fecal gas being incorrectly diagnosed as a fracture. In this dog, the fecal gas could be misdiagnosed as a fracture of the caudal aspect of the sacrum *(hollow black arrowhead)*.

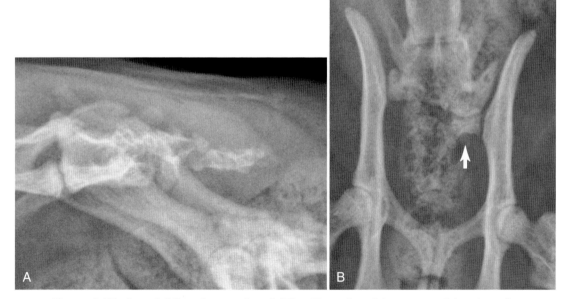

Figure 3-54. Lateral (**A**) and ventrodorsal (**B**) radiographs of the sacrocaudal region of a Manx cat. This cat had six lumbar vertebrae only. L6 is abnormally shaped, being shortened on the right side, and articulates with the pelvis at an angle. Note the pelvic asymmetry with respect to the lumbar spine. In **B**, a small bone fragment *(white arrow)*, a remnant of L7 or S1, also articulates with the left ilium. Fecal material precludes critical assessment of the caudal vertebrae in the ventrodorsal view. In the lateral view, it is apparent that the sacrum and caudal vertebrae are dysplastic/aplastic.

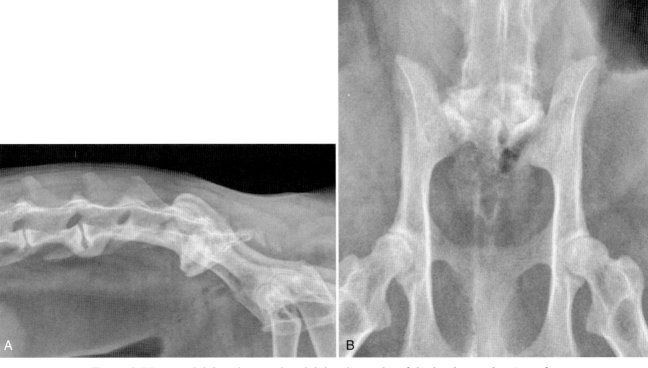

Figure 3-55. Lateral (A) and ventrodorsal (B) radiographs of the lumbosacral region of a 14-year-old Manx cat. L5, L6, and L7 are fused into a block vertebra. The sacrum is abbreviated, and there are only a few caudal vertebrae. These findings are typical of the breed and may influence pelvic and perineal innervation. There is significant spondylosis at L7-S1 with mild ventral displacement of S1.

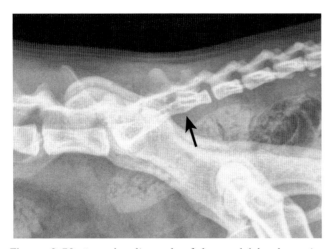

Figure 3-56. Lateral radiograph of the caudal lumbar spine and sacrum in a 6-year-old domestic shorthair cat. There is a smooth region of bone on the caudoventral aspect of the sacrum *(black arrow)* that can be misinterpreted as callus from prior trauma. This is a normal variant.

CAUDAL SPINE

The number of caudal vertebrae that are present varies according to breed and whether the tail has been docked. The vertebral arch is present on only the first few caudal vertebrae, quickly becoming progressively smaller until only a dorsal groove remains[3] (Figure 3-57). The caudal spine is also characterized by ventrally located hemal arches that protect the median caudal artery. The hemal arches are separate bones that articulate with the cranioventral aspect of Cd4, Cd5, and Cd6 (see Figure 3-57). Articular processes are present cranially but disappear by about Cd12.[3] Progressing farther caudally in the caudal spine, the vertebrae assume the shape of simple rods[3] (Figures 3-58 and 3-59).

The most common malformation of the caudal spine is the presence of hemivertebrae, which accounts for the morphology of the tail in screw-tail breeds, such as the Boston Terrier, Bulldog, and French Bulldog (Figure 3-60).

The caudal spine in the cat has no unique anatomic features that were not described for the dog.

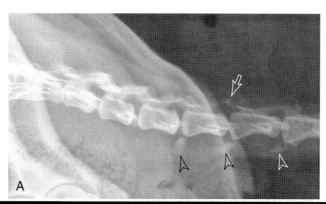
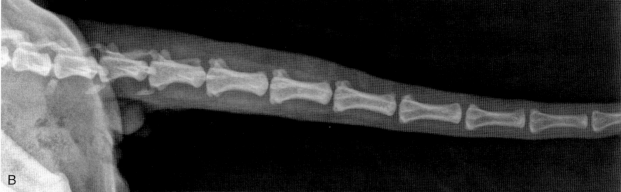

Figure 3-57. Lateral views of the most cranial (**A**) and middle (**B**) regions of the caudal spine of an 8-year-old Labrador Retriever. Note the disappearance of the vertebral arch at approximately Cd4 (*hollow white arrow* in **A**). Also note the cranioventrally located hemal arches on Cd4, Cd5, and Cd6 (*arrowheads* in **A**).

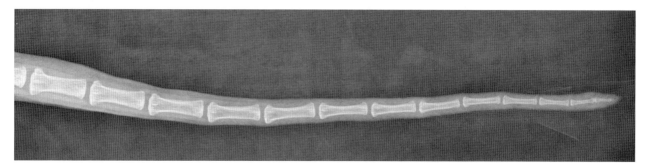

Figure 3-58. Lateral radiograph of the middle and caudal portions of the caudal spine of an 11-year-old German Shepherd. Individual vertebrae assume a simple rod shape progressing caudally in the tail.

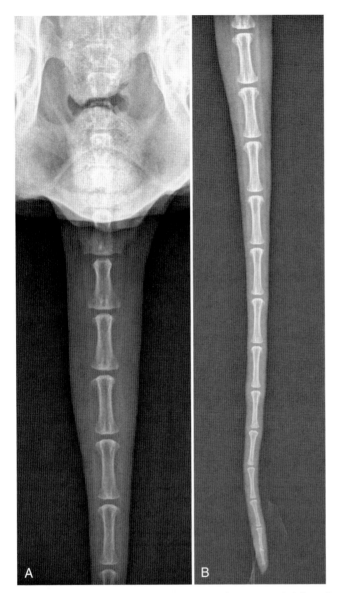

Figure 3-59. Ventrodorsal radiographs of the cranial (A) and caudal (B) regions of the caudal spine of an 11-year-old German Shepherd. Progressing caudally, note the change in size and shape of caudal vertebrae.

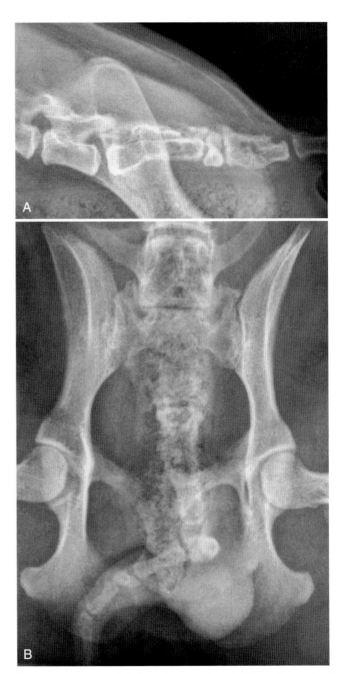

Figure 3-60. Lateral (A) and ventrodorsal (B) radiograph of the sacrocaudal region of a 10-year-old Boston Terrier. Multiple hemivertebrae account for the screw-tail morphology present in this breed and others.

REFERENCES

1. Evans H, de Lahunta A, editors: Prenatal development. *Miller's anatomy of the dog*, ed 4, St. Louis, 2013, Elsevier/Saunders.
2. Watson A, Evans H, de Lahunta A: Ossification of the atlas-axis complex in the dog. *Anat Histol Embryol* 15:122–138, 1986.
3. Evans H, de Lahunta A, editors: The skeleton. *Miller's anatomy of the dog*, ed 4, St. Louis, 2013, Elsevier/Saunders.
4. Watson A, Evans H: The development of the atlas-axis complex in the dog. *Anat Rec* 184:558, 1976.
5. Watson A: *The phylogeny and development of the occipito-atlas-axis complex in the dog*, Ithaca, NY, 1981, Cornell University.
6. Evans H, de Lahunta A: editors: Arthrology. *Miller's anatomy of the dog*, ed 4, St. Louis, 2013, Elsevier/Saunders.
7. Morgan J, Bailey C: *Exercises in veterinary radiology: spinal disease*, St Louis, 2000, Wiley Blackwell.
8. Baines E, Grandage J, Herrtage M, et al: Radiographic definition of the anticlinal vertebra in the dog. *Vet Radiol Ultrasound* 50:69–73, 2009.
9. Gutierrez-Quintana R, Guevar J, Stalin C, et al: A proposed radiographic classification scheme for congenital thoracic vertebral malformations in 'screw-tailed' dog breeds. *Vet Radiol Ultrasound* 55:585–591, 2014.
10. Fluckiger M, Damur-Djuric N, Hassig M, et al: A lumbosacral transitional vertebra in the dog predisposes to cauda equina syndrome. *Vet Radiol Ultrasound* 47:39–44, 2006.

CHAPTER 4

The Thoracic Limb

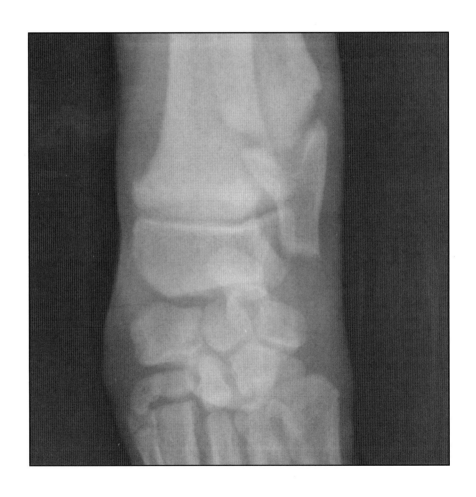

There is no direct osseous connection between the forelimbs and axial skeleton as occurs in the hindlimbs. The forelimbs support the trunk by a muscular sling known as the *pectoral girdle*, which extends from the medial aspect of the scapulae ventrally to the thoracic wall and sternum.

THE SCAPULA AND BRACHIUM

The *scapula* is the large flat bone of the shoulder (Figures 4-1 and 4-2). Because the trunk is supported by the pectoral girdle, which attaches to the scapulae, the normal position of each scapula with respect to the vertebral column varies considerably. The flat surface of the scapula is divided nearly equally into cranial and caudal halves by a protruding spine, which is more developed distally than proximally. The most distal aspect of the spine of the scapula is known as the *acromion* (see Figures 4-1 and 4-2). Distally, the scapular blade narrows to form a neck, the cranial margin of which is the *scapular notch*. The scapular notch extends into the glenoid cavity and articulates with the proximal humerus to form the *scapulohumeral* or *shoulder joint* (Figure 4-3). The *supraglenoid tubercle*, the site of origin of the biceps muscle, and the coracoid process are located at the most cranial aspect of the glenoid cavity. In the cat, the coracoid process is more developed than in the dog, and the acromion has a hamate process and a suprahamate process (Figure 4-4). Occasionally, a separate ossification center is visible radiographically at the caudal aspect of the glenoid cavity (Figure 4-5). This can be difficult to differentiate from an osteophyte. Infrequently, there is incomplete ossification at the distal aspect of the canine acromion (Figure 4-6).

In the dog, the clavicle is rudimentary, appearing as a very small opacity in the body of the brachiocephalicus muscle. In the cat, however, the clavicle is much more developed and is usually palpable clinically (Figures 4-7 and 4-8). In the dog, the clavicular remnant is seen only in ventrodorsal (or dorsoventral) radiographs; in the cat, the clavicle can be seen in both lateral and ventrodorsal (or dorsoventral) radiographs. Particularly when imaging one scapula or shoulder joint, there is a propensity to abduct the shoulder joint. This can give the false impression of disruption to the medial compartment of the joint (see Figure 4-8, *A*). In poorly positioned feline radiographs, the appearance of clavicles could be confused with an esophageal foreign object.

Lateral radiographs of the scapula and shoulder joint are best acquired with the limb of interest placed closest to the x-ray table and pulled craniodistally while the head and neck are extended dorsally and the uppermost limb pulled caudally. The orthogonal view of the scapula and shoulder is best acquired in a caudal-to-cranial direction (Cd-Cr) with the patient in dorsal recumbency and the limb extended cranially (see Figures 4-1 and 4-2).

Acquiring lateral and oblique radiographs of the scapula without superimposition of the spine is challenging. One method is to place the subject in lateral recumbency with the limb of interest uppermost (nondependent). The head and neck are flexed, and the elbow is directed dorsally to force the scapula dorsal to the cranial aspect of the thoracic spine. This results in a slightly oblique

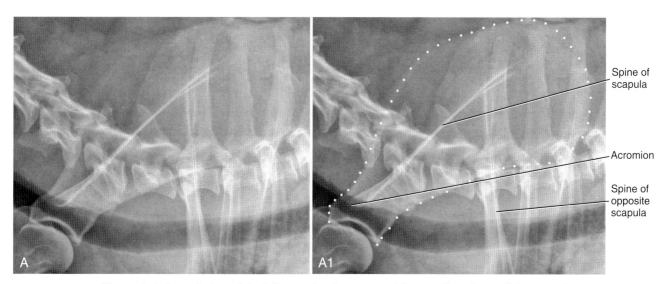

Figure 4-1. Lateral view of the left scapula of an 8-year-old Rottweiler. The caudal aspect of the cervical spine, the cranial aspect of the thoracic spine, and the contralateral right scapula are all superimposed on regions of the left scapula. Such superimposition is impossible to avoid using this positioning for a lateral view of the scapula. **A1,** The same image as in **A.** The border of the left scapula is outlined *(white dots)*.

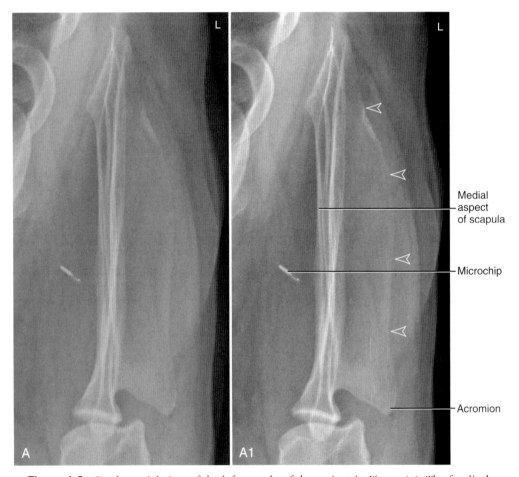

Figure 4-2. Caudocranial view of the left scapula of the patient in Figure 4-1. The forelimb has been abducted slightly from the thoracic wall to reduce superimposition. **A1,** The same image as in **A**. The *hollow white arrowheads* are the spine of the scapula.

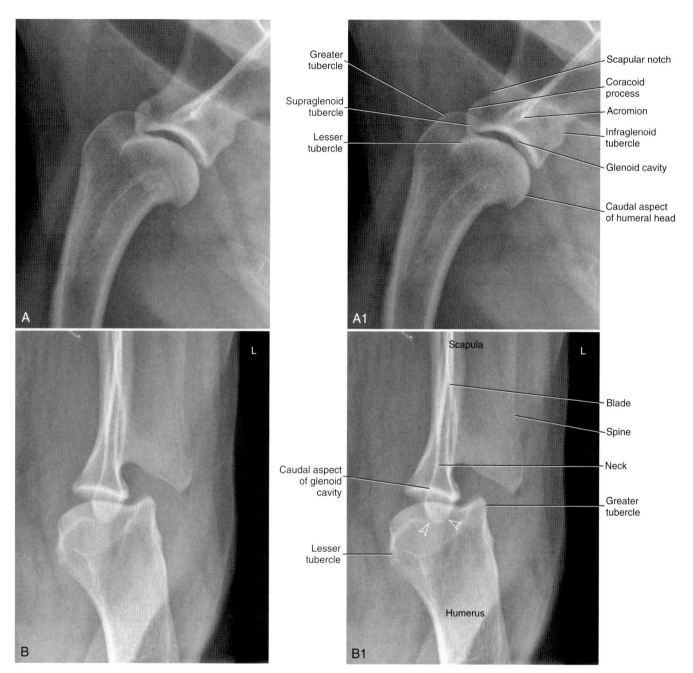

Figure 4-3. **A,** Lateral view of the left shoulder of a 7-year-old mixed breed dog. **A1,** The same view as in **A,** with labels. **B,** A caudocranial view of the left shoulder of an 8-year-old Rottweiler. **B1,** The same image as in **B,** with labels. The *hollow white arrowheads* are the most cranial aspect of the supraglenoid tubercle. An identification microchip is present dorsally.

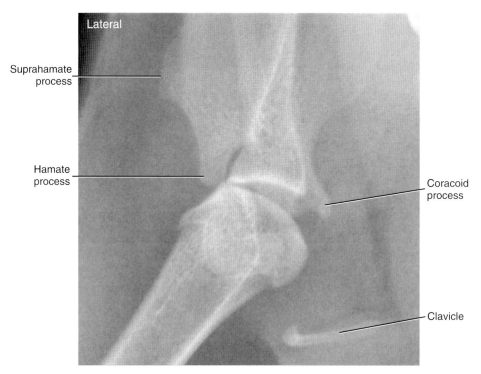

Figure 4-4. Caudocranial view of the distal aspect of the right scapula and shoulder joint of a 2-year-old Domestic Shorthair cat. The forelimb has been slightly abducted. This abduction causes the medial aspect of the joint to be increased in width; this is often misinterpreted as subluxation, or shoulder instability, but this is a normal finding.

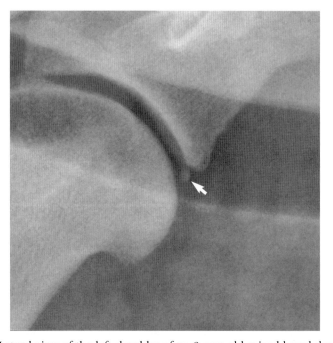

Figure 4-5. Lateral view of the left shoulder of an 8-year-old mixed breed dog. An osseous body at the caudal aspect of the glenoid cavity *(white arrow)* is considered an uncommonly occurring separate center of ossification. It is often difficult to differentiate between this apparent ossification anomaly and an osteophyte.

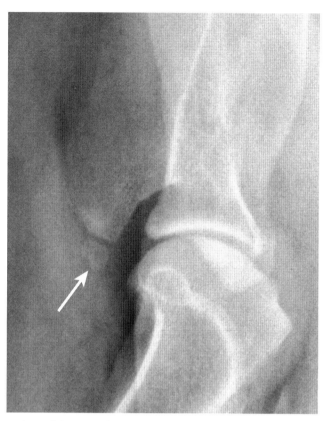

Figure 4-6. Craniocaudal view of the distal aspect of the right scapula. The *white arrow* highlights an incompletely fused ossification center at the distal aspect of the acromion.

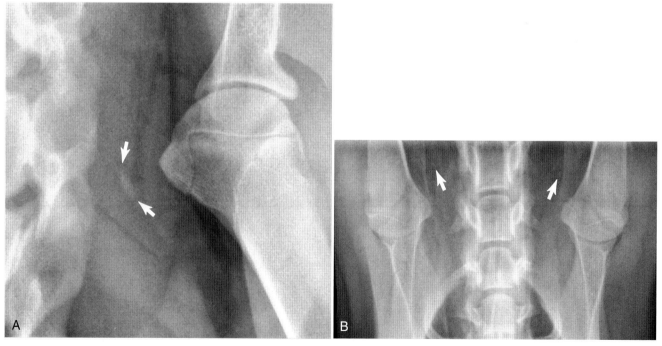

Figure 4-7. Craniocaudal view (**A**) of the left shoulder of a 10-month-old Bernese Mountain Dog. The brachium is slightly abducted. This is normal and does not indicate disruption to the medial aspect of the joint. The *solid white arrows* delineate the clavicle. **B**, Ventrodorsal view of the cervicothoracic junction of a 7-month-old Beagle with the forelimbs extended. The very faint mineral opacities delineated by the *solid white arrows* are the clavicular remnants.

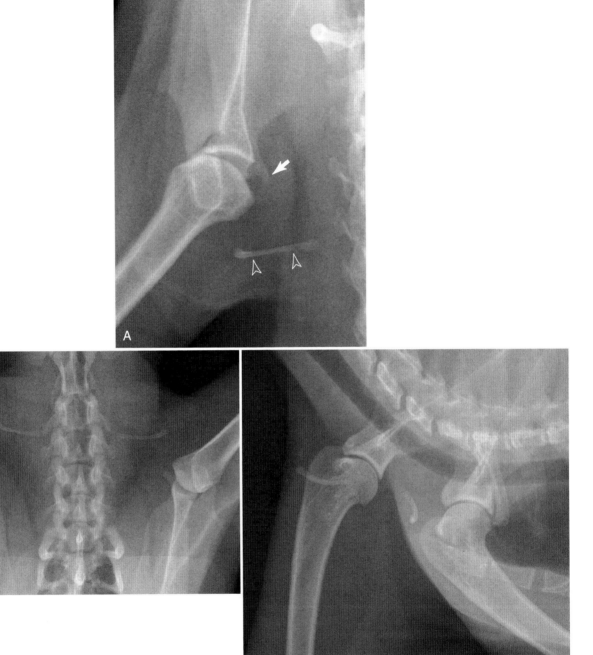

Figure 4-8. **A,** Caudocranial view of the right shoulder of a 20-year-old Domestic Shorthair cat. The *hollow white arrowheads* delineate the cranial margin of the clavicle. The *solid white arrow* is the coracoid process of the scapula. The coracoid process is well developed in cats and should not be confused with an osteophyte. Note the marked abduction of the humerus relative to the scapula. This is normal and should not be confused with medial instability. **B,** Ventrodorsal view showing the position of the clavicles relative to the spine. In **C,** the variable position of the clavicles relative to the shoulder joint is readily apparent in an 11-year-old Domestic Shorthair cat.

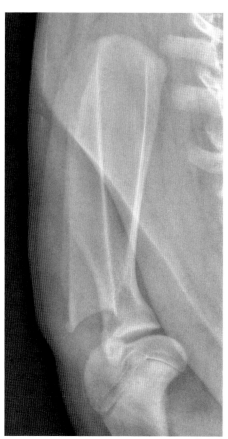

Figure 4-9. Oblique lateral view of the right scapula of a 1-year-old mixed breed dog. The patient is in lateral recumbency with the scapula of interest up (nondependent). The nondependent brachium is pushed dorsally while the neck is flexed. This eliminates superimposition of the scapula and vertebrae, and, in conjunction with the caudocranial (or craniocaudal) view, this view is useful in assessing body fractures.

view of the scapula, but without superimposition of the cervical and thoracic spine (Figure 4-9).

The shoulder joint is the articulation between the glenoid cavity of the scapula and the humeral head. Because shoulder lameness is common, the shoulder is one of the most frequently radiographed joints. Lateral views are best acquired as previously described for the lateral view of the scapula (Figure 4-10). A common fault is failure to pull the shoulder far enough cranially, resulting in superimposition of the proximal aspect of the humerus on the manubrium. This superimposition prevents accurate assessment of the humeral head (see Figure 4-10). When the limb is pulled craniodistally, the shoulder joint is usually cranioventral to the ventral aspect of the cervical trachea. Alternatively, intentionally using less traction so that the humeral head is superimposed on the lumen of the caudal aspect of the cervical trachea may facilitate the evaluation of intra-articular structures

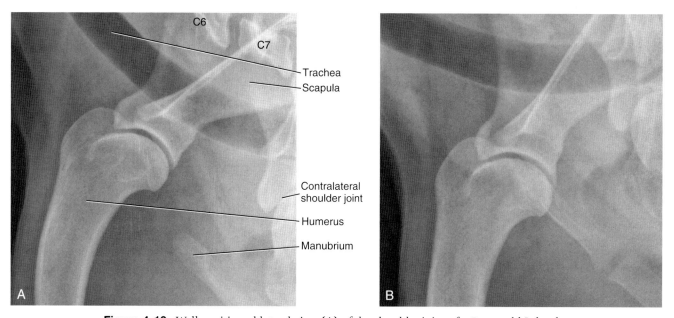

Figure 4-10. Well-positioned lateral view **(A)** of the shoulder joint of a 7-year-old Labrador Retriever. The shoulder is cranial and distal to the trachea and sternum, and the contralateral limb has been retracted caudally. This positioning allows optimal radiographic evaluation. In **B**, the shoulder is not pulled sufficiently cranially, and there is superimposition of the manubrium and the caudal aspect of the humeral head.

because of the additional contrast provided by the air in the trachea (Figure 4-11). However, it is impossible to control exactly which portion of the humeral head will be superimposed on the trachea and, in general, lateral radiographs should be made with the shoulder completely distant from the tracheal lumen. This is particularly true if the patient is under general anesthesia, because superimposition of the endotracheal tube over the joint will also compromise radiographic assessment (Figure 4-12).

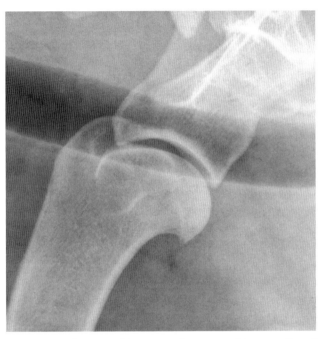

Figure 4-11. Lateral view of the shoulder joint of a 5-year-old Rottweiler. The forelimb has not been distracted maximally, and a portion of the humeral head is superimposed over the trachea. This sometimes provides additional contrast when evaluating the joint, but the portion of the humeral head that is superimposed on the trachea varies, creating inconsistency that can complicate image interpretation.

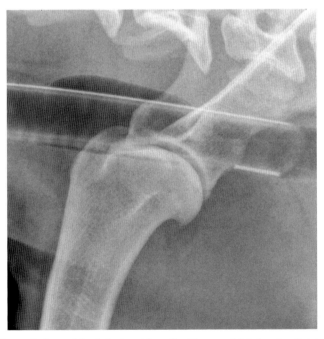

Figure 4-12. Lateral view of the left shoulder of a 10-year-old Labrador Retriever. The endotracheal tube is superimposed over the shoulder joint. This can complicate radiographic interpretation.

Osteochondrosis is a common disorder of the canine shoulder. Often the changes created by this abnormality are visible on conventional lateral views of the shoulder, where they are struck tangentially by the x-ray beam. However, sometimes the osteochondrosis lesion is located slightly medial to the true caudal aspect of the humeral head and is not struck tangentially by the x-ray beam. In this instance, additional views of the shoulder joint are needed to provide a more global assessment of the caudal aspect of the humeral head. With the patient in lateral recumbency as for the lateral view, the dorsum of the forelimb is rotated laterally (supination) to create the view in Figure 4-13, *B* (Md-LCrO). This supination profiles the caudolateral aspect of the humeral head. When

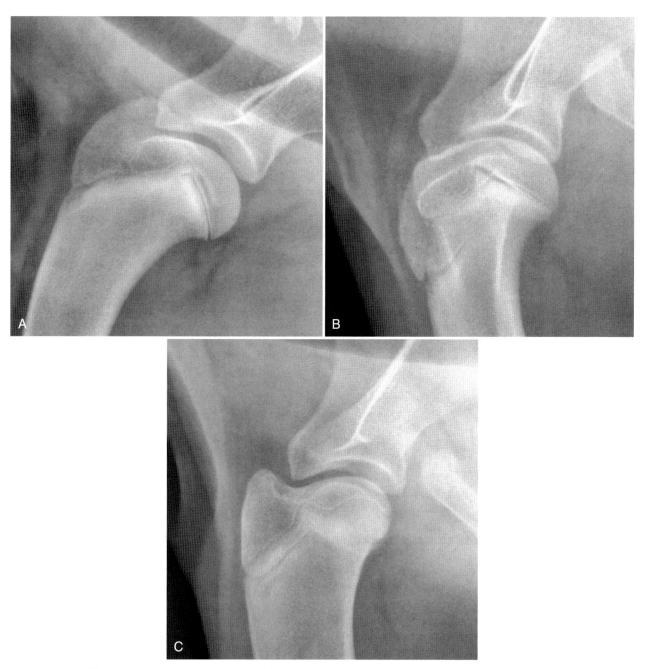

Figure 4-13. Lateral view (**A**) of the shoulder of a 9-month-old St. Bernard. This is the standard view when assessing for humeral head osteochondrosis. Lesions on the caudomedial or caudolateral aspect of the humeral head are often more easily detected by rotating the forelimb medially and laterally. With the patient in lateral recumbency, as for the lateral view, the dorsum of the forelimb is rotated laterally, as in **B**. This profiles the caudolateral aspect of the humeral head. When the dorsum of the limb is rotated medially, as in **C**, the caudomedial aspect of the humeral head is profiled.

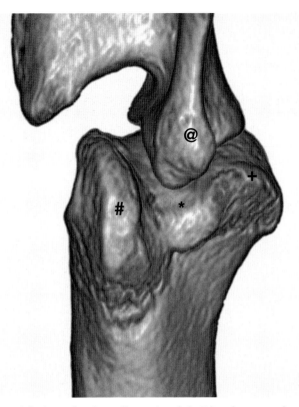

Figure 4-14. Cranial view of a three-dimensional (3-D) volume rendered CT of a canine shoulder joint. The symbol # is the greater tubercle, * is the bicipital groove, + is the lesser tubercle, @ is the supraglenoid tubercle. The biceps tendon originates from the supraglenoid tubercle and extends between the greater and lesser tubercles in the bicipital groove. The bicipital bursa, which is contiguous with the shoulder joint, provides lubrication for the tendon as it courses through the bicipital groove.

assessing the caudomedial aspect of the humeral head for medially located osteochrondrosis lesions, the dorsum of the limb is rotated medially as in Figure 4-13, C (pronation). In this position, the caudomedial aspect of the humeral head is profiled (Mr-LCdO).

The biceps tendon originates at the supraglenoid tubercle of the scapula and extends between the larger laterally located greater tubercle and the smaller medially located lesser tubercle (see Figure 4-14). In addition to a lateral and caudocranial view, a cranioproximal-to-craniodistal view of the cranial aspect of the humerus allows assessment of the biceps groove for changes that may develop secondary to biceps tenosynovitis or supraspinatus insertionopathy (Figure 4-15). This view is acquired by placing the patient in sternal recumbency on the x-ray table with the shoulders flexed and the forelimbs adjacent to the thoracic cage. The cassette or imaging plate is placed in the crook of the elbow and the x-ray beam directed vertically and centered on the proximal humerus.

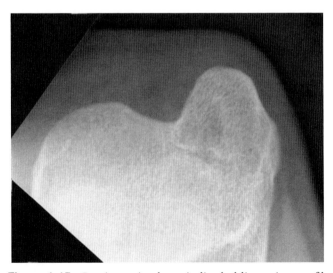

Figure 4-15. Cranioproximal-craniodistal oblique view profiling the greater and lesser tubercles in a 12-month-old Labrador Retriever. The biceps tendon is positioned in the groove between these two structures. This view is used in the assessment of secondary changes occurring as a consequence of bicipital tenosynovitis.

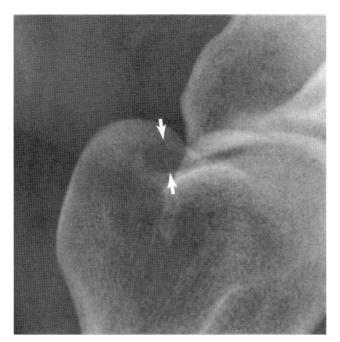

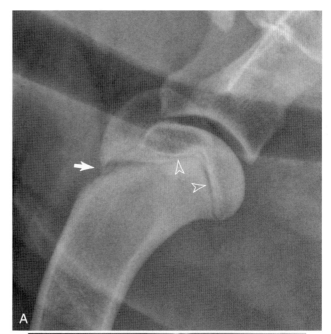

Figure 4-16. Lateral view of the proximal aspect of the humerus in an 8-year-old Greyhound. There is lucency at the base of the greater tubercle, delineated by the *solid white arrows*. This is relatively common. Although the cause has not been completely characterized, it might reflect a retained cartilage core, which can give the false impression of an aggressive lesion.

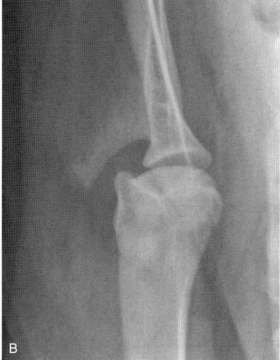

Figure 4-17. Lateral view (**A**) and caudocranial view (**B**) of the right shoulder of an 11½-month-old Labrador Retriever. The *solid white arrow* is the cranial margin of the proximal humeral physis. The zone of incomplete mineralization is normal and should not be confused with an aggressive process. The *hollow white arrowheads* are the caudal margin of the proximal humeral physis.

Occasionally, an ill-defined lucency can be seen associated with the base of the greater tubercle (Figure 4-16). The origin of this has not been completely characterized; it may reflect a retained cartilage core during early development. There are small vascular channels in this region and their presence probably contributes to the reduction in opacity present radiographically. Whether this is clinically significant is debated but most likely represents a normal variant.

The greater and lesser tubercles and the humeral head together make up the proximal humeral epiphysis. The associated physis is typically closed in dogs by 12 to 18 months, but the timing of fusion of the cranial aspect of the proximal humeral physis is variable. During the process of fusion, this area of the physis can look quite irregular and should not be confused with an aggressive lesion or a physeal fracture (Figure 4-17). In addition, the supraglenoid tubercle of the scapula develops as a separate ossification center, typically closing in the dog at 4 to 7 months (Figure 4-18).

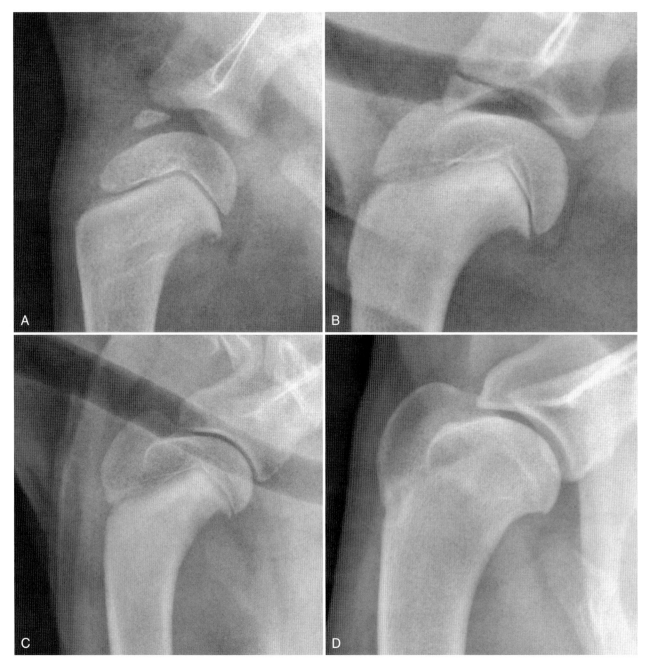

Figure 4-18. A, Lateral view of a 3-month-old Standard Poodle. The separate center of ossification of the supraglenoid tubercle is incompletely fused, and ossification of the greater tubercle has yet to occur. **B,** Lateral view of a 3½-month-old German Shorthaired Pointer. The ossification center associated with the supraglenoid tubercle of the scapula has yet to fuse; the normal fusion time is 4 to 7 months. **C,** Lateral view of a 5-month-old German Shepherd. Although compared with Figure 4-7, *A*, the lucency at the cranial aspect of the humeral physis is similar, the caudal metaphyseal region is mildly sclerotic, and there is a small metaphyseal lip at the caudal aspect of the physis. This is commonly seen in phases of rapid growth. **D,** Lateral view of the left shoulder of a 1½-year-old Boxer. Although the caudal aspect of the proximal humeral physis is closed, there is incomplete physeal closure cranially. The proximal humeral physis is typically closed by 12 to 18 months.

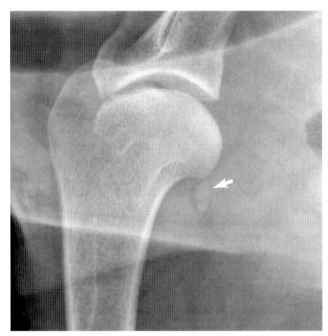

Figure 4-19. Lateral view of the caudal aspect of the shoulder joint in a 10-year-old Siberian Husky. The two circular opacities at the caudal aspect of the humeral head *(solid white arrow)* are the axillobrachial vein, the caudal circumflex humeral artery, and also a branch of the axillary nerve surrounded by fascial plane fat that provides adequate contrast to allow visibility. This opacity has been confused with a joint mouse.

In the mature patient, a small triangular region of fat is often present just caudal to the humeral head. Occasionally, the axillobrachial vein, caudal circumflex humeral artery, and branch of the axillary nerve are visible within that fat when they are struck end-on by the x-ray beam (Figure 4-19). The opacity created by these structures has been confused with a joint mouse.

The humeral shaft (Figure 4-20) is relatively smooth, with the deltoid tuberosity readily apparent at the lateral aspect of the mid-diaphysis (Figures 4-21 and 4-22). A linear margin of increased opacity extending distally from the caudal aspect of the greater tubercle distally, the *tricipital line,* is confluent with the deltoid tuberosity. The small tuberosity for the teres major appears as a slight irregularity on the medial aspect of the diaphysis at the level of the mid-aspect of the deltoid tuberosity. The nutrient canal penetrates the caudal cortex at approximately the junction of the middle and distal third of the diaphysis and is readily apparent on the lateral view (Figure 4-23). The nutrient canal should not be confused

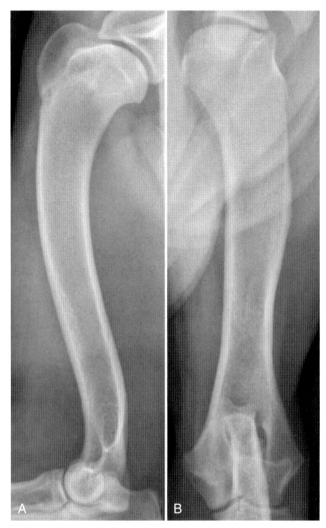

Figure 4-20. Lateral (A) and caudocranial (B) views of the humerus of a 1½-year-old Boxer. The canine humerus is more curvaceous than the feline humerus. The medullary cavity is of homogeneous opacity, with more coarse trabeculation apparent distally.

with a nondisplaced cortical fracture. In the cat, the nutrient canal is located at the same level but is in the medial cortex rather than being located caudally, and is usually not apparent radiographically. The feline humerus is straighter and more uniform in size throughout its length compared with the canine humerus (Figure 4-24). Conversely, in chondrodystrophoid breeds, the humeral head is typically flatter and appears relatively larger, and the diaphysis appears relatively shorter and more curved. The bulbous nature of the humeral head and lipping of the caudal humeral head in chondrodystrophic dogs can be misinterpreted as osteophytosis (Figure 4-25).

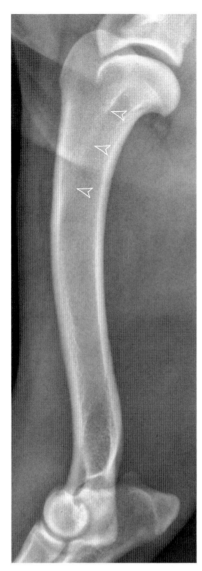

Figure 4-21. Lateral view of the right humerus of an 11-year-old mixed breed dog. The linear increase in opacity outlined by the *hollow white arrowheads* is the tricipital line proximally, extending into the deltoid tuberosity distally.

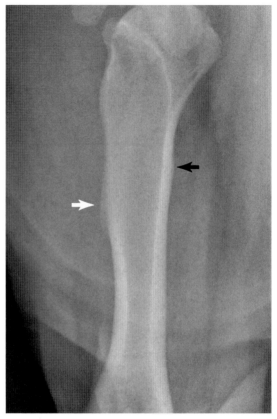

Figure 4-22. Caudocranial view of the right humerus of a 4-year-old Labrador Retriever. The *solid white arrow* is the distal aspect of the deltoid tuberosity. The *solid black arrow* is the tuberosity for the teres major muscle.

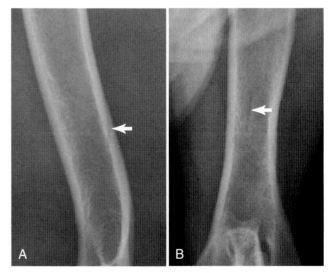

Figure 4-23. A, Close-up view of the distal humerus of a 10-year-old Labrador Retriever. The *solid white arrow* indicates the nutrient foramen penetrating the caudal cortex. This should not be confused with a fracture. B, Caudocranial view of the mid-shaft of the humerus of a 1½-year-old Boxer, the same patient as in Figures 4-18, **D**, and 4-20. The *solid white arrow* is the nutrient foramen obliquely penetrating the caudal cortex.

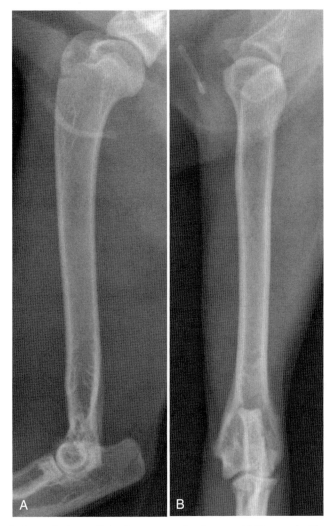

Figure 4-24. A, Lateral view of the humerus of a 2-year-old Domestic Shorthair cat. The clavicle is superimposed over the proximal humerus. The nutrient foramen in the mediodistal cortex is typically not visible radiographically. **B,** Caudocranial view of the humerus of a 16-year-old Domestic Shorthair cat. The clavicle is readily visible medial to the shoulder joint.

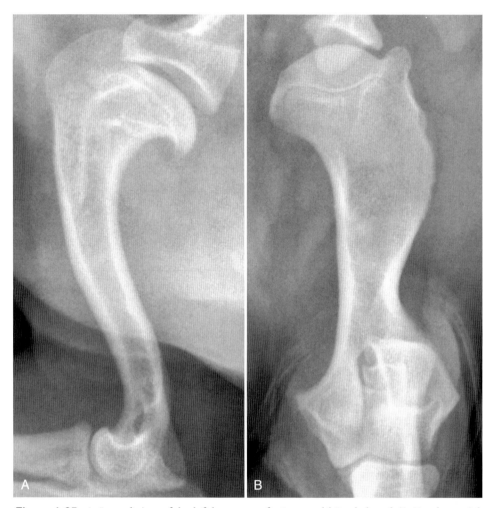

Figure 4-25. **A,** Lateral view of the left humerus of a 9-year-old Dachshund. **B,** Caudocranial view of the humerus of a 10-month-old Pembroke Welsh Corgi. The diaphyseal region of chondrodystrophoid breeds is typically more curved and the metaphyseal region more flared and bulbous than the non-chondrodystrophoid breeds. This lipping can be easily confused with osteophytosis.

THE ELBOW JOINT

The elbow joint is unique in that three long bones must grow in perfect synchrony to ensure normal joint congruency. Disruption in growth of either the ulna or the radius results in joint incongruency, which can lead to limb deformity, fragmented medial coronoid process, fractured anconeal process, and severe degenerative joint disease.

A standard radiographic examination of the elbow consists of a neutral lateral view with the brachium and antebrachium at an approximately 90-degree angle, a lateral view with the elbow highly flexed, and a craniocaudal (or caudocranial) view (Figures 4-26 and 4-27). Distally, the humerus consists of one condyle divided into lateral and medial parts, each of which has an associated epicondyle; the medial epicondyle is larger than the lateral epicondyle in both the dog and the cat. The lateral

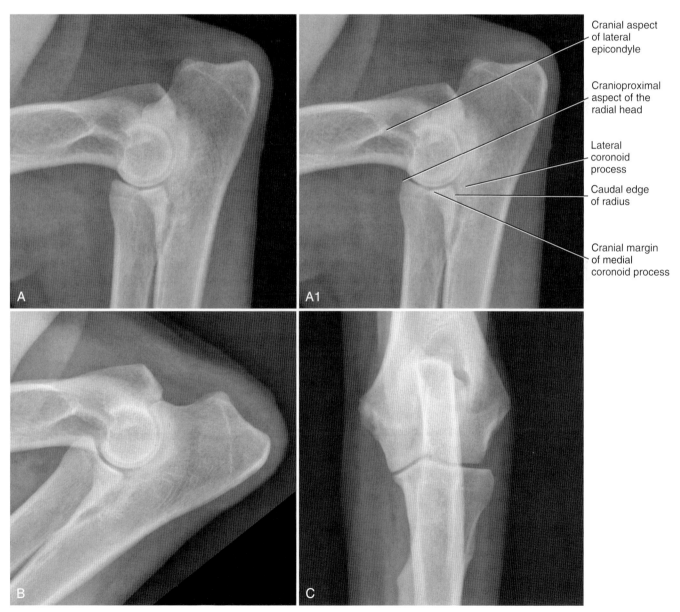

Figure 4-26. A standard radiographic examination of the elbow comprises a neutral lateral view (**A**), a flexed lateral view (**B**), and a craniocaudal view (**C**). A1, The same image as in A, with labels.

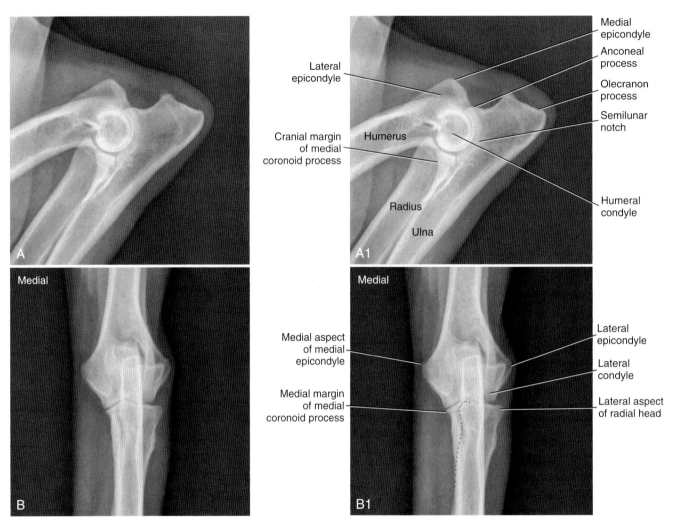

Figure 4-27. A, Flexed lateral (A1 with labels) and (B) craniocaudal (B1 with labels) views of the elbow of an 8-year-old Golden Retriever. The flexed view reduces superimposition of the humeral epicondyles over the anconeal process. The anconeal process is an early site for the development of degenerative joint disease. In **B1,** the *dotted black line* reflects the medial margin of the proximal radius. Note the lateral position of the radius, a landmark for distinguishing the lateral versus medial sides of the elbow joint in craniocaudal views. Note also the extent of the radius that does not articulate with the humerus in this view; this is often confused with elbow subluxation.

Chapter 4 ■ The Thoracic Limb 109

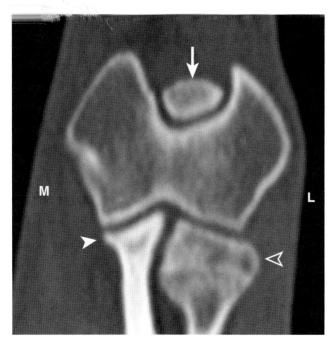

Figure 4-28. A dorsal plane CT image of a canine elbow, optimized for bone. M is medial, L is lateral; the *solid white arrowhead* is the medial-most aspect of the medial coronoid process. The *hollow white arrowhead* is the lateral aspect of the radial head, and the *white arrow* is the anconeal process of the ulna extending into the supratrochlear foramen of the humerus. Note how the lateral aspect of the radial head is lateral to the lateral aspect of the humeral condyle.

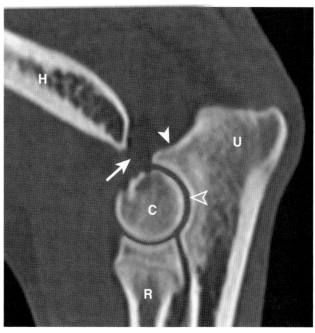

Figure 4-29. Sagittal plane CT image of a canine elbow optimized for bone. H is humeral shaft, C is humeral condyle, U is ulna, and R is radius. The *solid white arrow* is the anconeal process, the *hollow white arrow* is the trochlear notch, and the *white arrow* is the supratrochlear foramen. The apparent disconnect between the humeral shaft and the condyle is because the image plane (0.6 mm) is through the center of the condyle and therefore at the level of the supratrochlear foramen. In extension, the anconeal process extends into the olecranon fossa, the caudal aspect of the supratrochlear foramen.

aspect of the humeral condyle articulates with the head of the radius. The radius is the best anatomic landmark to use to determine medial versus lateral in craniocaudal views of the elbow; the radius is medial at the carpus, as discussed in the later section on the carpus, and lateral at the elbow. The olecranon process of the ulna cannot be used to distinguish medial from lateral in craniocaudal radiographs of the elbow because it is positioned near the midline in this view. Also important is the fact that, in the craniocaudal view, there is a relatively large portion of the lateral aspect of the proximal articular surface of the radius that does not articulate with the humerus; this is often confused with elbow subluxation (see Figures 4-27 and 4-28).

The medial aspect of the humeral condyle articulates with the trochlear notch of the ulna to form one of the most stable hinge joints in the body (Figure 4-29). The *olecranon fossa* is a deep cavity in the caudal part of the distal metaphysis of the humerus. The anconeal process of the ulna protrudes into this cavity when the elbow is in full extension (Figure 4-30). The olecranon fossa is contiguous with the *radial fossa*, a similar excavation on the cranial aspect of the distal metaphysis of the humerus, via the supratrochlear foramen. In the cat, the

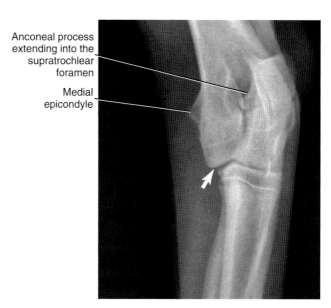

Figure 4-30. Craniocaudal oblique view (Cr15°L-CdMO) of the elbow joint of a 6½-month-old German Shepherd. This view enables a less obstructed view of the medial aspect of the humeral condyle and is useful in the medial humeral condylar lesions. The *solid white arrow* is the medial aspect of the articular margin of the medial part of the humeral condyle, a common location for osteochondrosis.

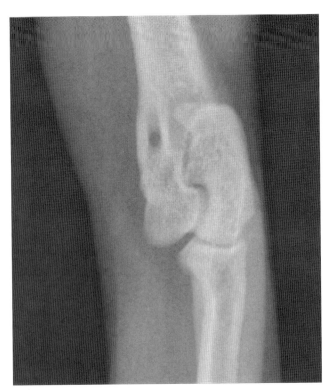

Figure 4-31. Craniolateral-caudomedial oblique view of the elbow of a mature cat. The supracondylar foramen is readily apparent, through which the brachial artery and median nerve pass.

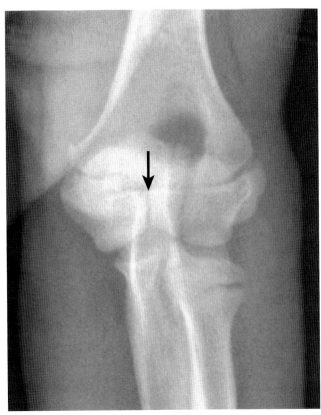

Figure 4-32. Craniocaudal view of the elbow of a 10-week-old Bulldog. Medial is to the left. There is incomplete fusion of the medial and lateral parts of the humeral condyle, readily evident by the sagittal plane linear lucency denoted by the *solid white arrow*. This is normal in a patient of this age. The humeral condyle is typically fused by 12 weeks.

humerus contains a supracondylar foramen located distomedially, through which the brachial artery and the median nerve pass (Figure 4-31). The feline humerus does not have a supratrochlear foramen.

Many orthopedic diseases of the elbow in young patients are particularly challenging to diagnose radiographically. Quality radiographs with optimal contrast, detail, and positioning are important with respect to imaging this joint. In addition, as with all complex joints, specifically positioned views are often required to highlight particular aspects of the joint. A craniocaudal view with the cranial surface of the antebrachium rotated medially 15 degrees (Cr15°L-CdMO) allows for better assessment of the medial part of the condyle. This is important with respect to assessing possible osteochondrosis of the medial part of the humeral condyle (see Figure 4-30). Also, superimposition of the ulna can confound assessment of the intercondylar region when evaluating for the presence of incomplete ossification of the humeral condyle. The lateral and medial parts of the humeral condyle normally fuse by 8 to 12 weeks,[1] and no radiographic evidence of a division between the two should be visible after this time (Figure 4-32). Ideally, the medullary cavity of the ulna is positioned centrally over the intercondylar region to reduce unwanted superimposition (Figure 4-33). This is often achieved by rotating the cranial surface of the antebrachium laterally 15 degrees and acquiring a Cr15°M-CdLO view.[2]

The anconeal process of the ulna is best evaluated in a fully flexed lateral view. This helps avoid superimposition with the large medial epicondyle of the humerus (Figure 4-34). There is some variation in contour of the anconeal process between breeds, which can make assessment of early remodeling associated with joint disease difficult to recognize (see Figures 4-26, 4-27, and 4-34). This is important because remodeling of the anconeal process is one of the earliest radiographically detectable changes associated with osteoarthritis. In a very small portion of medium and large breed dogs, the anconeal process arises as a separate center of ossification. This should not be confused with the disorder Ununited Anconeal Process (UAP) which manifests as a large wide radiolucent zone and large anconeal process fragment. The clinically important disorder UAP appears unrelated

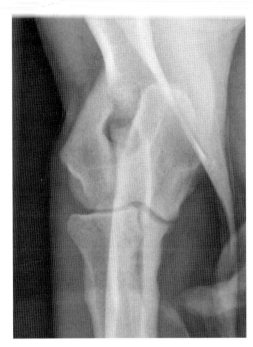

Figure 4-33. Cr15°M-CdLO view of a 9-year-old German Shorthaired Pointer. This view positions the medullary cavity of the ulna over the central aspect of the humeral condyle. This allows better assessment of the intercondylar region for ossification anomalies because, in a standard craniocaudal view, the ulnar cortex usually overlies the center of the humeral condyle.

to the presence of an anconeal process ossification center. It is possible the disorder UAP reflects abnormal maturation or mechanical failure of the anconeal process due to mild joint incongruency.

In the neutral lateral view, the medial coronoid process of the ulna is usually superimposed over the caudoproximal aspect of the radius. This prevents critical evaluation of the medial coronoid process of the ulna in the lateral view, but the medial coronoid process does create a curved triangular opacity that can be seen in many patients (Figure 4-35). Proper radiographic technique is mandatory for visualization of this triangular opacity in the lateral view (Figures 4-36 and 4-37). Joint congruency, particularly humeroulnar incongruency, is better evaluated in the craniocaudal view where the medial coronoid process of the ulna appears as a sharply marginated shelf of bone subjacent to the medial aspect of the humeral condyle (see Figure 4-27). Chondrodystrophoid breeds commonly have a smaller medial coronoid process that is most apparent in the lateral view (Figure 4-38). The physes associated with the distal humeral condyle and medial epicondyle can complicate radiographic interpretation in the young patient (Figure 4-39). These physes are typically closed by 7 to 8 months of age.

The triceps muscle group, the major extensor of the elbow joint, arises from the caudal aspects of the

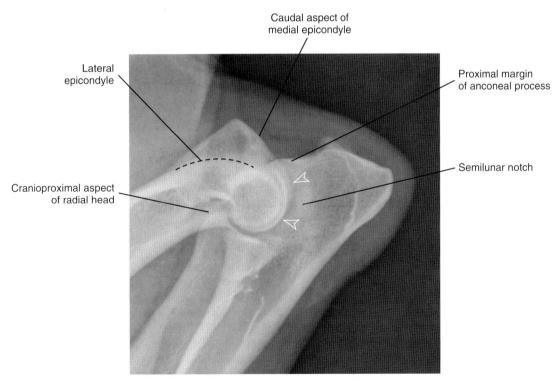

Figure 4-34. Flexed lateral view of the elbow of a 2-year-old German Shepherd. The *hollow white arrowheads* delineate the medial condyle extending from the epicondyle into the semilunar notch of the ulna.

Figure 4-35. A CT image of the level of the medial coronoid process of the ulna, optimized for bone. The image plane is transverse to the shaft of the radius and ulna. M is medial, R is radius, and U is ulna. The *hollow white arrowhead* is the most cranial aspect of the medial coronoid process. The *solid white arrow* is the radial notch or incisure of the ulna, which articulates with the caudomedial margin of the radius, known as the articular circumference. The articular circumference is larger than the radial incisure allowing some rotation of the antebrachium. The *black hollow arrow* is the smaller lateral coronoid process.

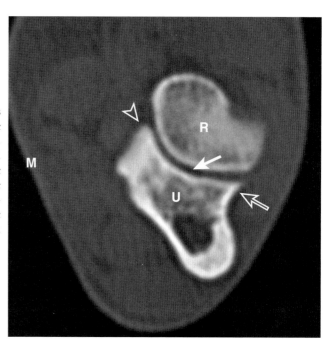

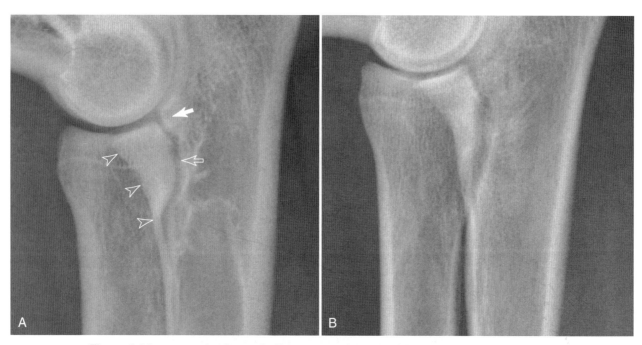

Figure 4-36. A, Lateral radiograph of the proximal aspect of the antebrachium of a 2-year-old German Shepherd. The cranial margin of the medial coronoid process is delineated by the *hollow white arrowheads*. Optimal radiographic technique is needed to identify this structure in this view. The lateral coronoid process is the *solid white arrow*. The radiolucent line outlined by the *hollow white arrow* is the caudal margin of the radial notch, between the medial and lateral coronoid processes, in which the caudoproximal aspect of the radius resides. This patient has a more coarse trabecular pattern in the proximal ulna when compared with patient **(B),** an 8-year-old Doberman Pinscher. Although this may be individual variation, a coarse trabecular pattern in this region is sometimes associated with degenerative joint disease.

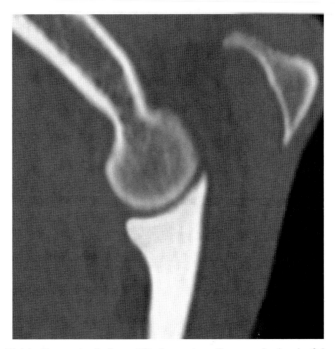

Figure 4-37. Thick slice sagittal plane CT of a canine elbow at the level of the medial coronoid process, optimized for bone. The *solid white arrow* is the medial coronoid process. Radiographically, this should be readily visible as a curved triangular opacity superimposed over the caudal aspect of the radial head. It should be congruent with the humeral condyle, and the humeroulnar joint space should be no smaller than the humeroulnar joint space radiographically.

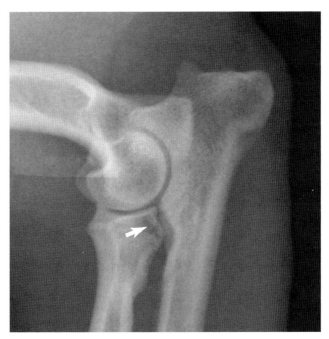

Figure 4-38. Lateral view of the elbow of a 10-month-old Pembroke Welsh Corgi. The medial coronoid process *(solid white arrow)* is typically smaller in chondrodystrophoid breeds. The humeroulnar joint appears relatively congruent.

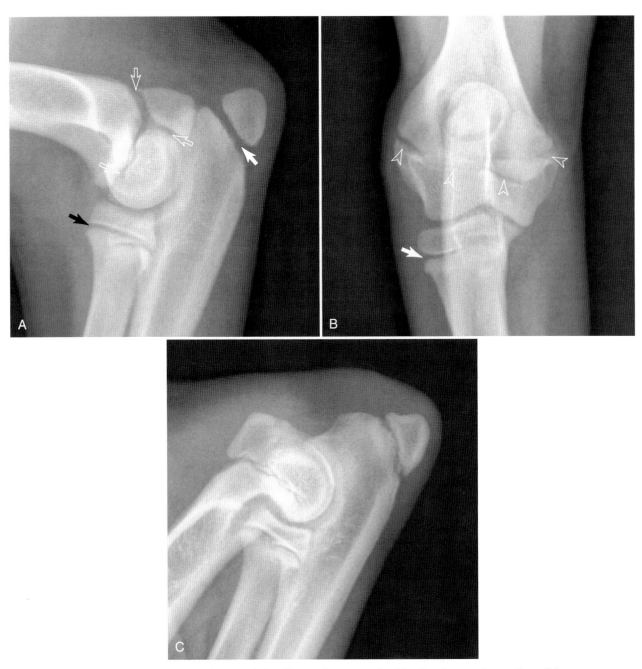

Figure 4-39. **A,** Lateral view of the elbow of a 3-month-old Labrador Retriever. The *solid white arrow* is the physis associated with the olecranon tuberosity. The *hollow white arrows* delineate the physis associated with the humeral condyle and medial epicondyle. The *solid black arrow* is the physis of the proximal radius. **B,** Craniocaudal view of the elbow of a 3½-month-old Boxer. The *hollow white arrowheads* delineate the distal humeral physis. It is common for both the medial and lateral aspects of the physis to be irregular with broad regions of lucency. The *solid white arrow* is the lateral aspect of the proximal radial physis. The cortical irregularity distal to this is normal, known as the *cutback zone*, which reflects a region of rapid osseous remodeling. **C,** Flexed lateral radiograph of the elbow of a 3½-month-old German Shorthaired Pointer. The physes are generally more mature than in the patient in **A.**

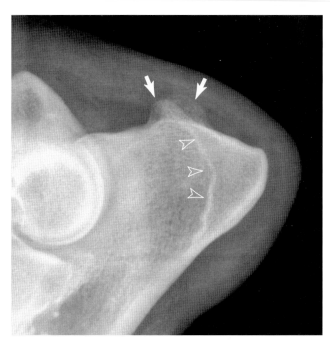

Figure 4-40. Lateral radiograph of the olecranon of a 2-year-old German Shepherd. The *hollow white arrowheads* are the scar associated with the physis of the olecranon. The *solid white arrows* delineate two protuberances associated with the tuberosity of the olecranon. Developmentally, the tuberosity of the olecranon is the fused olecranon apophysis.

scapula and the humeral shaft and attaches onto the olecranon process of the ulna. The olecranon process extends proximal to the trochlear notch and typically has two rounded unnamed prominences at its most proximal cranial margin. These should not be confused with enthesophytes due to stress remodeling (Figure 4-40).

Infrequently, a sesamoid bone is present in the supinator muscle at the craniolateral aspect of the proximal radius[4] in both dogs and cats. This should not be confused with an avulsion fragment or joint mouse (Figures 4-41 and 4-42). In cats, the sesamoid bone is rarely visible in the craniocaudal view and, when present, is most often visible in the lateral view (Figure 4-43, *A, B*).

ANTEBRACHIUM

The antebrachium extends from the elbow joint to the carpus and consists of the radius and ulna (Figure 4-44). As noted previously, although the radial head is lateral to the ulna at the elbow, the distal aspect of the radius is medial to the ulna at the carpus. At the elbow, the radial head interdigitates with the radial notch of the ulna and distally at the carpus, the styloid process of the ulna abuts the laterodistal aspect of the radius via the ulnar notch. Throughout the antebrachium, both the radius and the ulna are relatively smooth and regular. The radius has the larger diameter of the two and is the major weight bearer, whereas the ulna is highly attenuated distally. An important interosseous ligament is present at the proximal third of the radius at the level of the nutrient foramen. It is through this ligament that some of the forces borne by the radius at the antebrachiocarpal joint are transmitted to the ulna and then to the humerus. It is common to see an irregular cortical margination at the radial and ulnar attachment sites of the interosseous ligament. There is considerable variation in appearance of the interosseous ligament region, which complicates assessment of any disease that develops at this site (Figure 4-45).

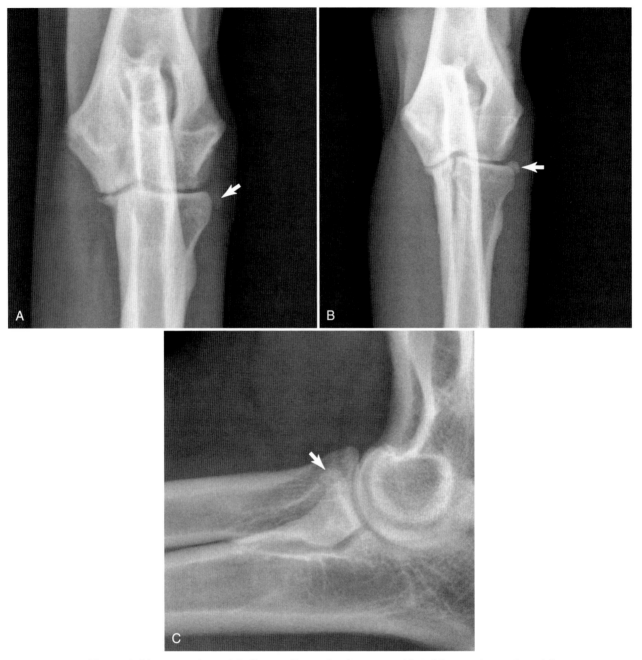

Figure 4-41. A, Craniocaudal elbow radiograph of a 3-year-old Golden Retriever. The *solid white arrow* indicates a sesamoid bone at the craniolateral aspect of the radial head. This sesamoid bone is in the supinator muscle and is reportedly radiographically apparent in approximately 30% of patients and should not be confused with a fracture fragment or joint mouse. It is seen less frequently in the lateral view. Craniocaudal view **(B)** and lateral view **(C)** of the elbow of a 7-year-old Rottweiler. In this patient, the sesamoid bone is apparent in both the craniocaudal and lateral views *(solid white arrows)*.

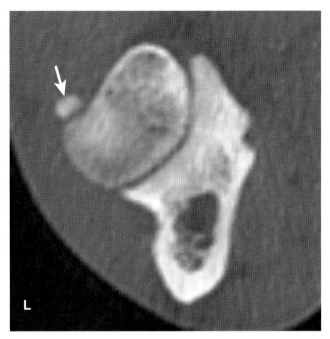

Figure 4-42. Transverse CT image of a canine elbow at the level of the radial head and medial coronoid process, optimized for bone. The osseous body at the craniolateral aspect of the humeral head is a sesamoid bone in the supinator muscle. This is an infrequent finding, occurring in about 30% of patients. It should not be confused with an osteochondral fragment that sometimes detaches from the medial coronoid process and migrates to the cranial aspect of the joint.

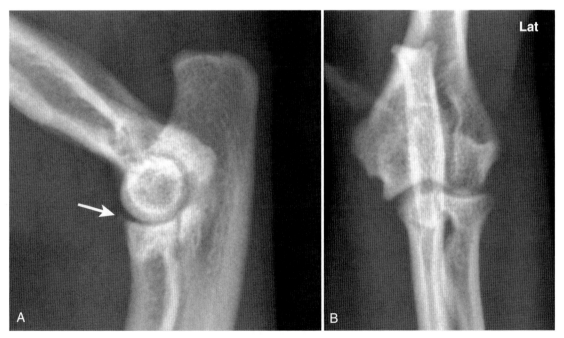

Figure 4-43. Lateral A, and caudocranial (B) views of the elbow of a 2-year-old Domestic Shorthair cat. The *white arrow* is the sesamoid associated with the supinator muscle. This is an infrequent finding in normal cats and should not be confused with a fracture fragment. The supinator sesamoid bone is lateral but is usually not evident in the craniocaudal view. The apparent widening of the medial aspect of the joint reflects normal joint laxity and is present in this view due to slight abduction of the joint during imaging.

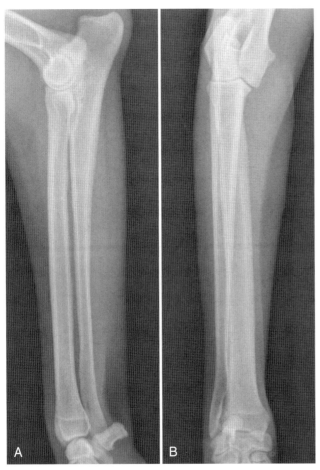

Figure 4-44. Lateral (**A**) and craniocaudal (**B**) views of the right antebrachium of a 7-year-old Labrador Retriever. The antebrachial muscles transition to tendons at the level of the distal antebrachium, resulting in reduced soft tissue mass at the carpus. The proximal aspect of the radius is lateral at the elbow, and the distal aspect of the radius is medial at the carpus. The proximal aspect of the ulna is larger at the elbow, and, in a well-positioned craniocaudal view, it will be located essentially on the midline. Distally, the ulna is lateral at the carpus.

The proximal radial physis typically closes at 7 to 10 months and the distal radial physis at 10 to 12 months. Approximately 70% of radial elongation occurs from growth at the distal physis, with the remaining 30% occurring from growth at the proximal physis[5] (Figure 4-46). The distal ulnar physis typically closes at approximately 9 to 12 months (Figure 4-47). All ulnar elongation distal to the elbow results from growth at the distal ulnar physis. In other words, the rate of growth from the distal ulnar physis must equal the rate of growth from the proximal and distal radial physes combined. The distal ulnar physis is V-shaped, which might increase its predisposition to growth retardation following concussion. During phases of rapid growth, especially in large dogs, cortical and periosteal remodeling can result in a highly

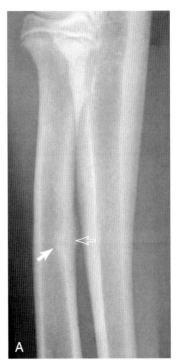
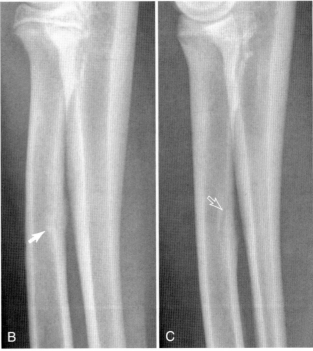

Figure 4-45. A, Lateral view of the proximal aspect of the antebrachium of a 2-year-old German Shepherd. Enthesopathy at the caudal aspect of the proximal radius in the region of the interosseous ligament *(hollow white arrow)* is common. There is marked variation in the appearance of this region. In this patient, there is also endosteal sclerosis *(solid white arrow)*. Lateral view of a 7-month-old Golden Retriever (**B**) and a similar view in a 9-month-old Labrador Retriever (**C**). In **B**, endosteal sclerosis, a common finding, is present *(solid white arrow)*, and in **C**, what is presumed to be the nutrient foramen is readily visible *(hollow white arrow)*.

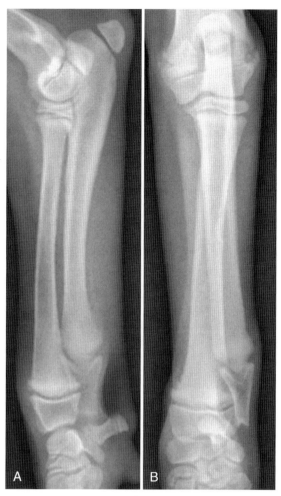

Figure 4-46. Lateral (**A**) and craniocaudal (**B**) views of the antebrachium of a 4-month-old Labrador Retriever. The physes associated with the distal humerus, proximal radius, distal radius, and ulna are readily apparent. The distal ulnar physis is typically V-shaped. The conical shape may predispose the physis to growth disruption following compression.

irregular appearance of the margin of the metaphysis; this is particularly true for the distal metaphysis of the ulna. The distal ulnar metaphyseal region is also where the diameter of the ulna decreases rapidly compared with the large immediately adjacent physis, further contributing to the marginal irregularity. This region of irregularity, called the *cutback zone*, is easy to misinterpret as an aggressive bone lesion, especially in the ulna (Figure 4-48). A derangement of physeal maturation in the distal antebrachium can result in the development of major conformational abnormalities and degenerative joint disease. Chondrodystrophoid dogs typically have a more curving radius and a mild valgus deformity of the manus. This may be associated with inherent asynchronous growth of the radius versus the ulna (Figure 4-49).

In the cat, long bones are typically straighter than in the dog, and the antebrachium is no exception. The feline ulna is larger distally compared with the canine ulna. Distally, it is lateral at the carpus and, as in the dog, does not contribute significantly to weight bearing (Figure 4-50).

The carpal pad lies palmar to the carpus. This can create superimposed soft tissue edges over the carpus, which can be radiographically misinterpreted as a fracture (Figure 4-51). In addition, lateral to the carpal pad there is a marked reduction in soft tissue volume immediately proximal to the antebrachiocarpal joint. This can result in the false impression of a destructive process in the distal lateral aspect of the radius and distal medial aspect of the ulna (Figure 4-52).

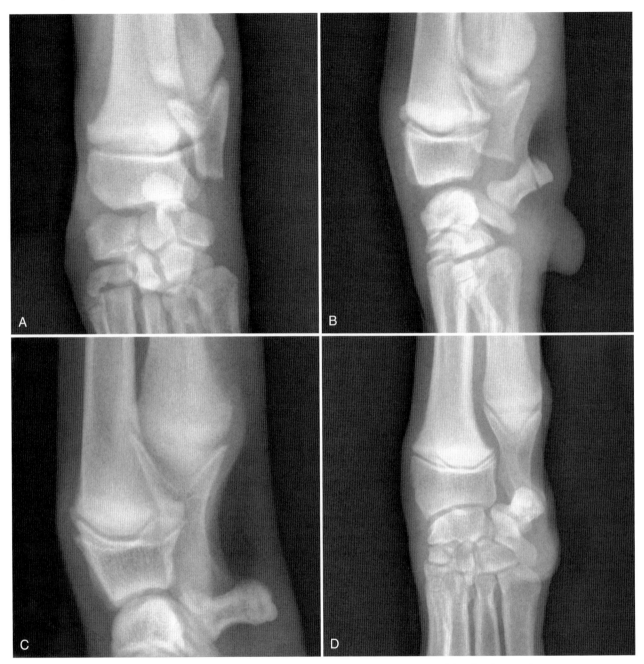

Figure 4-47. Dorsopalmar (**A**) and lateral (**B**) views of a 3-month-old Labrador Retriever. Note the blunted appearance to the styloid process of the ulna. This is normal at this age. In addition, in **B**, the accessory carpal bone physis is readily visible. This typically closes by 4 to 5 months. **C**, Lateral view of the distal antebrachium of a 3½-month-old German Shorthaired Pointer. The styloid process of the ulna is well developed compared with the patient in **A** and **B**, and the accessory carpal bone physis is fused. The caudal margin of the ulna is poorly defined. This is normal in young patients and is a manifestation of rapid growth and metaphyseal remodeling. **D**, Dorsolateral-palmaromedial oblique view of a 5-month-old Labrador Retriever. The obliquity enables unobstructed visualization of the distal ulnar physis.

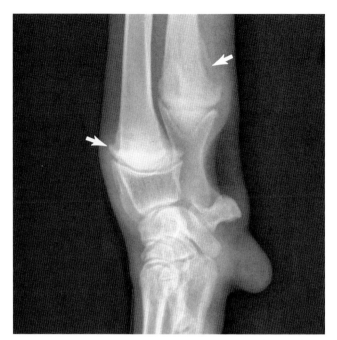

Figure 4-48. Lateral view of the distal antebrachium of a 5-month-old mixed breed dog. The metaphyseal cortex adjacent to the distal antebrachial physes is irregular *(solid white arrows)*. This is the normal cutback zone and is associated with rapid growth and remodeling.

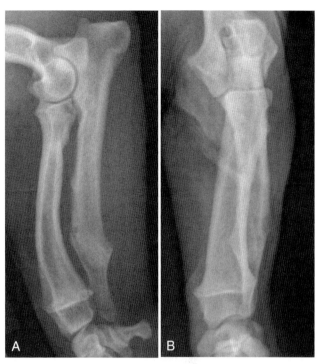

Figure 4-49. Lateral **(A)** and craniocaudal **(B)** views of a 10-month-old Pembroke Welsh Corgi. The antebrachium is short and the radius more bowed in chondrodystrophoid breeds. There is considerable variation in the appearance of the antebrachium in chondrodystrophoid breeds, and what denotes normal conformation is somewhat subjective. Assessing the elbow joint for congruency is an important component of this process. This patient has good elbow congruency, and, clinically, the limb had no obvious valgus or rotational deformity.

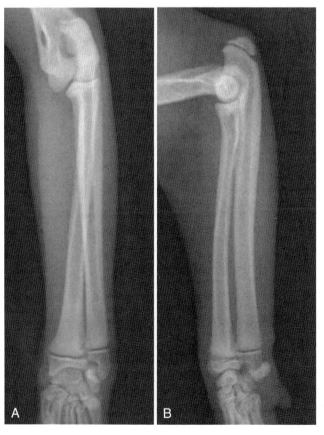

Figure 4-50. Craniocaudal **(A)** and lateral **(B)** views of a 4-month-old Domestic Shorthair cat. The feline distal ulna is larger than the canine. In this patient, the distal radial and ulnar physes are open and readily visible. The feline distal ulnar physis is transverse like the radius, unlike the canine, which is V-shaped. The supracondylar foramen at the mediodistal aspect of the humerus is readily visible.

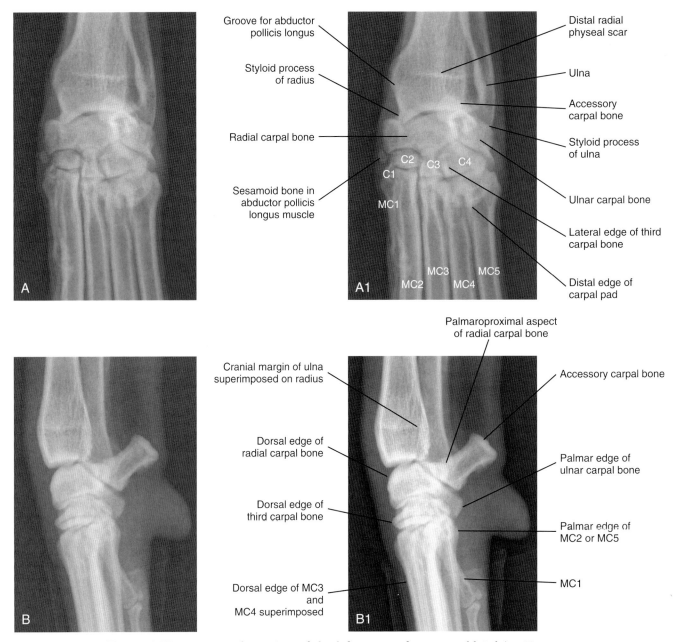

Figure 4-51. **A,** Dorsopalmar view of the left carpus of a 4-year-old Belgian Tervuren. **A1,** The same image as in **A,** with labels. The first, second, third, and fourth carpal bones are identified as such (*C1-C4*). The first, second, third, fourth, and fifth metacarpal bones are identified as such (*MC1-MC5*). Lateral view (**B**) and with labels (**B1**).

CARPUS

The compound joint known as the *carpus* is between the distal radius and ulna, and the proximal aspect of the metacarpal bones. It comprises the cuboidal carpal bones arranged in two rows and an occasional sesamoid bone. Carpal flexion occurs almost exclusively at the antebrachiocarpal joint. The carpus is a complex joint and, as with all such joints, a thorough radiographic assessment requires oblique projections in addition to the two standard orthogonal projections. Refer to the section on oblique projections in Chapter 1.

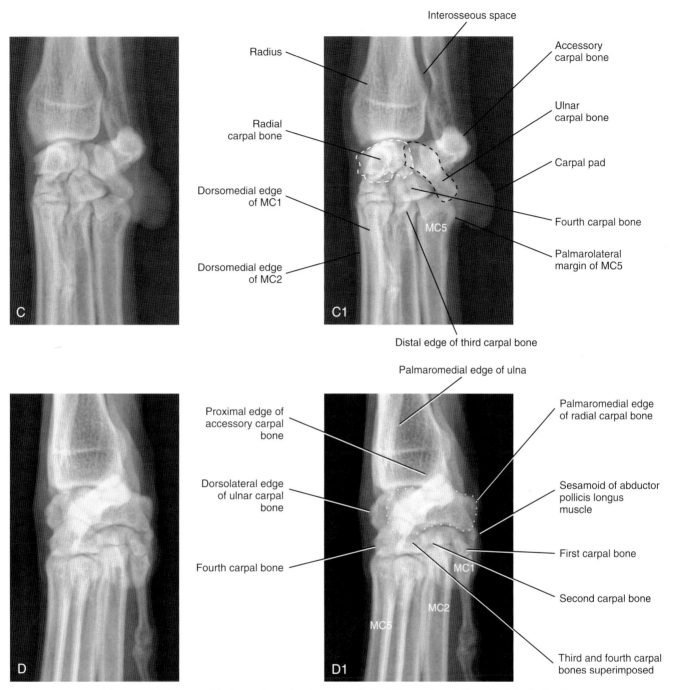

Figure 4-51, cont'd. Dorsolateral-palmaromedial oblique view (C), with labels (C1). In C1, the radial carpal bone is outlined by the *dashed white line;* the ulnar carpal bone is outlined by the *dashed black line.* Dorsomedial-palmarolateral view (D), with labels (D1). In D1, the radial carpal bone is outlined by the *dotted white line.*

The proximal row of carpal bones contains the radial, ulnar, and accessory carpal bones. The radial carpal bone, the largest cuboidal carpal bone, is primarily medial, and its proximal margin articulates with the distal aspect of the radius. The lateral margin articulates with the ulnar carpal bone, and the distal margin articulates with all four cuboidal carpal bones in the distal row. Medially, there is a portion of the radial carpal bone that is not articular; this is often confused with subluxation of the radial carpal bone (see Figures 4-51 and 4-53). The ulnar carpal

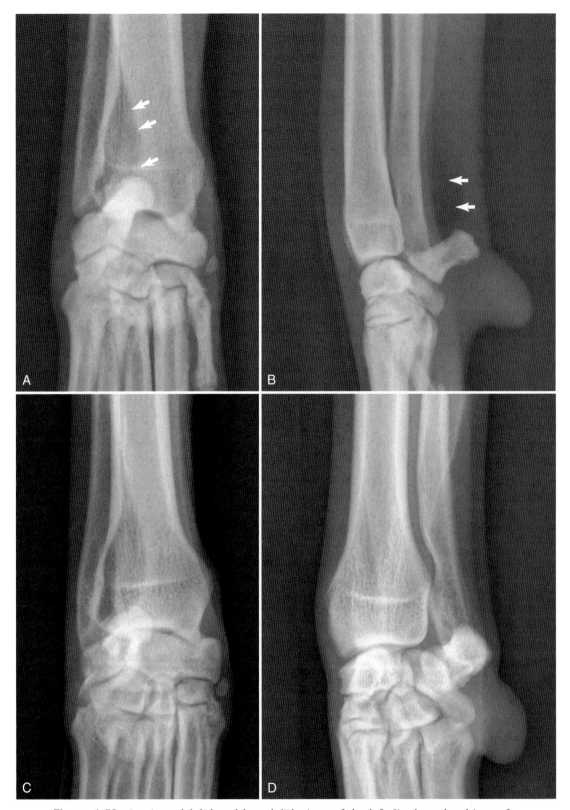

Figure 4-52. Craniocaudal (**A**) and lateral (**B**) views of the left distal antebrachium of a 7-year-old English Setter. There is a region of decreased opacity in the distal lateral aspect of the antebrachium in **A**, the medial margin of which is delineated by the *solid white arrows*. This is an artifact associated with the normal soft tissue void lateral and proximal to the carpal pad, the caudal margin of which is delineated in **B** by the *solid white arrows*. This should not be confused with an aggressive radial lesion. **C**, Craniocaudal view of the distal antebrachium of a 4-year-old Belgian Tervuren. The lucency over the distal lateral aspect of the radius is readily apparent. A craniolateral-caudomedial oblique view (**D**) is the optimal view to rule out this artifact.

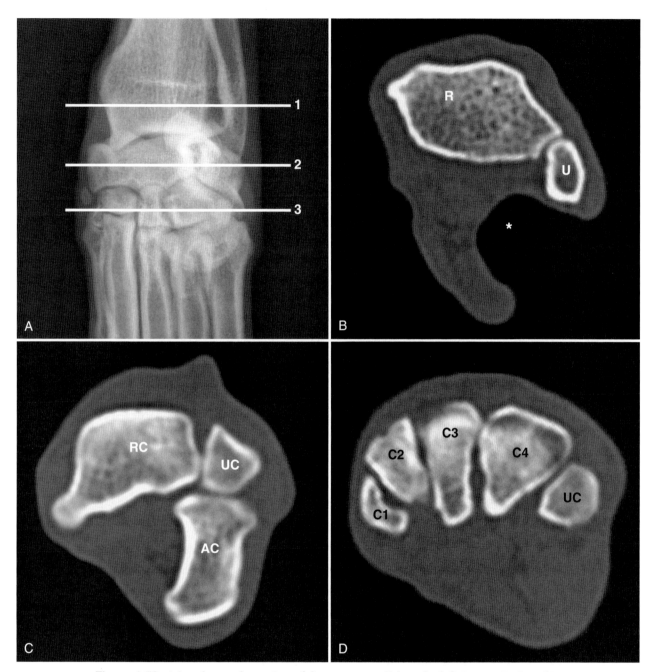

Figure 4-53. A is a dorsopalmar view of the carpus, medial on the left. Images **B, C,** and **D** are thin slice transverse CT images at the levels of 1, 2, and 3 respectively; dorsal is at the top of each image. In **B,** *R* is radius and *U* is ulna. Note the close association of the distal radius and ulna, immediately proximal to the antebrachiocarpal joint. Radiographically, the large soft tissue void palmar to the ulna *(white asterisk)* can create the false impression of a destructive process at the distal aspect of the radius and ulna. In **C,** *RC* is the radial carpal bone, *UC* is the ulnar carpal bone and *AC* is the accessory carpal bone. In **D,** each carpal bone is identified *C1-C5*. The most lateral bone, *UC,* is the distal aspect of the ulnar carpal bone. It articulates proximally with the radius, the ulna, the lateral aspect of the radial carpal bone, the accessory carpal bone, the fourth carpal bone, and the fifth metacarpal bone; hence, it is visible at the level of the distal row of carpal bones.

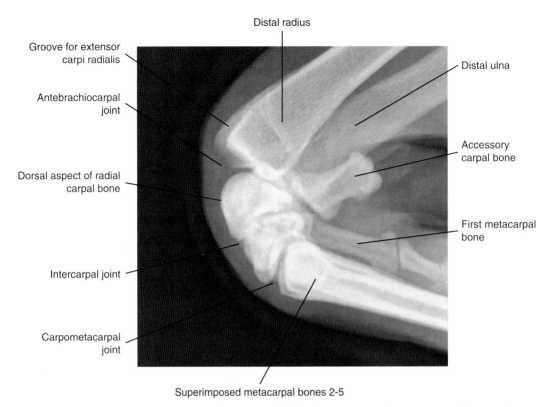

Figure 4-54. Flexed lateral view of a 6-year-old Labrador Retriever. The majority of the carpal range of motion occurs at the antebrachiocarpal joint. The flexed lateral view is particularly useful in the assessment of the proximal margin of the radial carpal bone.

bone articulates proximally with the radius, the ulna, the lateral aspect of the radial carpal bone, the accessory carpal bone, the fourth carpal bone, and the fifth metacarpal bone. Some dogs have a small sesamoid bone located at the distal medial aspect of the radial carpal bone. This sesamoid bone is located in the tendon of the abductor pollicis longus muscle and should not be confused with a chip fracture or joint mouse (see Figure 4-51). The distal row of carpal bones contains the first, second, third, and fourth carpal bones, which articulate with the radial and ulnar carpal bones proximally and their respective metacarpal bones distally (see Figure 4-51). The articulation between the antebrachium and proximal row of carpal bones is the antebrachiocarpal joint. The articulation between the proximal and distal rows of carpal bones is the intercarpal joint. The articulation between the distal carpal and metacarpal bones is the carpometacarpal joint. Almost all of the range of motion in flexion occurs at the antebrachiocarpal joint (Figure 4-54).

The feline distal antebrachium and carpus is similar to the canine with the exception of the ulna. Distally at the carpus, the feline ulna is more bulbous and rounded (Figure 4-55).

Stress radiography is often needed to definitively diagnose the nature of soft tissue injuries a joint may have sustained; this is particularly important in the carpus. When a soft tissue injury is suspected, pressure should be applied to the antebrachium and manus in a way to apply tension to the appropriate soft tissue structures (Figure 4-56). Appropriate recognition of carpal instability can markedly alter patient outcome.

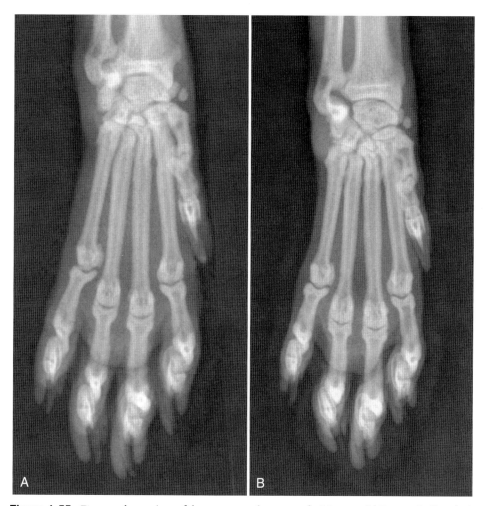

Figure 4-55. Dorsopalmar view of the carpus and manus of a 20-year-old Domestic Shorthair cat (**A**) and an 11-year-old Domestic Shorthair cat (**B**). The distal aspect of the ulna is larger than in the canine. A small radiolucent region is present in the distal aspect of the ulna. This is commonly present in the cat and does not reflect an aggressive lesion.

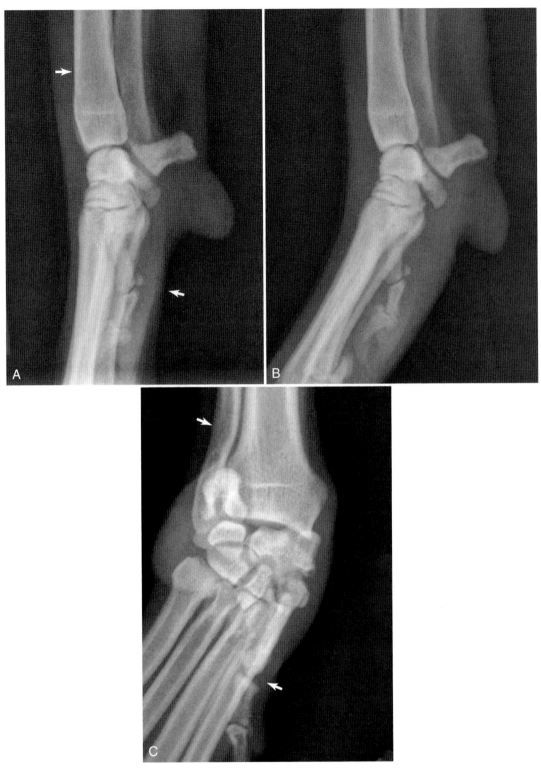

Figure 4-56. To examine the carpus radiographically while under stress, pressure should be applied such that the soft tissues under investigation are placed under tension. In the case of damage to the flexor retinaculum, either pressure can be applied in the direction of the *arrows* (**A**) or the limb can be positioned and loaded to mimic weight bearing. **B**, Stressed lateral view of the right forelimb of a 7-year-old English Setter. This patient sustained injury to the carpus. There is carpal hyperextension centered at the level of the intercarpal joint as a result of disruption to palmar soft tissues. Soft tissue swelling is evident palmar to the metacarpal bones. **C**, Dorsopalmar view of a 5-year-old Border Collie. Force has been applied in the direction of the *arrows,* stressing the medial collateral ligament and associated soft tissue structures. Excessive laxity is confirmed. This important finding may not be apparent on nonstressed views.

MANUS

The manus consists of five metacarpal bones and, most commonly, four digits, each of which has three phalanges. There is variability in the presence of the first metacarpal bone and associated phalanges, the dewclaw, secondary to either hypoplasia or removal at birth. The second through fifth metacarpophalangeal joints are each characterized by two sesamoid bones at the palmar aspect of each joint and one sesamoid bone at the dorsal aspect of each joint (Figures 4-57 and 4-58). The palmar sesamoid bones are numbered from medial to lateral, with the most medial sesamoid bone of the second metacarpophalangeal joint being sesamoid bone #1 and the most lateral sesamoid of the fifth metacarpophalangeal joint being sesamoid bone #8 (Figure 4-58). Occasionally, the

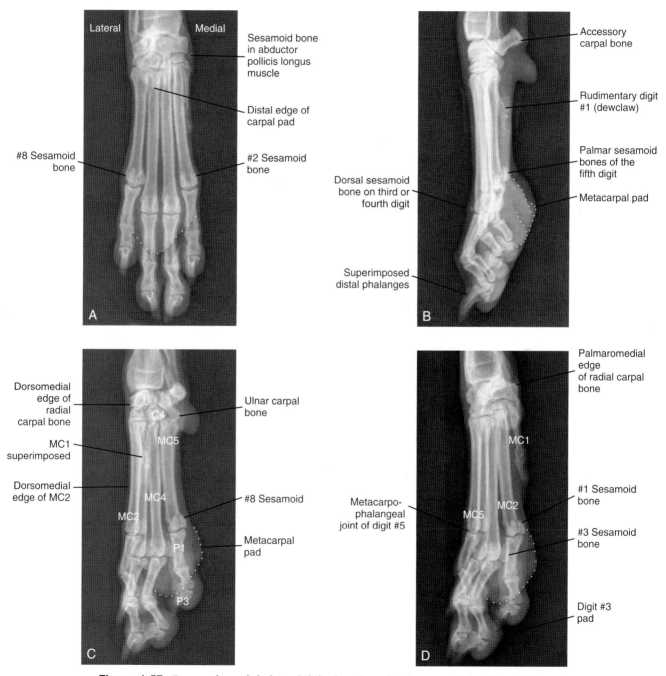

Figure 4-57. Dorsopalmar (**A**), lateral (**B**), dorsolateral-palmaromedial oblique (**C**), and dorsomedial-palmarolateral oblique (**D**) views of the carpus and manus of a 4-year-old Belgian Tervuren. Note in **A** how the medial aspect of the radial carpal bone does not articulate with the radius; this is often misinterpreted as a subluxation. The *dotted white line* in **A-D** is the metacarpal pad. Sometimes the opacity created by the carpal pad superimposed on the digits can give the false impression of an oblique phalangeal fracture.

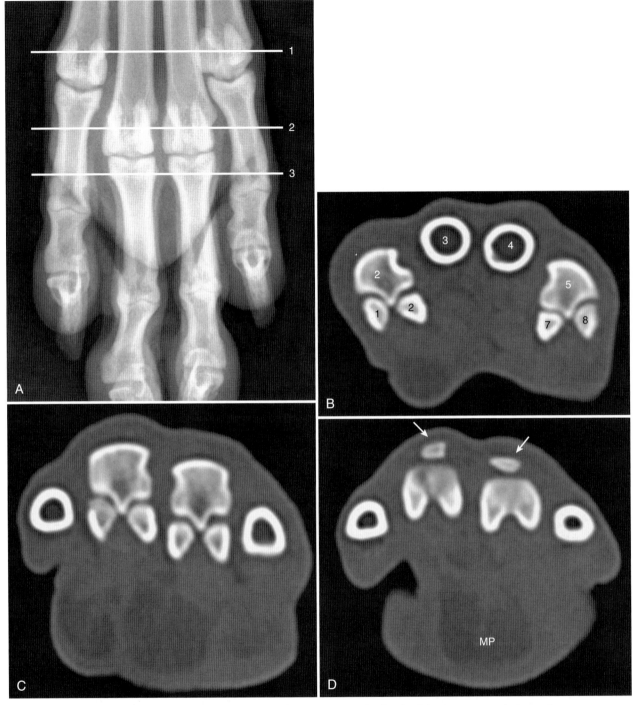

Figure 4-58. A is dorsopalmar view of the manus at the level of the metacarpophalangeal joints. Images **B**, **C**, and **D** are thin slice transverse CT images at levels 1, 2, and 3, respectively; dorsal is at the top of each image. In **B**, the second through fifth digits are identified as palmar sesamoid bones 1, 2, 7, and 8. The palmar sesamoid bones articulating with the distal metacarpal bones are readily visible in both **B** and **C**. In **D**, the small sesamoid bones at the dorsoproximal aspect of the proximal phalanx are readily visible *(solid white arrows)*. In **D**, the metacarpal pad (MP) is evident.

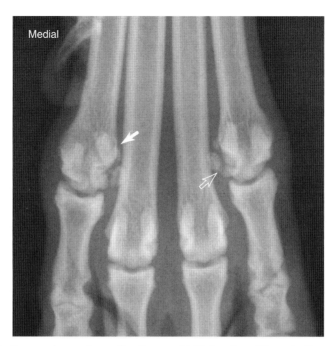

Figure 4-59. Dorsopalmar view of a 9-year-old Dalmatian. There is fragmentation and remodeling of the #2 palmar sesamoid bone *(solid white arrow)* and of the #7 palmar sesamoid bone *(hollow white arrow)*. These changes are often associated with incomplete sesamoid bone ossification, are often exacerbated by trauma, and may or may not cause lameness.

palmar sesamoid bones, particularly second and seventh, will be multipartite (Figure 4-59). This can result in lameness in some patients.[6]

The large metacarpal pad is palmar to the metacarpophalangeal joints. The shape of this pad is such that superimposition can result in linear oblique radiolucencies over the digits, which can be mistaken for fractures (see Figure 4-57). Each metacarpal bone consists of a base proximally and a head distally; the head articulates with the base of the associated proximal phalanx (see Figure 4-57). The proximal and middle phalanges consist of a base, body, and trochlea. The distal phalanges consist of a base proximally and an ungual process and associated unguis, or nail, distally (Figure 4-60).

Postnatally, there is a visible physis on the distal aspect of the second through fifth metacarpal (and metatarsal) bones that closes at approximately 6 to 7 months of age. There is no visible physis of the proximal aspect of the second through fifth metacarpal (or metatarsal) bones. In the proximal and middle phalanges of digits two through five, the visible physes are located proximally and these close earlier (Figures 4-61 and 4-62). There are no visible physes on the distal aspect of the proximal or middle phalanges of digits two through five. There is never a radiographically visible physis on the distal phalanx in the dog or cat.

Superimposition of the phalanges can sometimes be alleviated by applying traction to the digit of interest (Figure 4-63). The distal phalanx will be absent in cats that have been declawed at this level (Figure 4-64). Polydactyly is uncommon in both the dog and the cat but can be more common in the cat. It usually manifests as additional vestigial digits medially. These patients are typically asymptomatic (Figure 4-65).

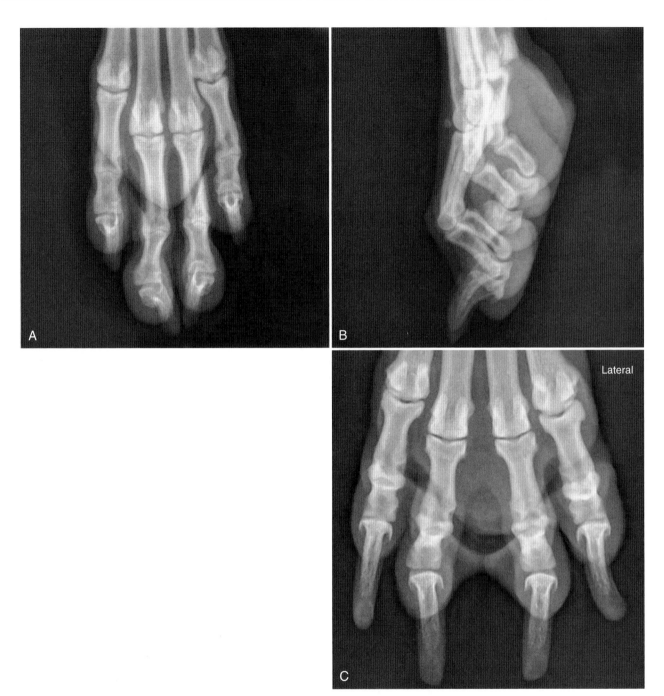

Figure 4-60. Dorsopalmar (**A**) and lateral (**B**) views of the digits of a 4-year-old Belgian Tervuren. In **A**, it is not possible to differentiate medial from lateral. Although the second and fifth digits are shorter than the third and fourth, and, typically, the fifth digit is marginally shorter than the second, this is unreliable; using radiographic markers or including the distal aspect of the carpus is necessary. There are paired sesamoid bones at the palmar aspect of each metacarpophalangeal joint and one sesamoid bone dorsally. **C**, Dorsopalmar view of the digits of a 4-year-old Cardigan Welsh Corgi. The distal border of the metacarpal pad is readily visible. There is some divergence between the third and fourth phalanges; this is commonly seen.

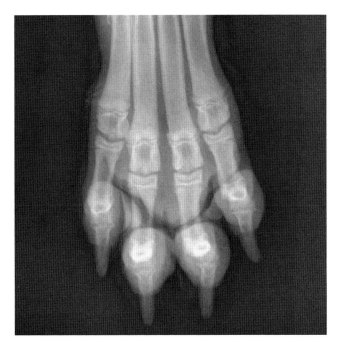

Figure 4-61. Dorsopalmar radiograph of a 4-month-old Labrador Retriever. Open distal metacarpal and proximal phalangeal physes are readily apparent.

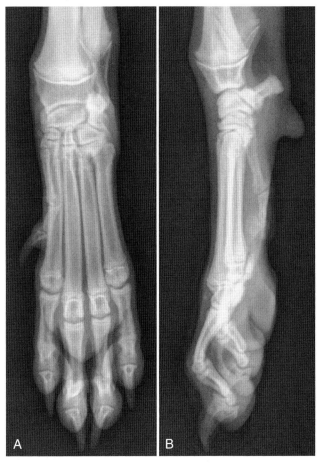

Figure 4-62. Dorsopalmar (**A**) and lateral (**B**) views of a 6-month-old mixed breed dog. The distal radial and ulnar physes are readily apparent. In addition, physes associated with the distal metacarpal bones have not yet closed.

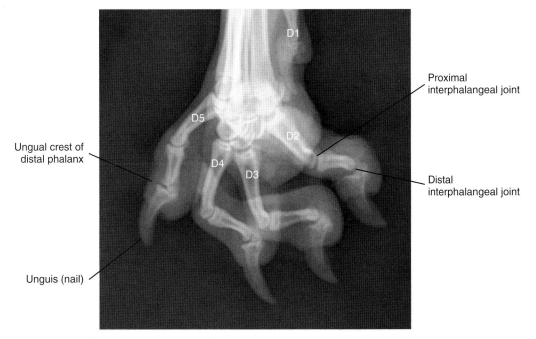

Figure 4-63. Lateral view of the digits of the manus, distracted apart (the *splayed toe view*). Each digit, with the exception of the first digit, also has an associated digital pad. The medial and laterally located second and fifth digits are typically shorter than the centrally located third and fourth digits.

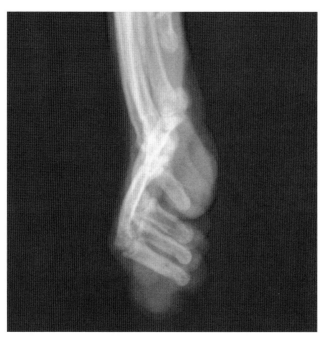

Figure 4-64. Lateral radiograph of the manus of a 9-year-old Siamese cat. The patient has been declawed. The distal phalanx is absent from all digits.

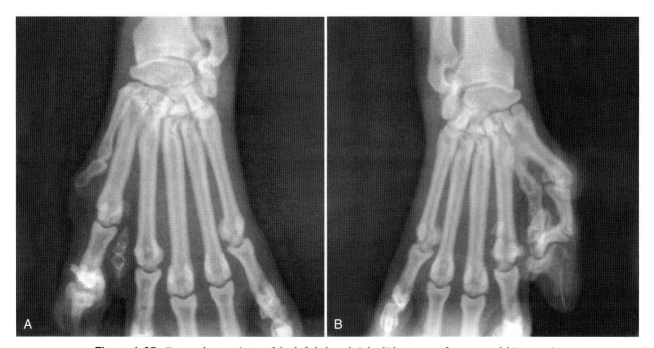

Figure 4-65. Dorsoplamar views of the left (**A**) and right (**B**) manus of a 6-year-old Domestic Shorthair cat. Polydactyly is present bilaterally. Supernumerary digits are most commonly medial as in this patient.

REFERENCES

1. Ticer J, editor: General principles. *Radiographic techniques in small animal practice*, Philadelphia, 1975, Saunders, pp 97-102.
2. Marcellin-Little D, DeYoung D, Ferris K, et al: Incomplete ossification of the humeral condyle in spaniels. *Vet Surg* 23(6):475-487, 1994.
3. Frazho J, Graham J, Peck J, et al: Radiographic evaluation of the anconeal process in skeletally immature dogs. *Vet Surg* 39(7):829-832, 2010.
4. Wood A, McCarthy P, Howlett C: Anatomic and radiographic appearance of a sesamoid bone in the tendon of origin of the supinator muscle of dogs. *Am J Vet Res* 46(10):2043-2047, 1985.
5. Riser W: The dog: his varied biologic makeup and its relationship to orthopedic diseases. *J Am Anim Hosp Assoc* 1985. Monograph.
6. Read R, Black A, Armstrong S, et al: Incidence and clinical significance of sesamoid disease in Rottweilers. *Vet Rec* 130(24):533-535, 1992.

CHAPTER 5

The Pelvic Limb

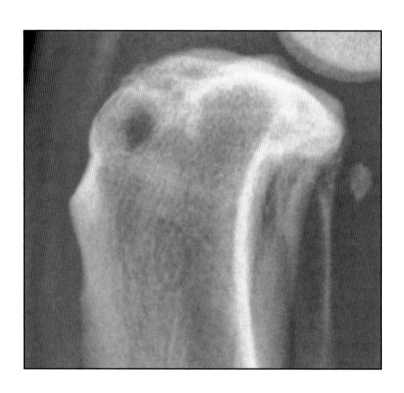

PELVIS

When radiographing the pelvis, it is typical to obtain ventrodorsal and lateral views. For the ventrodorsal view, the pelvic limbs can be either extended or flexed. The projection with the pelvic limbs extended is referred to as the *extended hip projection* and the projection with the pelvic limbs flexed the *frog-leg projection*. For the lateral view, it is typical for the pelvic limb closest to the imaging plate to be pushed cranially and for the opposite, nondependent, limb to be pulled caudally to avoid overlapping the femurs.

The pelvis is composed of four paired bones: ilium, ischium, pubic bone, and acetabular bone. The *acetabular bone* is a small triangular bone that lies between the caudoventral aspect of the ilium and the cranioventral aspect of the ischium (Figures 5-1 and 5-2). The acetabular bone fuses with adjacent bones at approximately 12 weeks of age to form the ventral portion of the acetabulum.[1] Before fusion of the acetabular bone, there are many radiolucent junctions in the region of the acetabulum. These are between the cranial aspect of the acetabular bone and the ilium; between the caudal aspect of the acetabular bone and the ischium; and dorsal to the acetabular bone between the ilium and the ischium (see Figure 5-2). Thus, superimposition of all of these radiolucent junctions makes it difficult to identify the acetabular bone radiographically in its entirety, although portions of it can be seen before complete fusion occurs (Figure 5-3). Care should be taken to avoid misinterpreting the numerous cartilaginous junctions in the acetabular region of young animals as fractures (Figure 5-4). Superimposition of fecal material with the pelvis can also lead to an incorrect diagnosis of a pelvic fracture, especially in lateral radiographs (Figure 5-5).

The ilium and sacrum join at the *sacroiliac joint*, a combined synovial and cartilaginous joint that is united by a thin joint capsule. The union between the ilium and sacrum is a *synchondrosis*.[2] The cartilage between the ilium and sacrum creates a radiolucent line separating the two structures. This radiolucency persists throughout life and is commonly confused with a sacroiliac fracture or subluxation in patients of all ages (Figures 5-6 and 5-7). In a ventrodorsal radiograph, the sacroiliac junction appears to be composed of multiple longitudinal radiolucent lines rather than a single line because the articulation is not perfectly opposing. The interdigitations create alternating lines of opacity and lucency (Figures 5-8 and 5-9). The caudal margin of the normal sacroiliac joint is characterized by a smooth transition at the junction of the ilium and the sacrum (see Figure 5-7); the character of this transition is the key to detecting sacroiliac subluxation. If there is subluxation of the ilium with respect to the sacrum, a *malalignment*, or step, will be formed at this transition.

The ilium is divided subjectively into a cranially located wing, the most cranial portion of which is the *iliac crest*, and a caudally located body (Figure 5-10). The iliac crest is fused with the wing of the ilium in most dogs by 2 years of age, but a permanently incomplete fusion occurs in more than 10% of dogs[3] (Figure 5-11). This incomplete fusion can be misinterpreted easily as a

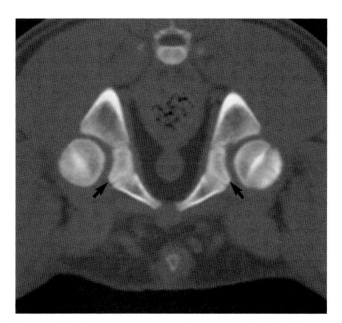

Figure 5-1. Transverse computed tomography (CT) image through the acetabular region of a 3-month-old Golden Retriever. The acetabular bone is visible at the central region of the acetabulum *(arrows)*.

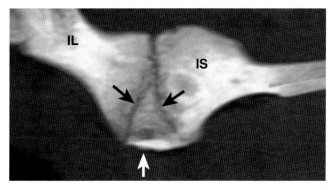

Figure 5-2. Maximum Intensity Projection (MIP) CT image of the medial aspect of the acetabulum from a 4-month-old Golden Retriever. The triangular segment of bone at the ventromedial aspect of the acetabulum *(black arrows)* is the acetabular bone prior to fusion. Cranial is to the left. The cartilaginous junction between the various components of the pelvis is clear from this image. Note that a Maximum Intensity Projection image is a composite, or summation, of 2-dimensional CT slice images to create a volumetric image. The thickness of the volume is controlled by the number of individual slices that are summed. *Il*, Ileum; *Is*, ischium.

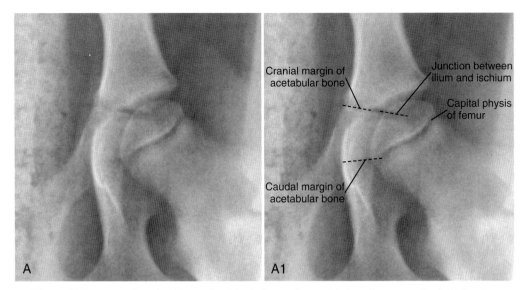

Figure 5-3. Ventrodorsal radiograph of the left coxofemoral joint of a 4-month-old Labrador Retriever (A) and corresponding labeled radiograph (A1). The acetabular bone is the triangular opacity that makes up the central portion of the acetabulum.

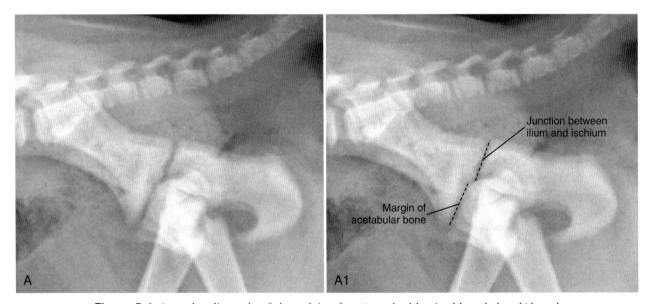

Figure 5-4. Lateral radiograph of the pelvis of a 10-week-old mixed-breed dog (A) and corresponding labeled radiograph (A1). The cartilaginous junction between the ilium and the ischium, and at the cranial margin of the acetabular bone, is quite conspicuous and can be misdiagnosed as a fracture. In A1, the margin of the acetabular bone that is identified is the cranial margin; the caudal margin of the acetabular bone cannot be seen in this radiograph.

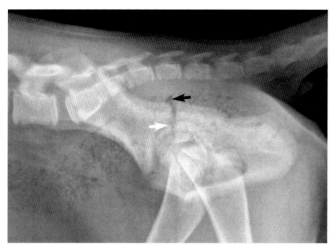

Figure 5-5. Lateral radiograph of the pelvis of a 4-month-old Labrador Retriever. There is a large amount of feces in the rectum. A vertical linear gas collection in the fecal matter superimposed on the ilium and the acetabulum *(white arrow)* could be misinterpreted as a fracture. However, upon careful inspection, the radiolucent line extends peripheral to the pelvis *(black arrow)*, indicating that it cannot be part of the pelvis.

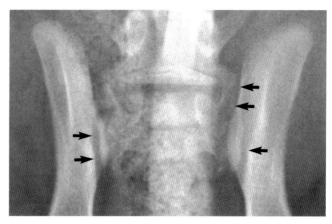

Figure 5-6. Ventrodorsal radiograph of the sacroiliac region of a 4-month-old Labrador Retriever. The cartilaginous junction between the ilium and the sacrum is of soft tissue opacity, and therefore appears more lucent than adjacent bones *(arrows)*, which can be confused with a fracture or subluxation. Fecal material is superimposed on the right sacroiliac joint, making a portion of the joint indistinguishable.

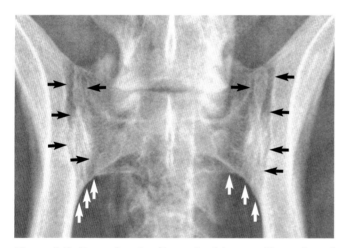

Figure 5-7. Ventrodorsal radiograph of the sacroiliac region of a 5-year-old Australian Shepherd. The cartilaginous junction between the ilium and sacrum persists throughout life, meaning there will always be a radiolucent region between these two bones *(black arrows)*. This should not be confused with an iliosacral subluxation or fracture. The smooth transition between the ilium and the sacrum *(white arrows)* is evidence that there is no malalignment or subluxation. This transition will have a slight malalignment, or a *step*, when there is a subluxation of the sacroiliac joint.

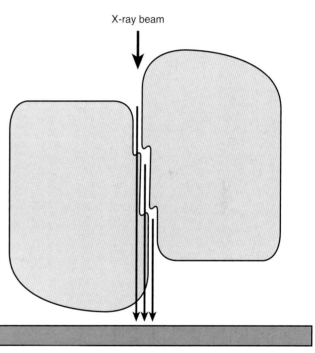

Figure 5-8. Drawing illustrating why multiple radiolucent lines will be created from the junction of two structures that have an irregular edge at the junctional point. Here, two objects are touching but the junction is composed of curving surfaces. In every location where the curving surface deviates, a separate region of decreased x-ray attenuation will be created *(vertical black arrows)*. This will create a series of radiolucent lines and the appearance of the junction of multiple structures, when in fact there is only one structure creating the junction. The sacroiliac junction is an example of this type of junction but any joining of structures of this general shape will be subject to this effect.

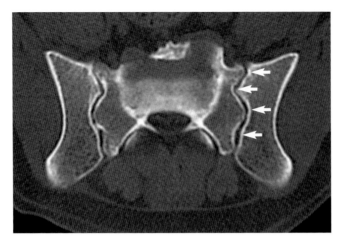

Figure 5-9. Transverse CT image through the sacroiliac joint in a 12-year-old Labrador Retriever; the ventral aspect of the dog is at the top of the image. Note the interdigitations and undulating nature of the sacroiliac junction *(arrows)*. The x-ray beam striking the ventral aspect of the subject cannot project a structure with this configuration as a single radiolucent line; rather, a series of adjacent radiolucencies are projected.

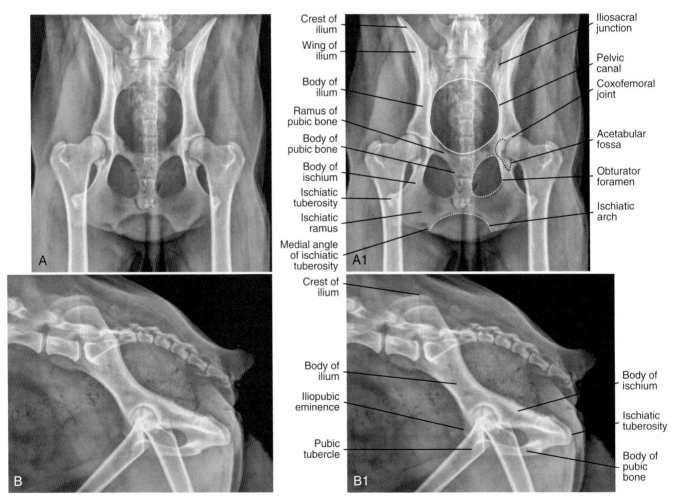

Figure 5-10. Ventrodorsal **(A)** and lateral **(B)** radiographs of the pelvis of a 7-year-old Australian Shepherd and corresponding labeled radiographs **(A1, B1)**.

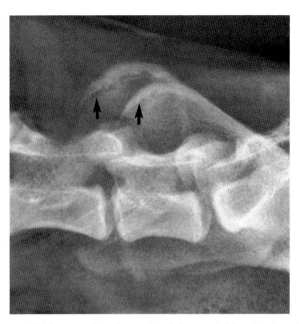

Figure 5-11. Lateral radiograph of the caudal lumbar/cranial pelvic region of a 6-year-old Labrador Retriever. The crest of the ilium has not fused to the wing *(arrows)*. This is commonly encountered in dogs and should not be misinterpreted as a fracture.

fracture. Usually the nonfused iliac crest is seen in the lateral view. In the ventrodorsal view, the iliac crest is superimposed on the wing of the ilium and any incomplete union is typically not visible. However, if there is slight obliquity to the pelvis in the ventrodorsal view, the radiolucent junction between the iliac crest and wing of the ilium can occasionally be seen in that view as well (Figure 5-12).

The acetabulum is formed at the confluence of the ilium, the ischium, and the pubis. It is a major articular component of the coxofemoral joint. The acetabulum is characterized by a nonarticular depression, the *acetabular fossa*, that occupies the central portion of the articular surface in the ventrodorsal view (see Figure 5-10, *A1* and Figures 5-13 and 5-14). Radiographic assessment of joint laxity, as for canine hip dysplasia detection, is commonly done using the extended hip ventrodorsal view. In this view, congruency should be evaluated cranial to the acetabular fossa, which corresponds roughly to the cranial third of the joint space (see the region of the *white arrowheads* in Figure 5-14). Since the acetabular fossa disrupts the smooth articulation between the femoral head and the acetabulum, congruency cannot be assessed in this area. In some young dogs, there is a small separate center

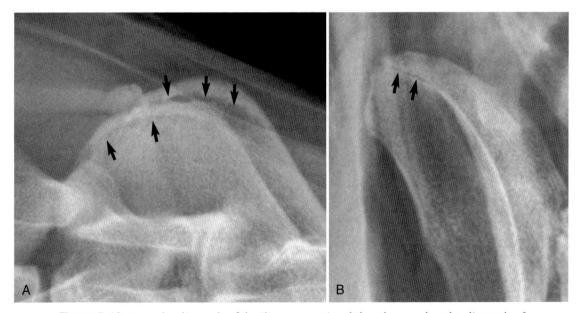

Figure 5-12. Lateral radiograph of the iliac crest region **(A)** and ventrodorsal radiograph of the right iliac crest **(B)** of a 10-year-old German Shepherd. The iliac crest has not fused to the wing of the ilium, creating the expected radiolucency in the lateral view (*black arrows* in **A**). In this dog, the pelvis was slightly oblique when the ventrodorsal view was made, so the separation between the iliac crest and the iliac wing can also be seen in this view (*black arrows* in **B**).

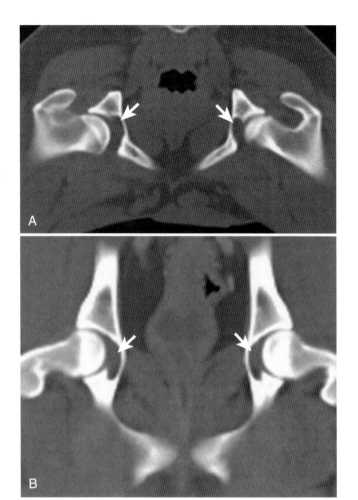

Figure 5-13. Transverse (A) and dorsal (B) CT images of the acetabular region from a 1-year-old Golden Retriever showing the characteristics of the acetabular fossa *(black arrows)*.

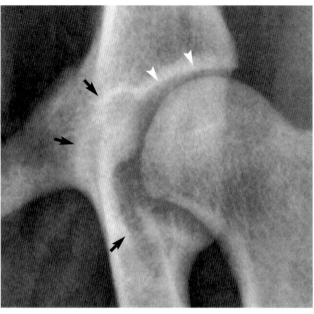

Figure 5-14. Radiograph of the left acetabulum of a 7-year-old Australian Shepherd. The acetabular fossa *(black arrows)* occupies the central portion of the acetabulum. Congruency of the coxofemoral joint should be assessed cranial to the acetabular fossa *(white arrowheads)* but not in the region of the acetabular fossa. The coxofemoral joint in this dog is very mildly incongruent, based on the joint space being slightly wider just cranial to the acetabular fossa than at the most craniolateral aspect of the joint.

of ossification located at the craniolateral margin of the acetabulum; this will eventually fuse with the ilium and should not be confused with a fracture fragment (Figures 5-15 and 5-16).

Two additional, supplementary, radiographic projections are sometimes used to assess the acetabulum. One is the *dorsal acetabular rim (DAR)* view, which, obvious from the name, is designed to assess the configuration of the dorsal acetabular rim. To acquire the DAR view, the anesthetized patient is placed in sternal recumbency and the pelvic limbs are pulled cranially. If correctly positioned, the center of the x-ray beam will pass through the long axis of the ilial shaft and result in the superimposition of the wings of the ilium, the body of the ilium, the acetabulum, and the tuber ischii. This will result in an unobstructed projection of the dorsal rim of the acetabulum[4] (Figure 5-17). In the DAR view of a normal dog, the dorsal aspect of the coxofemoral joint space should be oriented horizontally, and the rim of the acetabulum should be relatively pointed (Figure 5-18).

The second supplementary acetabular view is an *open-leg lateral* view of the acetabulum, which provides an oblique, unobstructed lateral view of the acetabulum and femoral head. To acquire this view, the subject is

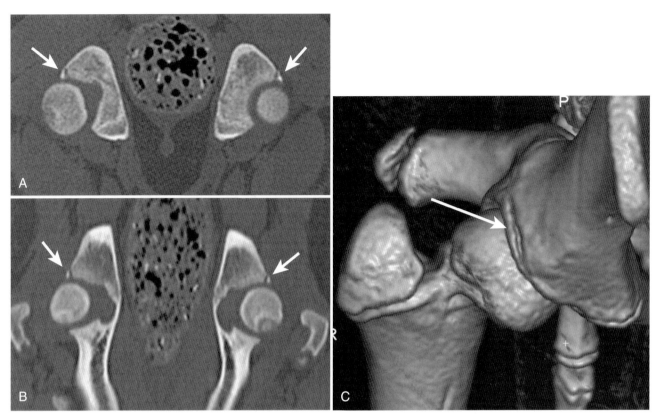

Figure 5-15. Transverse (**A**) and dorsal (**B**) CT images, and a 3-dimensional CT volume rendering (**C**) of the acetabular region of the pelvis from a 4-month-old Golden Retriever. Note the small separate center of ossification at the craniolateral aspect of the acetabulum *(arrows)*. This can be misinterpreted as a fracture in radiographs. This center will fuse with the acetabulum during the subsequent few months and be inconspicuous by 1 year of age.

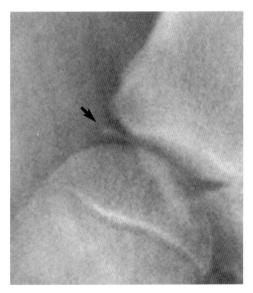

Figure 5-16. Ventrodorsal radiograph of the right acetabular region of a 4-month-old Weimaraner. There is a small opacity adjacent to the craniolateral aspect of the acetabulum due to a separate ossification center *(arrow)*. This should not be confused with a fracture fragment.

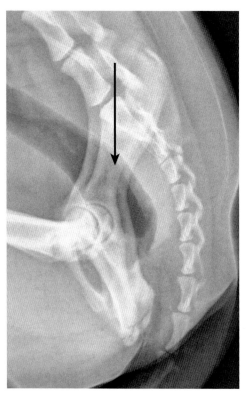

Figure 5-17. Representation of the orientation of the x-ray beam *(black arrow)* and pelvis for acquisition of a dorsal acetabular rim (DAR) view. The x-ray beam strikes the dorsal surface of the acetabulum tangentially, allowing its configuration to be assessed.

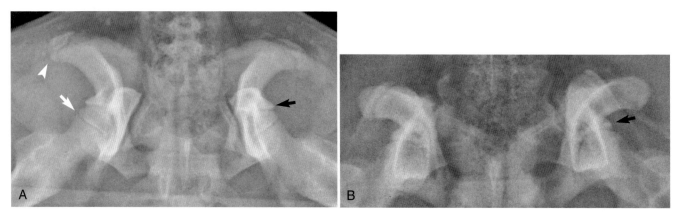

Figure 5-18. Dorsal acetabular rim views of a 7-month-old Vizsla **(A)** and an 11-month-old Mastiff **(B)**. In **A**, the dorsal acetabular rim is indicated by the *black arrow*. Note how the coxofemoral joint space just beneath the acetabular rim is oriented horizontally and that the acetabular rim is relatively pointed. In this young dog, the capital physis is also open *(white arrow)*. The large protuberance *(white arrowhead)* is the tuber ischii; its physis is also visible. The dog in **B** is slightly older and was provided for comparison. The acetabular rim is indicated by the *black arrow*.

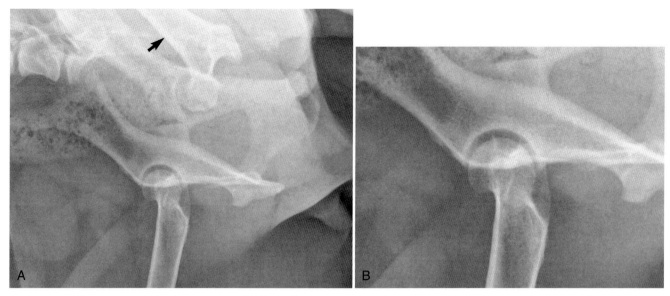

Figure 5-19. Open-leg lateral radiograph **(A)** and close-up view **(B)** of the right coxofemoral joint of a 4-year-old mixed breed dog. Note the unobstructed view of the coxofemoral joint. In **A**, note the position of the left femur *(arrow)*, having been abducted laterally and dorsally.

positioned in lateral recumbency with the coxofemoral joint of interest placed dependently (down leg), against the x-ray table. The nondependent (up) leg is abducted laterally and dorsally to the extent that its final position is actually dorsal to the lumbar spine. The x-ray beam is then centered on the dependent hip joint of interest. The open-leg lateral view provides an unobstructed lateral view of the acetabulum and proximal femur (Figure 5-19).

The ischium, which is composed of a body, ramus, and tuberosity, contributes to the formation of the acetabulum, the obturator foramen, and the pubic symphysis (see Figure 5-10).[1] The *pubic symphysis* is characterized by a fibrous union in most dogs younger than 5 years of age; after that time there is usually complete bony fusion (Figure 5-20).[2] Before fusion, the relative radiolucency of the fibrous union of the pubic symphysis can be misinterpreted radiographically as a fracture (Figure 5-21). In male dogs, superimposition of fecal material, the caudal vertebrae, and the prepuce and os penis decreases the conspicuity of the pubic symphysis in many ventrodorsal radiographs by creating confusing summation opacities. The pubic symphysis is not visualized in lateral radiographs because the primary x-ray beam does not strike the junction head-on in this projection.

The caudal portion of the ischium, the *ischiatic tuberosity*, develops from a separate ossification center

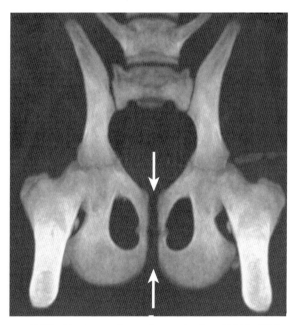

Figure 5-20. Maximum intensity projection (MIP) CT image of the pelvis of a 4-month-old Labrador Retriever from the ventral perspective. Note the large fibrous union at the pelvic symphysis *(black arrows)*. (A Maximum Intensity Projection, or MIP, image is a composite, or summation, of 2-dimensional CT slice images to create a volumetric image. The thickness of the volume is controlled by the number of individual slices that are summed.)

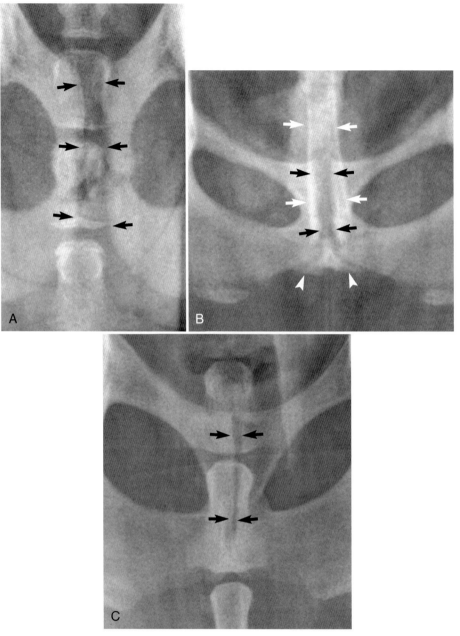

Figure 5-21. Ventrodorsal radiographs of the pubic symphysis region of an 11-week-old Golden Retriever (**A**), an 18-month-old Welsh Corgi (**B**), and a 1-year-old Golden Retriever (**C**). Various stages of pubic symphysis (*black arrows* in each illustration) fusion can be seen in these radiographs. In **A** and **C**, the caudal vertebrae are superimposed on the pubic symphysis. In **B**, the os penis (*white arrows*) is superimposed on the pubic symphysis. In **B**, there is a triangular segment of bone at the caudal aspect of the pubic symphysis (*white arrowheads*); this is a separate center of ossification of the medial portion of the ischium at the ischiatic arch.

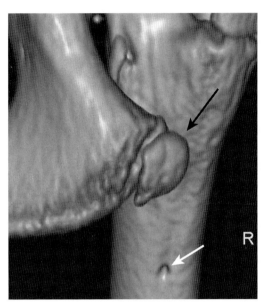

Figure 5-22. Three-dimensional CT volume rendering of the dorsal aspect of the right ischium from a 4-month-old Golden Retriever. Note the separate center of ossification for the tuber ischii *(black arrow)*. The defect in the caudal cortex of the femur *(white arrow)* is the nutrient canal.

(Figures 5-22 and 5-23). Typically, ossification of the ischiatic tuberosity begins at the most lateral aspect of the ischium (see Figure 5-23, *A*). With age, the lateral aspect of the ossification center fuses with the ischium, and ossification of the cartilaginous tuber ischii continues medially (see Figure 5-23 *B, C,* and *E*). As ossification of the tuber ischii proceeds, some dogs will also develop a secondary center of ossification at the caudal aspect of the pubic symphysis, associated with the ischiatic arch (see Figures 5-21, *C*, and 5-23, *C* and *E*). In lateral radiographs of the pelvis, the secondary ossification centers of the caudal aspect of the pelvis are easily confused with fractures (see Figure 5-23, *D, F,* and *G*).

The pubic bone is composed of two rami and a body (see Figure 5-10). The pubic bone and the ischium bound the obturator foramen (see Figure 5-10). Assessing the symmetry of the obturator foramina in ventrodorsal radiographs is a good method to determine whether any rotation of the pelvis was present during radiography. Assessing the degree of rotation is important because rotation influences the apparent radiographic depth of the acetabulum and radiographic congruity of the coxofemoral joint. With the dog in dorsal recumbency, the obturator foramina are not normally parallel with the dorsal plane; the plane of each foramen is angled from ventromedially to dorsolaterally. As a result, when the long axis of the body is rotated, the obturator foramen on the side opposite of the direction of rotation appears larger, and the coxofemoral joint on that side appears to have greater depth and congruency than the opposite coxofemoral joint (Figures 5-24 and 5-25). In addition, the iliac crest on the side of the direction of rotation appears wider because it is struck more *en face* by the primary x-ray beam (see Figure 5-25).

In male dogs, the os penis and prepuce superimpose on the pelvis in ventrodorsal radiographs. The opacity

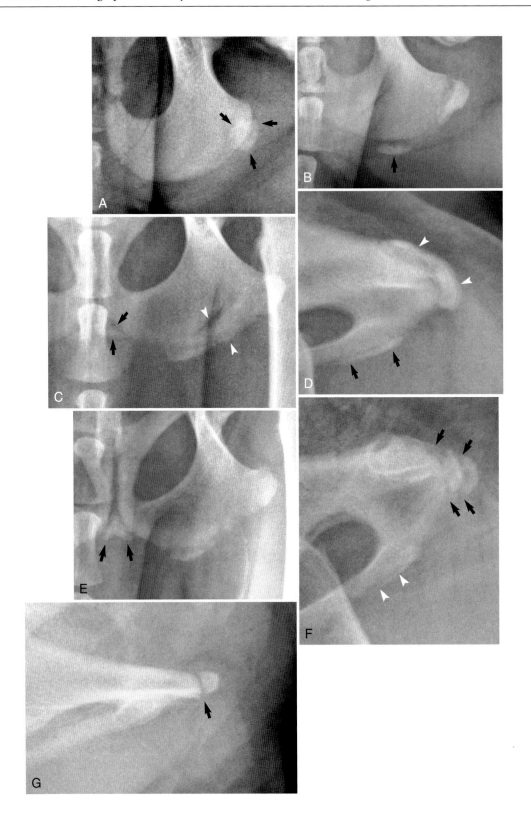

Figure 5-23. A, Ventrodorsal radiograph of the left tuber ischii of a 14-week-old Golden Retriever. B, Ventrodorsal radiograph of the left tuber ischii of a 7-month-old Labrador Retriever. Ventrodorsal radiograph of the left tuber ischii (C) and lateral radiograph of the tuber ischii (D) of an 8-month-old Labrador Retriever. Ventrodorsal radiograph of the left tuber ischii (E), lateral radiograph of the tuber ischii (F), and open-leg lateral (G) radiographs of the left tuber ischii of a 1-year-old Labrador Retriever. In the 14-week-old dog (A), the secondary ossification center of the tuber ischii appears as a focal opacity at the lateral margin of the ischium *(arrows)*. In the 7-month-old dog (B), the development of the ossification center of the lateral aspect of the tuber ischii has progressed and is nearly fused, but there is now ossification of the more medial aspect of the ossification center *(arrow)*. In the 8-month-old dog (C-D), the lateral aspect of the ossification center of the tuber ischii has fused and ossification of the medial portion of the secondary center has progressed *(black arrows* in C). Also in this dog is a triangular opacity (the patient's left side marked with *white arrowheads* in C) that represents a secondary center of ossification associated with the ischial arch. The incompletely fused tuber ischii *(black arrows* in D) and incompletely developed ischiatic arch *(white arrowheads* in D) could be confused with fractures in the lateral view. The 1-year-old dog (E-G) has tuber ischii ossification similar to the 8-month-old, but the separate center of ossification associated with the ischiatic arch is more conspicuous *(arrows* in E) because there is not overlap of feces or the caudal vertebrae. As in the 8-month-old dog, the incompletely developed tuber ischii *(black arrows* in F) and ischiatic arch *(white arrowheads* in F) could be mistaken for fractures. In the open-leg lateral view of the pelvis, the x-ray beam strikes the cleavage plane between the ossification center of the tuber ischii directly, leading to a very conspicuous radiolucent region *(arrow* in G) that could even more likely be confused with a fracture.

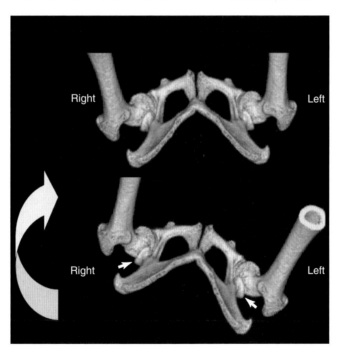

Figure 5-24. Caudal view of a canine pelvis. The *top* depiction represents a dog positioned symmetrically for a pelvic radiograph. The obturator foramina slope dorsally (remember, the dog is in dorsal recumbency) from medial to lateral. The congruity of the coxofemoral joints will appear equal with this positioning. In the *bottom* depiction, the pelvis is rotated in a clockwise direction when viewed from its caudal aspect; this equates to a dog in dorsal recumbency having the sternum rotated to its left. As a result, the right obturator foramen will appear larger in the radiograph and the left obturator foramen will appear smaller. In the bottom depiction, when looking at the relative position of the dorsal rim of each acetabulum *(arrows)*, it is clear that the coverage of the right femoral head by the acetabulum appears greater than the coverage of the left femoral head; the right acetabulum looks deeper and the joint more congruent. This type of rotation can create a false assessment of joint laxity and/or depth.

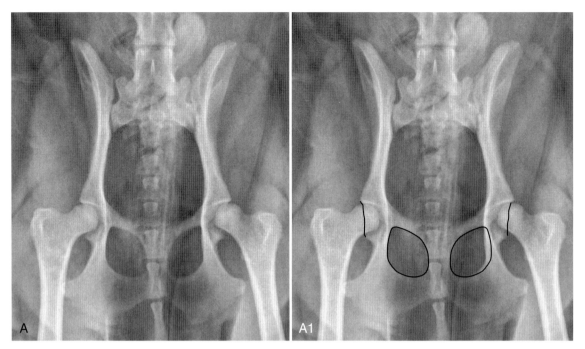

Figure 5-25. **A,** Ventrodorsal radiograph of the pelvis of a 2-year-old Shetland Sheepdog. The pelvis was rotated slightly during radiography, with the sternum rotated slightly to the dog's left. This creates the appearance of a deeper coxofemoral joint on the right and a shallower coxofemoral joint on the left. In the labeled radiograph **(A1),** the *outline* of the right obturator foramen has been duplicated and superimposed on the left obturator foramen, and it is clear that the right obturator foramen is larger. The dorsal acetabular margin has been *outlined with lines;* note the appearance of increased coverage of the femoral head on the right due to the rotation, and the apparent reduced coverage on the left. Also note the apparent increased width of the left iliac crest; this will occur on the side toward which the longitudinal rotation occurred.

created by these structures can be misinterpreted as an abdominal mass, and can also interfere with critical evaluation of the pelvis (Figure 5-26). Anal sacs are generally not visible radiographically, but occasionally one or both contain gas, rendering them visible. This gas can be superimposed on the pelvis in ventrodorsal radiographs, creating a focal radiolucency that can be misinterpreted as a lytic bone lesion[5] (Figure 5-27). Anal sac gas usually superimposes on the ischium; its craniocaudal location can range from the medial aspect of the acetabulum to the region of the tuber ischii.[5]

In general, the overall shape of the pelvis of most dogs appears similar radiographically, regardless of breed. In cats, the overall shape of the pelvis is more rectangular than in the dog, with a proportionally longer ischium with respect to the length of the ilium (Figure 5-28). The general anatomic principles noted for the canine pelvis also apply to the feline pelvis.

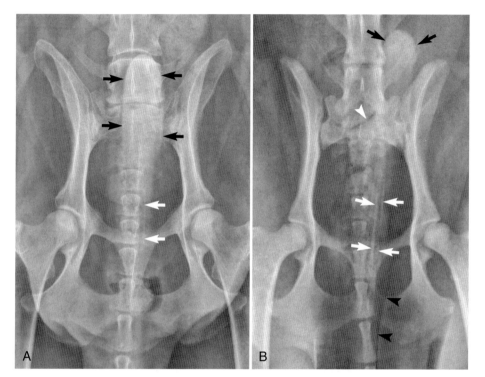

Figure 5-26. Ventrodorsal radiographs of the pelvis of a 14-year-old mixed breed dog (**A**) and a 2-year-old Shetland Sheepdog (**B**). In **A** and **B**, the prepuce *(black arrows)* creates an obvious opacity superimposed on the cranial aspect of the pelvic region. Superficial masses or objects that have a margin that is struck tangentially by the primary x-ray beam have a conspicuous edge; this is the rationale for the noticeable opacity created by the prepuce in ventrodorsal radiographs in male dogs. The os penis is more conspicuous in **B** than in **A** but can be seen in both dogs *(white arrows)*. In **B**, other confusing opacities are present. Gas trapped in feces *(white arrowhead)* and the margin of the tail base *(black arrowheads)* could be misinterpreted as fractures. Visualization of the margin of the tail base follows the same principle that accounts for increased conspicuity of the margin of the prepuce. The margin of the tail base clearly extends beyond the bone margin; this criterion is helpful in distinguishing a pseudofracture from a real fracture.

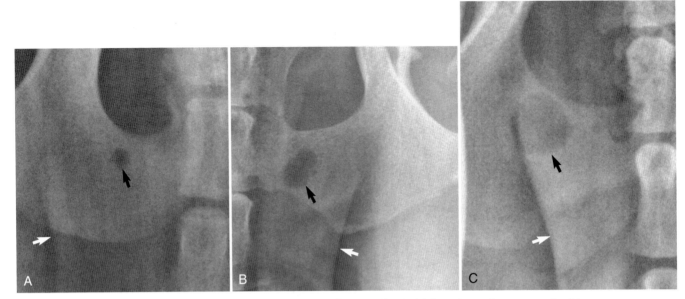

Figure 5-27. Ventrodorsal radiographs of the right ischial region of a 4-month-old Weimaraner (**A**), of the left ischial region of a 2-year-old German Shepherd (**B**), and of the right ischial region of a 1-year-old Labrador Retriever (**C**). In each dog, anal sac gas superimposed on the ischium creates the appearance of a lytic lesion *(black arrow)*. The margin of the base of the tail is visible in each dog *(white arrow)*.

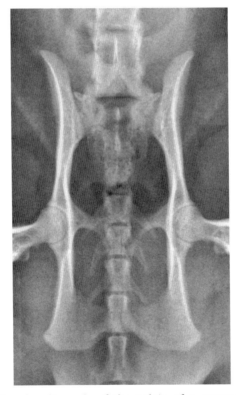

Figure 5-28. Ventrodorsal radiograph of the pelvis of a 7-year-old domestic cat. The pelvis is generally more rectangular than in the dog, and the ischium is slightly longer proportionally.

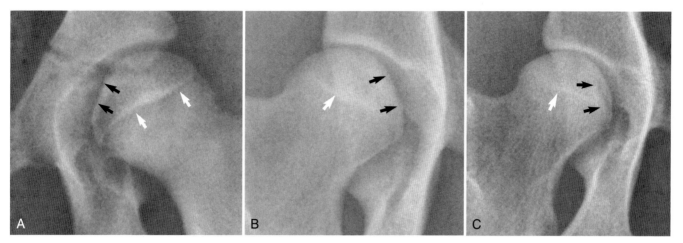

Figure 5-29. Ventrodorsal radiographs of the left coxofemoral joint of a 4-month-old Golden Retriever (**A**), the right coxofemoral joint of a 3-year-old Labrador Retriever (**B**), and the right coxofemoral joint of a 10-year-old Labrador Retriever (**C**). In **A** and **B**, the fovea capitis creates a flattened appearance to the medial aspect of the femoral head *(black arrows)*. In **C**, the fovea capitis is not struck tangentially by the primary x-ray beam due to the position of the femur during radiography. As a result, the fovea capitis creates a generalized radiolucent region rather than a change in shape of the femoral head *(black arrows)*. The physis of the femoral head epiphysis, also called the capital epiphysis, is visible in **A** *(white arrows)*. In **B** and **C**, a thin sclerotic line is present at the site of the closed physis of the femoral head *(white arrows)*.

FEMUR AND STIFLE

The most commonly acquired radiographic projections of the femur and stifle are lateral and craniocaudal (or caudocranial) views, and these are the projections illustrated herein.

The proximal aspect of the femur, the *femoral head*, articulates with the acetabulum to form the coxofemoral joint. The femoral head is essentially round with a central depression, the *fovea capitis*, that is the femoral attachment site of the *ligament of the head of the femur*, formally called the *round ligament*[1] (Figure 5-29). If the fovea capitis is not struck tangentially by the primary x-ray beam, it can appear as a region of decreased opacity in the femoral head rather than as a flattening of the margin (see Figure 5-29, *C*). The fovea capitis is often misinterpreted as femoral head flattening that occurs as a consequence of joint incongruity. The flattening that does occur following chronic incongruity is located more proximally on the weight-bearing aspect of the femoral head, and not medially where the fovea capitis is located.

The femoral head originates from a separate center of ossification, the *capital epiphysis,* and, in young dogs, the physis of the femoral head is visible (see Figure 5-29, *A*). This physis should be fused by 1 year of age, joining the head with the neck. A thin radiopaque line may persist throughout life at the site of the closed physis of the capital epiphysis (Figure 5-30; see Figure 5-29, *B*).

In addition to the femoral head and neck, the other major morphologic feature of the proximal femur is the *greater trochanter*. The greater trochanter is just lateral to the femoral head and neck and the depression between these structures is the *trochanteric fossa* (Figure 5-31). The gluteus medius, gluteus profundus, and piriformis insert on the greater trochanter.[1] The *lesser trochanter* is a smaller protuberance on the proximal aspect of the femur, located

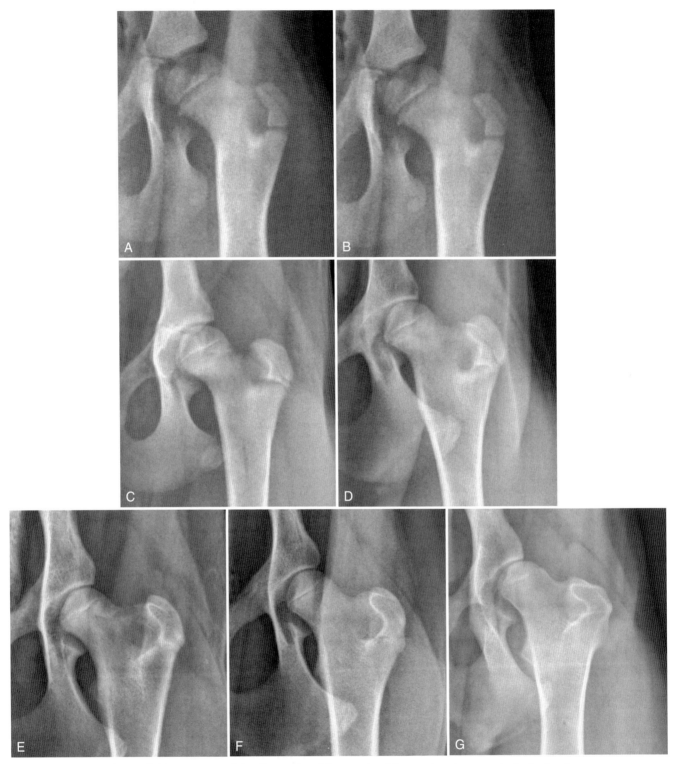

Figure 5-30. Craniocaudal radiographs of the proximal femur showing morphologic developmental changes that occur with time. **(A)** Two-month-old German Shepherd. **(B)** Three-month-old Golden Retriever. **(C)** Four-month-old Labrador Retriever. **(D)** Seven-month-old Labrador Retriever. **(E)** Eight-month-old German Shepherd. **(F)** Eleven-month-old German Shepherd. **(G)** Fourteen-month-old German Shepherd.

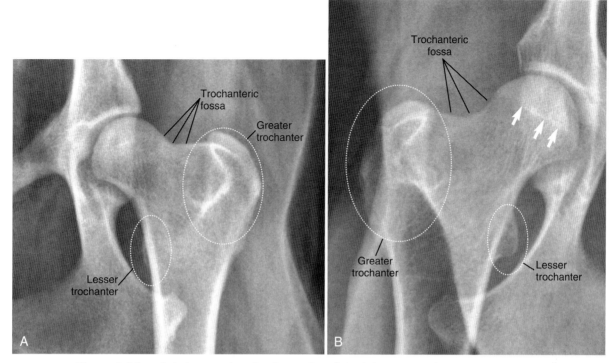

Figure 5-31. Labeled ventrodorsal radiographs of the left proximal femur of a 7-year-old Australian Shepherd (**A**) and the right proximal femur of a 14-year-old mixed breed dog (**B**). The size of the lesser trochanter varies between these dogs. In **B**, the thin sclerotic line marking the location of the physis of the femoral head, now closed, can be seen *(arrows)*.

medial and distal to the greater trochanter (see Figure 5-31). The size of the lesser trochanter is more variable than the size of the greater trochanter. The iliopsoas muscle attaches to the lesser trochanter.

The greater and lesser trochanter each originates from its own center of ossification. The physis of the greater trochanter is commonly visible in radiographs of young dogs. The conspicuity of this physis diminishes with age and should not be confused with a fracture (see Figure 5-30 and Figure 5-32). The physis of the lesser trochanter is much smaller, and its visibility depends on the relationship of the primary x-ray beam and the plane of the physis. Usually, the physis of the lesser trochanter is superimposed on the femoral cortex and is not visible. Occasionally, the femur is positioned such that the physis of the lesser trochanter is visible, and, in this instance, the lesser trochanter appears relatively isolated. The physis of the lesser trochanter is misinterpreted as a fracture more often than is the physis of the greater trochanter (Figure 5-33). The physes of the greater and lesser trochanters should be closed by approximately 9-12 months of age (see Figure 5-30).

A large artery penetrates the caudal cortex of the femur through a nutrient canal located in the proximal aspect of the diaphysis. In lateral views, this canal can be confused with a fracture, and, in craniocaudal projections, it can appear as a focal lytic lesion surrounded by a region of increased opacity. The foramen may not be visible, may be visible in one projection only, or may be visible in both the lateral and craniocaudal projection, depending on its relationship with the primary x-ray beam (see Figures 5-11 and 5-34).

The distal end of the femur is characterized by two condyles, the *medial condyle* and the *lateral condyle*. These condyles articulate with the proximal tibia, with interposed menisci, to form the *stifle joint*. Cranially, there is a bony ridge at the axial margin of each femoral condyle; these ridges bound the centrally located femoral trochlea (Figure 5-35). Caudally, there is a depression between the femoral condyles, the *intercondylar fossa*. Many small vessels enter the femur through this fossa, creating distinct radiolucencies, which in a craniocaudal radiograph can be confused with permeative lysis (Figures 5-36 and 5-37).

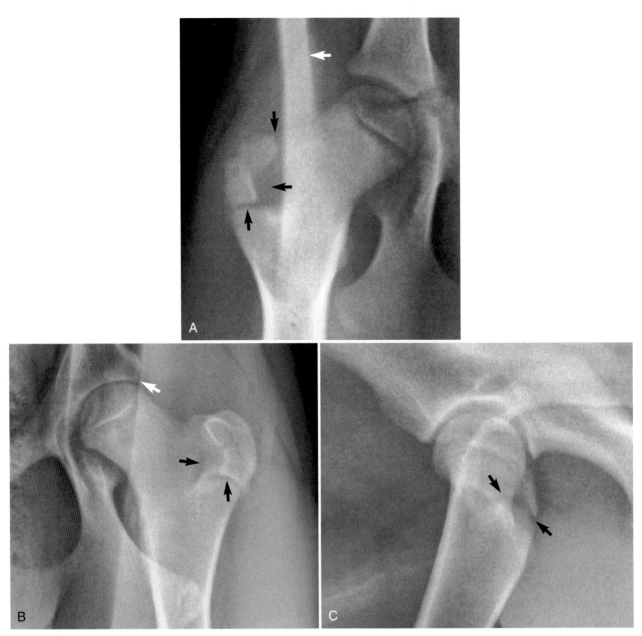

Figure 5-32. Ventrodorsal radiographs of the right proximal femur of a 4-month-old Labrador Retriever (**A**) and the left proximal femur of a 6-month-old Greyhound (**B**) and an open-lateral radiograph of the right proximal femur of an 8-month-old Labrador Retriever (**C**). The physis of the greater trochanter is visible in each dog *(arrows)*. In the younger dog (**A**), the greater trochanter appears as a more isolated structure, whereas, in **B**, the physis is narrowing in the process of closure. In **A**, the thick opaque region is a skin fold *(white arrow)*, and, in **B**, the thin line *(white arrow)* is an artifact created by a positioning device.

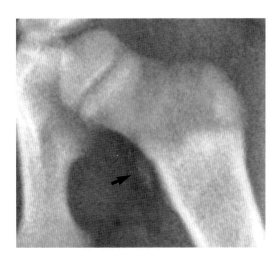

Figure 5-33. Ventrodorsal radiograph of the left proximal femur of a 9-week-old mixed breed dog. The femur is positioned such that the lesser trochanter *(arrow)* is not superimposed on the cortex; this appearance is easily confused with a fracture.

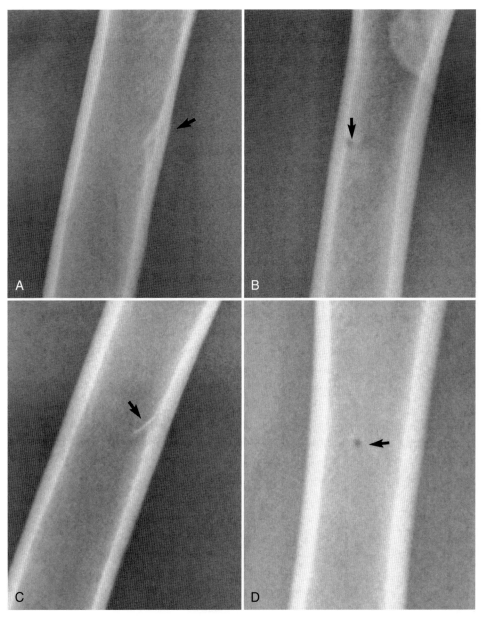

Figure 5-34. A, Lateral radiograph of the mid-portion of the femur of a 6-year-old Vizsla. **B,** Craniocaudal radiograph of the mid-portion of the femur of a 6-month-old Greyhound. Lateral **(C)** and craniocaudal **(D)** radiographs of the mid-portion of the femur of an 8-month-old Labrador Retriever. The nutrient canal of the femur is visible in each image *(arrows)*.

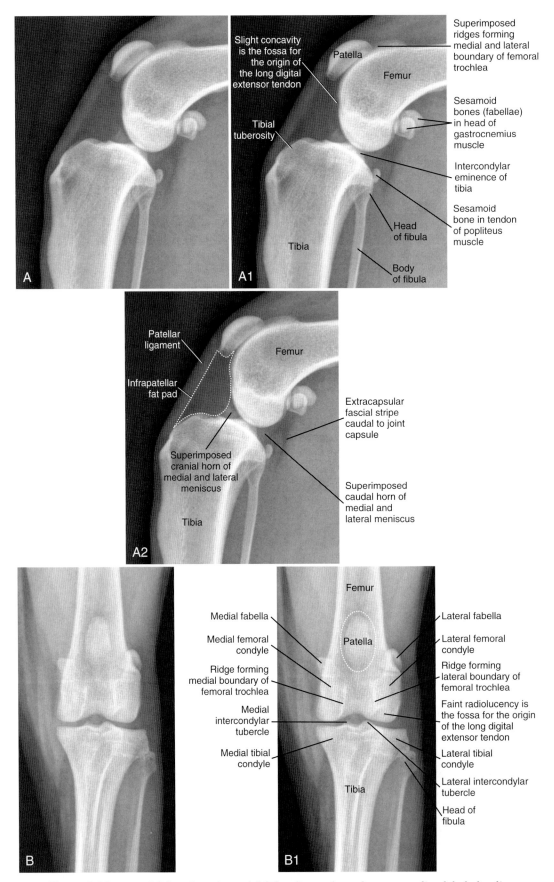

Figure 5-35. Lateral (A) and craniocaudal (B) radiographs and corresponding labeled radiographs (A1, A2, B1) of the distal femoral and proximal tibial region of a 5-year-old English Setter.

Chapter 5 ■ The Pelvic Limb   159

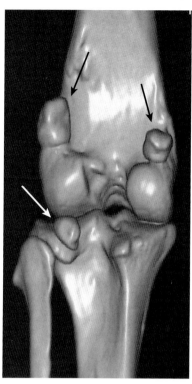

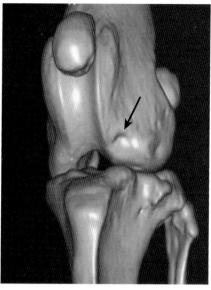

Figure 5-38. Three-dimensional volume rendered CT image of the stifle joint of a 6-year-old Belgian Malinois. The small focal depression in the distolateral aspect of the femur is the site of origin of the long digital extensor tendon *(black arrow)*. This depression is termed the *extensor fossa*.

Figure 5-36. Three-dimensional volume rendered CT image of a canine stifle joint from the caudal perspective. There is a sesamoid bone in each head of the gastrocnemius muscle *(black arrows)* and a sesamoid bone in the origin of the popliteus muscle *(white arrow)*. Note the larger size of the lateral gastrocnemius sesamoid compared to the medial; this is normal. The multiple small depressions in the intercondylar fossa of the femur serve as vascular access ports and can appear in craniocaudal radiographs as regions of bone lysis; this should not be confused with an aggressive lesion.

The long digital extensor tendon originates from a depression that lies at the cranial and lateral aspects of the lateral femoral condyle (Figure 5-38). This depression is termed the *extensor fossa*. The extensor fossa appears in lateral radiographs as a concave defect just distal to the distal aspect of the patella and in craniocaudal radiographs as an ill-defined radiolucent region in the distal aspect of the lateral condyle (see Figures 5-35 and 5-39). The difference in appearance of the extensor fossa in lateral versus craniocaudal radiographs relates to the orientation of the fossa with regard to the primary x-ray beam. The extensor fossa has been confused with an osteochondrosis lesion. The extensor fossa becomes more conspicuous radiographically in dogs with osteoarthritis due to new bone formation at the periphery of the fossa.

The femoral condyles originate from a separate center of ossification, and the distal femoral physis is visible in radiographs of dogs younger than 1 year of age. This physis is undulating, which creates interdigitating radiolucent regions (see Figure 5-39). In very young animals during the process of rapid ossification, the margin of the distal femoral epiphysis and the patella can appear irregular, due to rapid irregular ossification. This should not be misinterpreted as lysis due to infection (Figure 5-40). The appearance of subchondral bone flattening can also be seen; this is also usually a manifestation of the rapid osteogenesis that is occurring (see Figure 5-40, *B*). Also in young dogs, the stifle joint space can appear wide due to the increased thickness of the articular cartilage at this stage; this appearance is not a sign of joint effusion (see Figure 5-40).

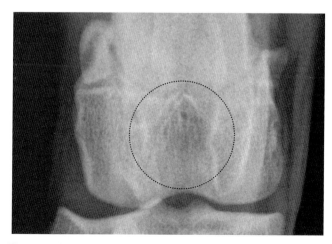

Figure 5-37. Craniocaudal radiograph of the distal femur of a 6-month-old Greyhound. The multiple focal radiolucencies in the intercondylar fossa represent vascular foramina *(dotted circle)*. These can be confused with permeative lysis.

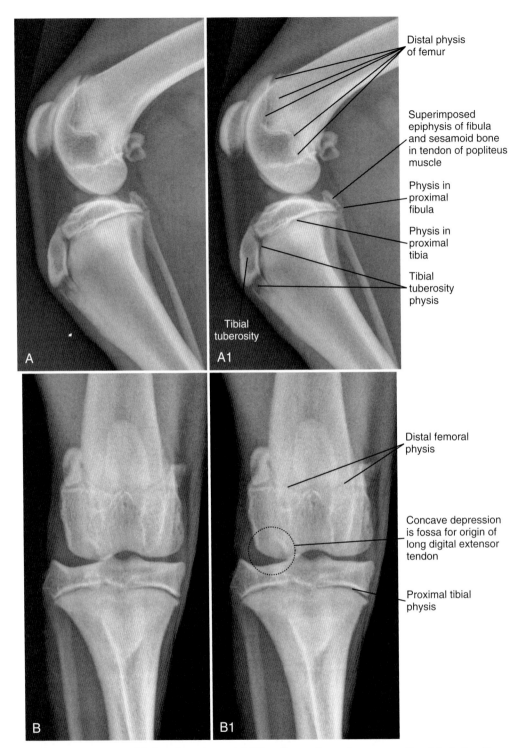

Figure 5-39. Lateral (**A**) and craniocaudal (**B**) radiographs and corresponding labeled radiographs (**A1, B1**) of the stifle region of a 6-month-old Greyhound. The distal femoral and proximal tibial growth plates are open and therefore radiolucent. The distal femoral physis is undulating, creating interdigitating radiolucent lines at the physis rather than a single continuous radiolucency.

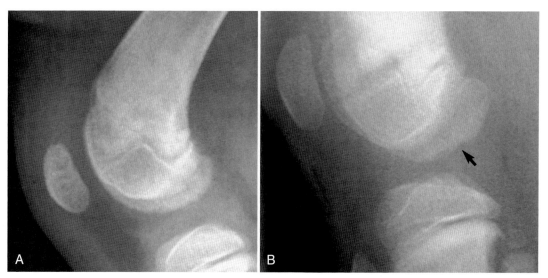

Figure 5-40. **A,** Lateral radiograph of the stifle of a 10-week-old Labrador Retriever. The patella and the distal aspect of the femur have mottled appearances. This is commonly seen in young subjects during periods of rapid ossification and has been confused with lysis due to infection. The space between the distal aspect of the femur and the proximal aspect of the tibia appears wide. This is due to the increased thickness of the articular cartilage present in young animals and should not be misinterpreted as joint effusion. **B,** Lateral radiograph of the stifle of an 8-week-old English Setter. Note the irregular margin and flattened appearance of the femoral condyle *(arrow)*. This is a normal appearance during this phase of rapid skeletal maturation and is not evidence of an epiphyseal lesion, such as osteochondrosis or infection.

Four sesamoid bones can be visible in the stifle joint, the largest of which is the *patella* (see Figure 5-35). The patella is in the tendon of insertion of the quadriceps femoris muscle. The continuation of the tendon of the quadriceps femoris muscle distal to the patella is the *patellar ligament*[1] (see Figure 5-35, *B1*). The exact position of the patella in a lateral view relative to the distal end of the femur depends on the amount of flexion of the joint at the time of radiography. In a well-positioned craniocaudal view, the patella should be centered in the femoral trochlea. The distal end of the patella might have an irregular shape that creates a focal radiolucency; this should not be confused with geographic lysis (Figure 5-41).

Each head of the gastrocnemius muscle has a sesamoid bone, the *medial and lateral gastrocnemius fabellae*. The lateral gastrocnemius fabella is usually slightly larger than the medial gastrocnemius fabella (see Figures 5-35 and 5-36). The medial gastrocnemius fabella is occasionally positioned more distal than the lateral gastrocnemius fabella as a normal variant.[6] This is most common in West Highland white terriers, occurring in approximately 70% of examined dogs. Distal positioning of the medial gastrocnemius fabella is also found in approximately 9% of other small breed dogs, mostly terriers, but rarely, if ever, as a normal variant in large dogs (Figures 5-42 and 5-43). Observation of multipartite or fragmented gastrocnemius fabellae has also been reported.[7] This can involve

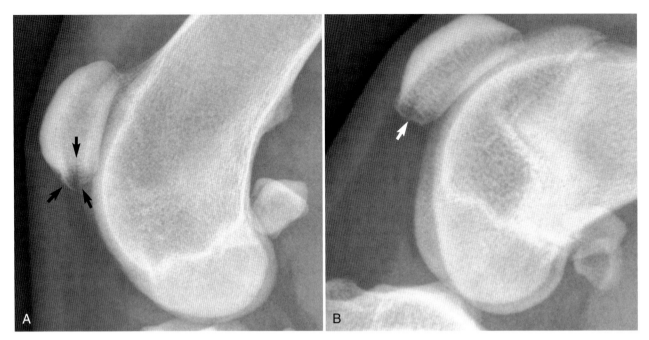

Figure 5-41. A, Lateral radiograph of the distal femur of a 7-year-old Rhodesian Ridgeback. B, Lateral radiograph of the distal femur of an 11-month-old Golden Retriever. There is decreased opacity of the distal aspect of the patella *(arrows)* in both dogs. This is created by the shape of the patella in these individuals and is not evidence of geographic lysis.

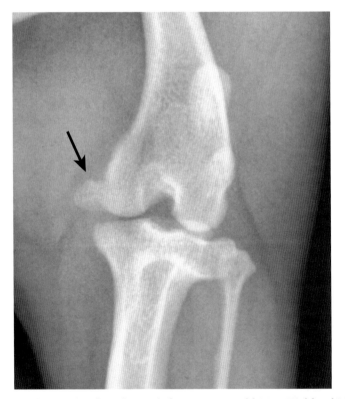

Figure 5-42. Caudocranial stifle radiograph from a 6-year-old West Highland White Terrier. Note the distal positioning of the medial gastrocnemius sesamoid bone *(black arrow)*. This is a normal variant.

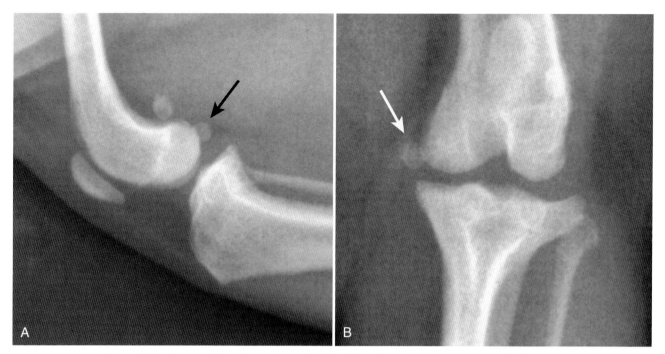

Figure 5-43. Lateral (**A**) and caudocranial (**B**) stifle radiographs from a 3-year-old Shih Tzu. Note the distal positioning of the medial gastrocnemius sesamoid bone *(black arrow)*. This is a normal variant.

either the medial or lateral gastrocnemius fabella and is not thought to be clinically significant (Figures 5-44 and 5-45). Finally, the medial gastrocnemius fabella is occasionally absent and no clinical significance is attached to this finding.

The fourth sesamoid bone of the stifle joint, the popliteal sesamoid bone, lies in the tendon of origin of the popliteus muscle (see Figures 5-36 and 5-46). The popliteal sesamoid bone is not present in every subject and its absence is not clinically significant. Sometimes, the popliteal sesamoid bone can be seen only in the lateral view due to being superimposed with other structures in the craniocaudal view (see Figure 5-35, *A* and *B*).

There is an accumulation of fat, the *infrapatellar fat body*, or *infrapatellar fat pad*, between the patellar ligament, the craniodistal aspect of the femur, and the cranioproximal aspect of the tibia. The infrapatellar fat pad is visible radiographically as a relatively less opaque, distinct structure because of the inherently lower physical density and effective atomic number of fat compared with muscle, tendons and ligaments (see Figure 5-35). The infrapatellar fat pad develops in the fibrous layer of the joint capsule. Fluid, or less commonly a mass, within the stifle joint can compress the caudal aspect of the fat pad causing it to appear distorted and displaced cranially. In addition, the extracapsular fascial stripe caudal to the stifle can become displaced caudally with moderate or severe stifle joint effusion (see Figure 5-35).

Neither the medial nor the lateral meniscus can be evaluated radiographically. The menisci contribute to the soft tissue opacity interposed between the distal femur and proximal tibia (see Figure 5-35) but the menisci are not large enough, or surrounded by tissue of sufficiently different opacity, to be assessed radiographically.

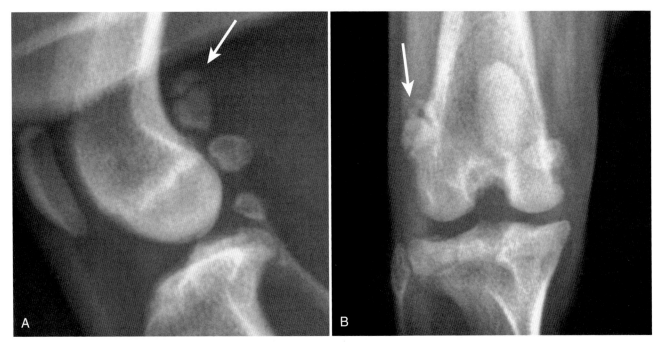

Figure 5-44. Lateral (A) and caudocranial (B) radiographs of the stifle of a 5-year-old Miniature Pinscher. The lateral gastrocnemius sesamoid bone is multipartite *(black arrows)*. This is a clinically insignificant normal variation.

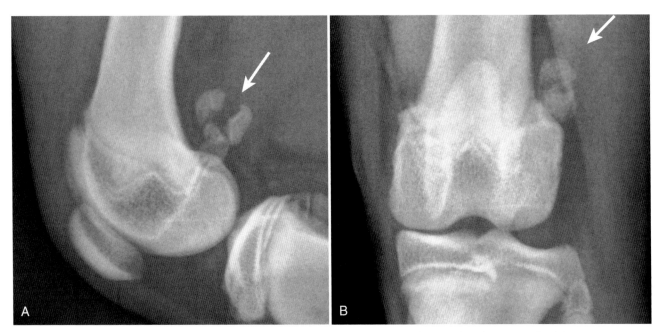

Figure 5-45. Lateral (A) and caudocranial (B) radiographs of the stifle of a 5-month-old Labrador Retriever. The lateral gastrocnemius sesamoid bone is multipartite *(black arrows)*. This is a clinically insignificant normal variation.

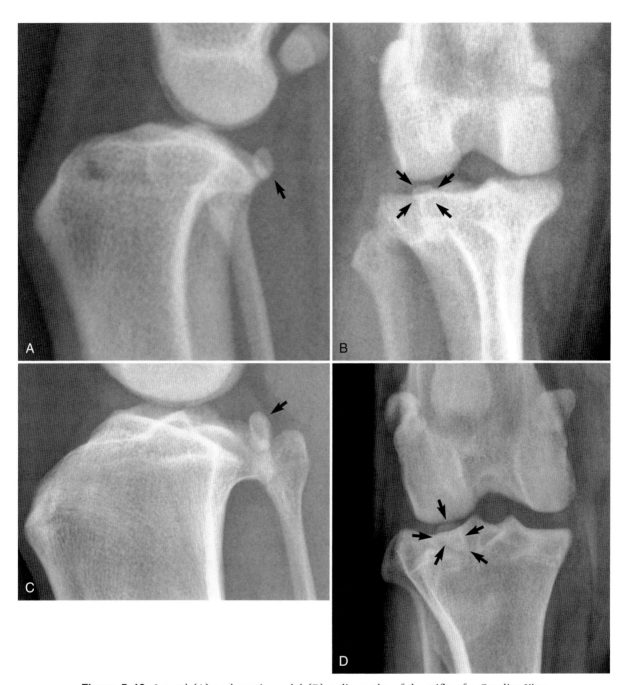

Figure 5-46. Lateral (**A**) and craniocaudal (**B**) radiographs of the stifle of a Cavalier King Charles Spaniel. Lateral (**C**) and craniocaudal (**D**) radiographs of the stifle of a 7-year-old Rhodesian Ridgeback. Each dog has a sesamoid bone *(arrows)* in the tendon of origin of the popliteus muscle that is visible in both views.

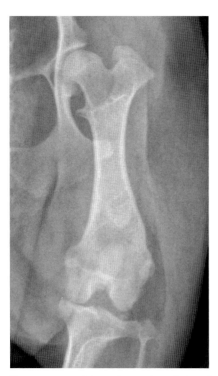

Figure 5-47. Craniocaudal radiograph of the femur of a 12-year-old Dachshund. Note the proportionally shorter femur, characteristic of chondrodystrophic dogs. The greater and lesser trochanters are also proportionally larger than in nonchondrodystrophic dogs. The femoral head is also slightly larger than the neck, giving an impression of osteophytosis at the margin of the femoral head.

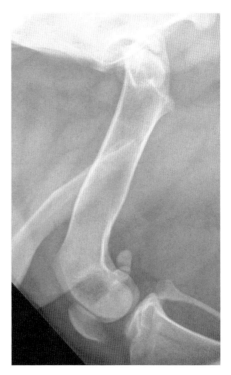

Figure 5-48. Lateral radiograph of the femur of a 1½-year-old Basset Hound. The distal aspect of the femur is curved, with caudal displacement of the femoral condyles. This is typical in chondrodystrophic dogs.

The femur in chondrodystrophic breeds has a different appearance than in other types of dogs, being proportionally shorter with larger protuberances, such as larger greater and lesser trochanters. The character of the junction of the femoral head with the femoral neck in chondrodystrophoid dogs gives the false impression of osteophyte formation. The femoral condyles are also proportionally larger in chondrodystrophoid dogs, and the distal end of the femur is curved, with caudal displacement of the femoral condyles (Figures 5-47 and 5-48).

With the advent of digital radiography, contrast resolution has improved dramatically compared to film-screen (analog) images. This leads to greater conspicuity of some soft tissue structures that were seen only rarely in analog images. For example, the popliteal lymphocenter is commonly seen in lateral digital radiographs of the canine stifle. The popliteal lymphocenter is usually composed of one lymph node, but occasionally multiple lymph nodes are seen (Figure 5-49).

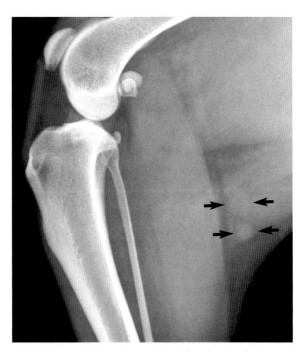

Figure 5-49. Lateral radiograph of the thigh of a 5-year-old English Setter. The contrast resolution in this digital radiograph is excellent, and multiple muscles and fascial planes are visible. Two nodular opacities in the caudal aspect of the image represent lymph nodes *(arrows)* in the popliteal lymphocenter.

Chapter 5 ■ The Pelvic Limb **167**

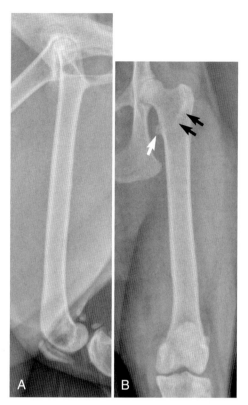

Figure 5-50. Lateral (**A**) and craniocaudal (**B**) radiographs of the femur of a 13-year-old domestic cat. The morphologic features of the feline femur are the same as in the dog. In general, the feline femur is more linear than the canine femur. The lesser trochanter *(white arrow)* may be proportionally larger than in many dogs. The linear opacity in the proximal femur on the craniocaudal view *(black arrows)* is an edge of the greater trochanter.

The feline femur is characterized by the same morphologic features as described for the dog. In general, the femur in the cat is more linear than in the dog, and the lesser trochanter may be proportionally larger than in many dogs (Figure 5-50). There are two differences in the radiographic appearance of the stifle joint in the cat versus the dog. First, the feline patella has a tapered and longer distal end (Figure 5-51). This should not be confused with patellar osteophyte formation. Second, small mineralized foci are commonly seen in the cranial aspect of the joint in lateral radiographs of the feline stifle, being found in 37 of 100 cats in one survey[8] (Figure 5-52). These mineralized foci represent regions of ossification or chondro-osseous metaplasia of the medial meniscus[8,9] and may be clinically insignificant when small but when larger are associated with cartilage erosion on the medial femoral condyle. Thus, though they have been considered to be a normal variant, they are pathologic and a potential cause of feline stifle pain.

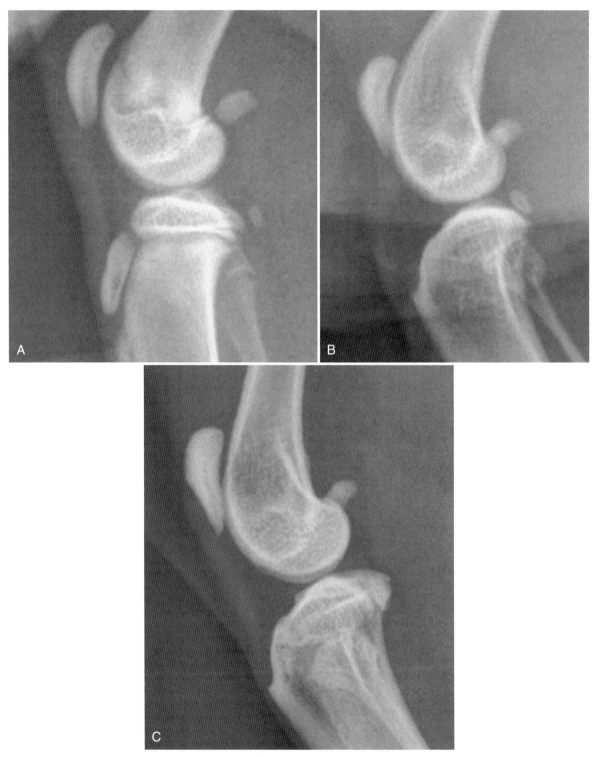

Figure 5-51. Lateral radiographs of the stifle of a 9-month-old domestic cat (**A**), a 5-year-old domestic cat (**B**), and a 14-year-old domestic cat (**C**). Note the elongated distal aspect of the patella in these cats. This is a normal patellar configuration.

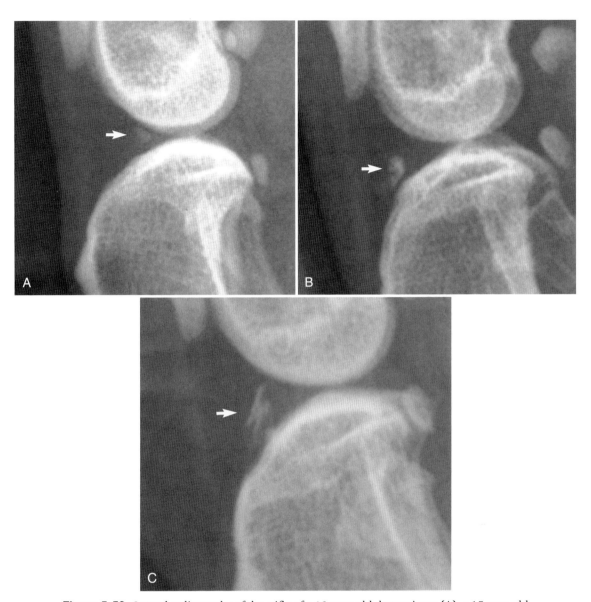

Figure 5-52. Lateral radiographs of the stifle of a 12-year-old domestic cat (**A**), a 15-year-old domestic cat (**B**), and a 9-year-old domestic cat (**C**). Focal opacities in the cranial aspect of each stifle joint *(arrows)* represent ossification of the cranial aspect of a meniscus; these are termed *meniscal ossicles*. The clinical significance of this finding has not been characterized completely, but, when large, these ossicles cause cartilage erosion on the medial femoral condyle.

TIBIA AND FIBULA

As with the femur and the stifle, the most commonly acquired radiographic projections of the tibia and fibula, collectively termed the *crus*, are lateral and craniocaudal (or caudocranial) views. These projections are illustrated herein.

The proximal surface of the tibia that articulates with the mescii is relatively flat and is characterized by a medial and lateral condyle (see Figure 5-35). Between the condyles is an intercondylar eminence that is composed of medial and lateral intercondylar tubercles. The tibial condyles arise from a separate center of ossification, the *proximal tibial epiphysis* (see Figure 5-39).

The *tibial tuberosity* is located on the cranioproximal aspect of the tibia and is the insertion site of the quadriceps femoris muscle and parts of the biceps femoris and sartorius muscles[1] (see Figures 5-35, 5-39, and 5-53). The tibial tuberosity arises from a separate center of ossification, and in young dogs the entire tuberosity is surrounded by cartilage, often leading to a misdiagnosis of avulsion of the tibial tuberosity (Figure 5-54). The

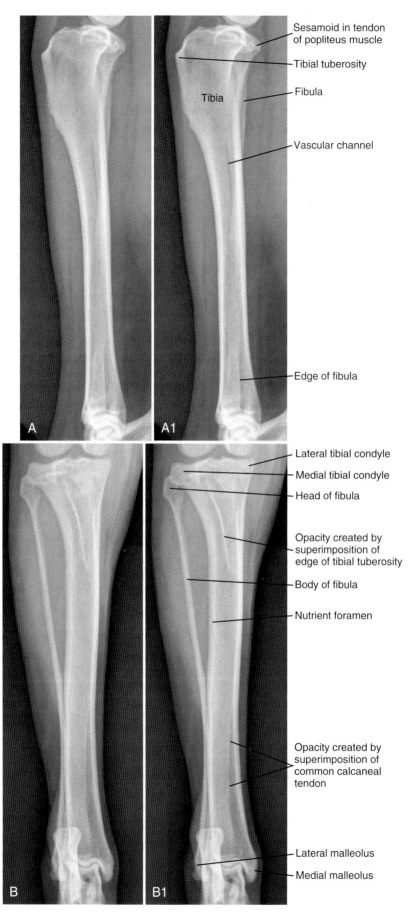

Figure 5-53. Lateral (**A**) and craniocaudal (**B**) radiographs, and corresponding labeled radiographs (**A1, B1**), of the tibia of a 2-year-old Labrador Retriever.

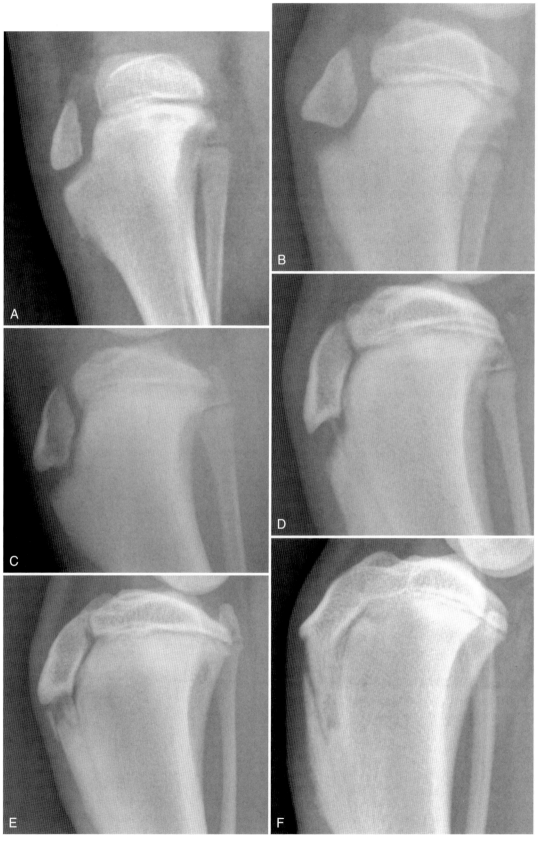

Figure 5-54. Lateral radiographs of the stifle of a 2½-month-old Labrador Retriever (**A**), a 3-month-old Border Collie (**B**), a 4-month-old Golden Retriever (**C**), a 5-month-old Labrador Retriever (**D**), a 6-month-old Greyhound (**E**), and a 10-month-old Golden Retriever (**F**). These illustrations depict the maturation and fusion of the tibial tuberosity.

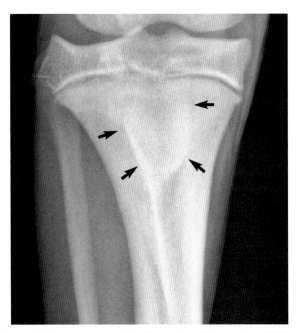

Figure 5-55. Craniocaudal radiograph of the proximal tibia of a 6-month-old Greyhound. Note the opacity *(arrows)* created by superimposition of the incompletely fused tibial tuberosity.

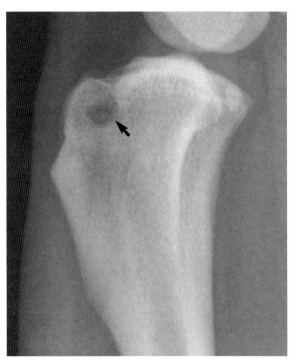

Figure 5-56. Lateral radiograph of the proximal tibia of a 3-year-old Boston Terrier. Note the focal radiolucent region in the proximal-cranial aspect of the tibia *(arrow)*. This is a normal variant found in some dogs and should not be misinterpreted as an aggressive lesion.

tibial tuberosity eventually fuses with the proximal tibial epiphysis and with the cranial aspect of the proximal tibial metaphysis (see Figure 5-53). An incompletely united tibial tuberosity creates a circular opacity in craniocaudal radiographs of the stifle joint (Figure 5-55).

In some mature dogs with a fused tibial tuberosity, a focal radiolucency is present in the proximal-cranial aspect of the tibia. The prevalence of this finding in one survey was approximately 21% and in 2 dogs where the region was examined histologically it was found to be caused by a hyaline cartilage core.[10] This radiolucency was more common in toy, small and medium breeds compared to large and giant breeds, and dogs with this focal radiolucency had higher odds of developing medial patellar luxation but a cause and effect was not proven. At this time, this radiolucent region can be considered a normal variant and should not be misinterpreted as an aggressive bone lesion, though dogs having this appearance appear to be at greater risk for developing medial patellar luxation (Figures 5-56 and 5-57).

The caudal aspect of the tibial cortex is penetrated by a nutrient canal at the approximate middle of the diaphysis (Figure 5-58). The canal is located laterally; thus it is

Chapter 5 ■ The Pelvic Limb   **173**

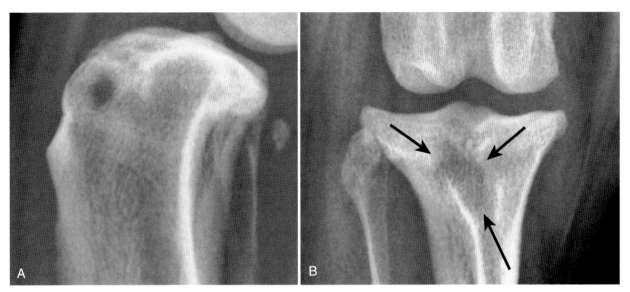

Figure 5-57. Lateral (A) and caudocranial (B) radiographs of the stifle joint of a 2-year-old Bulldog. Note the focal radiolucent region in the proximal-cranial aspect of the tibia. In the caudocranial view this variation creates an ill-defined region of radiolucency *(black arrow)* that could be confused with bone effacement. This is a normal variant found in some dogs and should not be misinterpreted as an aggressive lesion.

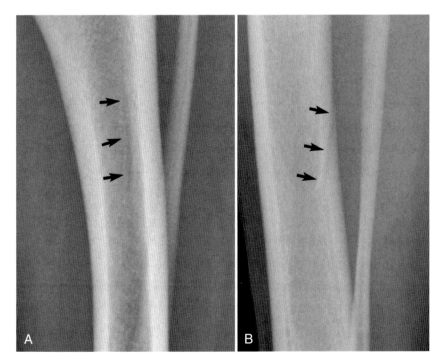

Figure 5-58. Lateral (A) and craniocaudal (B) radiographs of the proximal tibia of an 8-month-old Labrador Retriever. The nutrient canal of the tibia is evident *(arrows)*. As the canal is lateral, it is best seen in the craniocaudal view (B). In the lateral view, the tunnel in the cortex is projected as a linear radiolucent region superimposed on the medullary cavity (A).

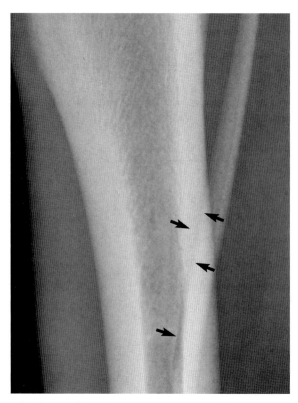

Figure 5-59. Lateral radiograph of the crus in the region of overlap between the tibial and fibular cortices. The eye interprets a black line at the edge of this overlap *(arrows)* due to the mach effect. This optical illusion can be misinterpreted as a fracture or the nutrient canal. In a lateral radiograph, the nutrient canal typically is oriented more vertically than the line created by the mach effect.

best seen in a craniocaudal view where it appears as a radiolucent canal through the lateral cortex. In the lateral view, the nutrient canal is still often visible but as a linear radiolucent region superimposed on the medullary cavity (see Figure 5-53, A and A1). In the lateral view, the fibula can be superimposed on the tibia in approximately the same location as the nutrient artery. When two structures cross each other, an artifactual change in opacity is created at the margin; this is due to the *mach effect*.[11,12] The mach effect is a visual artifact, creating a black line at the edge of the overlapping tibial and fibular cortices; this line can be misinterpreted as a fracture or as the nutrient foramen (Figure 5-59).

The distal end of the tibia articulates with the talus to form part of the tarsocrural joint. There are two concave depressions in the distal tibial articular surface, the *tibial cochleae*, which articulate with the trochlea of the talus. The medial part of the distal end of the tibia is the *medial malleolus*, which provides tarsocrural joint stabilization on the medial side (Figure 5-60).

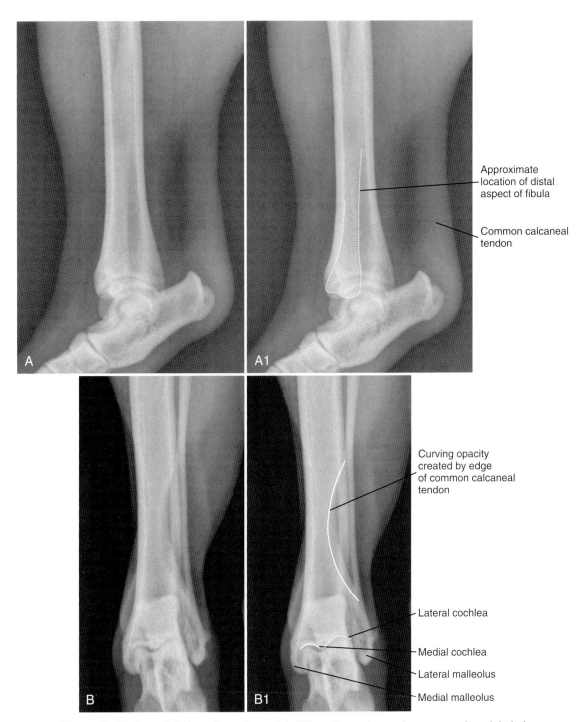

Figure 5-60. Lateral (A) and craniocaudal (B) radiographs, and corresponding labeled radiographs (A1, B1), of the distal aspect of the crus of a 10-month-old American Staffordshire Terrier.

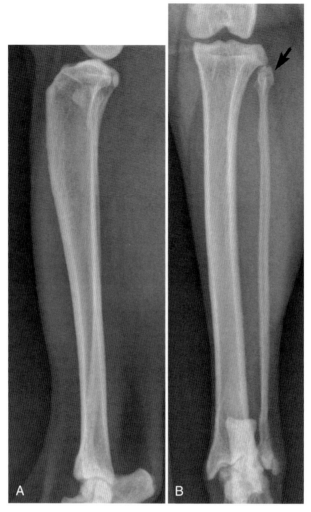

Figure 5-61. Lateral (**A**) and craniocaudal (**B**) radiographs of the tibia of a 3-year-old domestic cat. The tibia in the cat is characterized by the same morphologic features described for the dog. The fibula does tend to articulate farther distally with the lateral tibial condyle than in the dog *(arrow)*.

The fibula articulates with the caudal aspect of the lateral tibial condyle and extends distally to end at the lateral malleolus, which provides tarsal joint stabilization on the lateral side (see Figures 5-53 and 5-60).

The feline crus is characterized by the same morphologic features as described for the dog (Figure 5-61). Some minor differences exist. The feline fibula tends to articulate farther distally on the lateral tibial condyle (see Figure 5-61), and the lateral malleolus has a smooth lateral osseous protuberance not found in the dog (Figure 5-62).

PES

The pes is composed of the tarsal bones, the metatarsal bones, and the phalanges. In addition to lateral and dorsoplantar views, oblique views of the pes are commonly acquired; these views are included in the illustrations for this section. Oblique views are needed due to the multiple small bones composing the pes and the complicated opacities created by superimposition of these structures. Chapter 1 covers the principle of oblique radiography. It is important to remember that the topographic term for the front surface of the pelvic limb changes from cranial to dorsal, and the term for the rear surface changes from caudal to plantar, at the level of the tarsocrural joint.[13]

The tarsus is composed of seven individual bones: the talus; the calcaneus; the central tarsal bone; and the first, second, third, and fourth tarsal bones (Figure 5-63). The first tarsal bone is the most variable; it might not be present as an individual bone, in which instance it is

Chapter 5 ■ The Pelvic Limb **177**

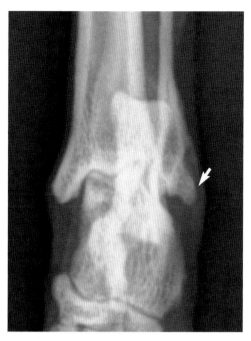

Figure 5-62. Craniocaudal view of the distal aspect of the crus of a 4-year-old Manx. The lateral malleolus in cats is characterized by a relatively large, laterally extending smooth protuberance *(arrow)* not found in dogs.

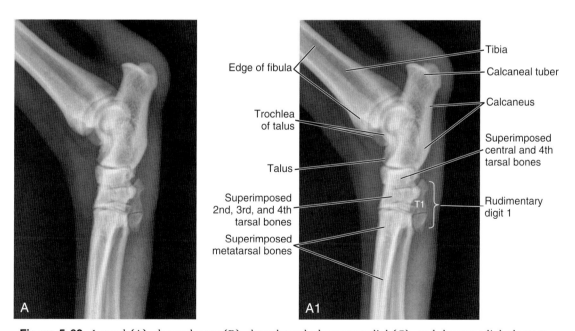

Figure 5-63. Lateral (**A**), dorsoplantar (**B**), dorsolateral plantaromedial (**C**), and dorsomedial plantarolateral (**D**) radiographs, and corresponding labeled radiographs (**A1, B1, C1, D1**), of the tarsus of a 1-year-old American Staffordshire Terrier. *MT,* Metatarsal; *T,* tarsal.

Continued

178 Atlas of Normal Radiographic Anatomy and Anatomic Variants in the Dog and Cat

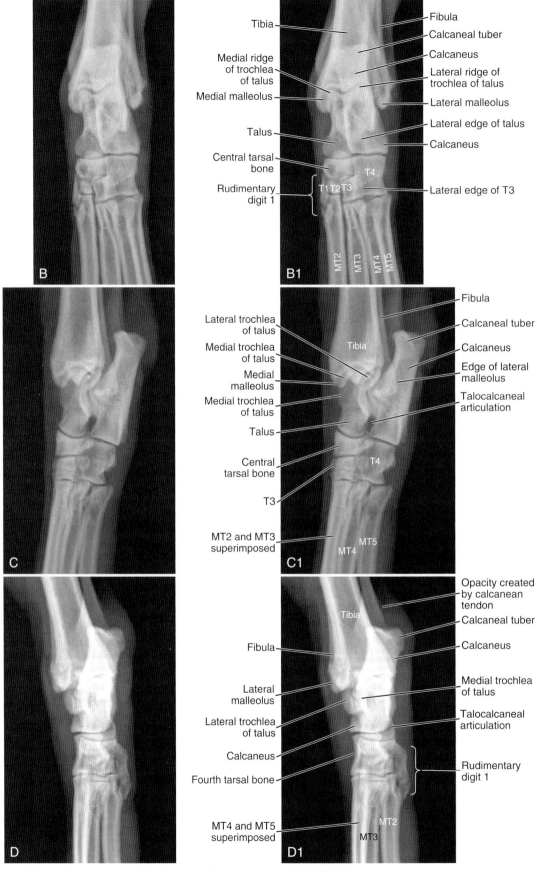

Figure 5-63, cont'd.

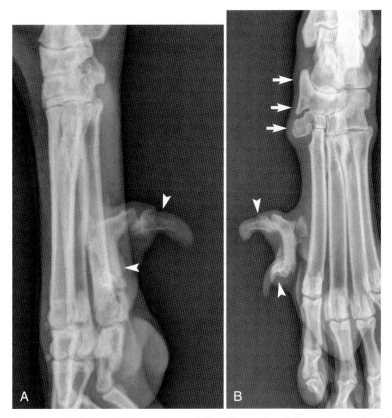

Figure 5-64. Lateral (A) and dorsoplantar (B) radiograph of the tarsocrural region of a 4-year-old Beauceron. There are double dewclaws *(white arrowheads)*, a breed standard. There is also an anomaly of the central and first tarsal bones *(white arrows)*. It is unclear exactly what anomaly has resulted in the abnormal shape of the tarsal bones, but this appearance is typical of Beaucerons with double dewclaws, and this tarsal anomaly is occasionally seen in other breeds not associated with a double dewclaw.

fused with the proximal portion of the first metatarsal bone.[1] Most dogs do not have development of the first digit on the pes to the extent that a distal phalanx and nail, termed the *dewclaw*, is present. A dewclaw is more commonly found in the manus. In breeds that do have a dewclaw on the pes, they are often removed shortly after birth as they have no function, are subject to injury, and are not required by breed standards. However, certain breed standards, including the Beauceron, require the presence of a double rear dewclaw (Figures 5-64 and 5-65). The double dewclaw phenotype in the Beauceron is also associated with an anomalous shape of the central and first tarsal bones (see Figure 5-64).

As noted previously, the complexity of the tarsus makes careful scrutiny of all bones and surfaces challenging, even when oblique projections are acquired. Given the importance of assessing the trochlea of the talus in young dogs, a special dorsoplantar view of the tarsocrural joint with the joint in flexion can be useful. In this view, there are fewer structures superimposed on the medial and lateral trochlea of the talus, allowing for more accurate assessment (Figure 5-66).

In general, neither chondrodystrophic dogs nor cats have significantly different tarsal morphology from other dog types. In chondrodystrophic dogs, the calcaneus appears proportionally larger (Figure 5-67), and in cats there is a smooth projection from the plantar-proximal aspect of the fifth metatarsal bone that is not present in dogs (Figure 5-68).

Because the metatarsal and phalangeal anatomy is not different between metatarsus and metacarpus, the description of the metacarpal (metatarsal) bones and phalanges and their sesamoid bones is included with the forelimb and not repeated here.

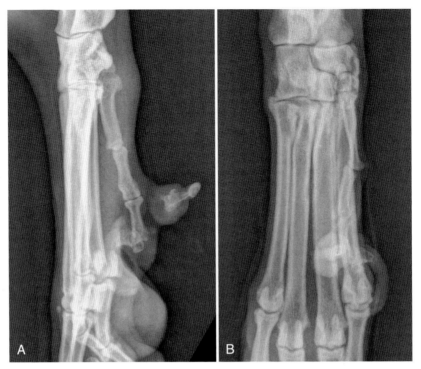

Figure 5-65. Lateral (A) and dorsoplantar (B) radiographs of the tarsal region of a 5-year-old Great Pyrenees. There is a completely developed first digit, with metatarsal bone I and phalanges I and II. There is duplication of the distal phalanx, forming a double dewclaw.

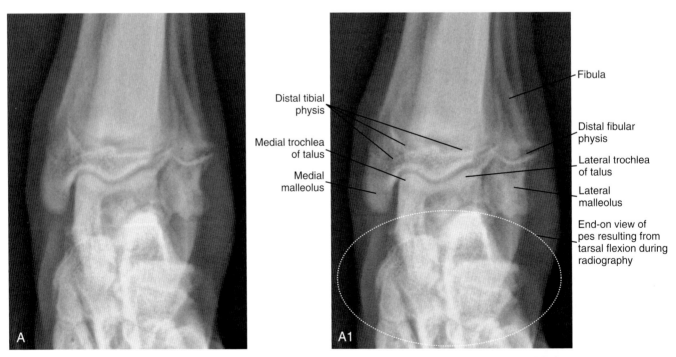

Figure 5-66. Dorsoplantar radiograph with tarsal flexion (A), and corresponding labeled radiograph (A1), of the left tarsus of a 7-month-old Fila Brasileiro. Flexion during radiography prevents other bones of the tarsal joint from being superimposed on the trochlea of the talus, allowing for more critical assessment of the trochlea.

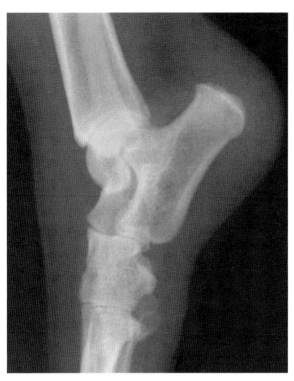

Figure 5-67. Lateral radiograph of the tarsus of a 6-year-old Basset Hound. The calcaneus is proportionally larger in chondrodystrophic dogs.

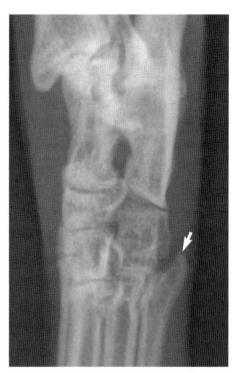

Figure 5-68. Dorsolateral-plantaromedial radiograph of the tarsus of a 10-year-old domestic cat. The smooth projection from the proximal aspect of the fifth metatarsal bone *(arrow)* is a normal finding; this is not present in dogs.

REFERENCES

1. Evans H, editor: The skeleton. *Miller's anatomy of the dog,* ed 3, Philadelphia, 1993, Saunders.
2. Evans H, editor: Arthrology. *Miller's anatomy of the dog,* ed 3, Philadelphia, 1993, Saunders.
3. Fagin B, Aronson E, Gutzmer M: Closure of the iliac crest ossification center in dogs: 750 cases. *J Am Vet Med Assoc* 200:1709-1711, 1992.
4. Slocum B, Devine T: Dorsal acetabular radiographic view for evaluation of the canine hip. *J Am Anim Hosp Assoc* 26:280-296, 1990.
5. Dennis R, Penderis J: Radiology corner—anal sac gas appearing as an osteolytic pelvic lesion. *Vet Radiol Ultrasound* 43:552-553, 2002.
6. Störk CK, Petite AF, Norrie RA, et al: Variation in position of the medial fabella in West Highland white terriers and other dogs. *J Small Anim Pract* 50:236-240, 2009.
7. Comerford E: The stifle joint. In Barr FJ, Kirberger RM, editors: *BSAVA Manual of canine and feline musculoskeletal imaging,* UK, 2006, BSAVA, pp 135-149.
8. Freire M, Brown J, Robertson ID, et al: Meniscal mineralization in domestic cats. *Vet Surg* 39:545-552, 2010.
9. Whiting P, Pool R: Intrameniscal calcification and ossification in the stifle joints of three domestic cats. *J Am Anim Hosp Assoc* 21:579-584, 1984.
10. Paek M, Engiles JB, Mai W: Prevalence, association with stifle conditions, and histopathologic characteristics of tibial tuberosity radiolucencies in dogs. *Vet Radiol Ultrasound* 54:453-458, 2013.
11. Papageorges M: How the mach phenomenon and shape affect the radiographic appearance of skeletal structures. *Vet Radiol Ultrasound* 32:191-195, 1991.
12. Papageorges M, Sande R: The mach phenomenon. *Vet Radiol Ultrasound* 31:274-280, 1990.
13. Smallwood J, Shivley M, Rendano V, et al: A standardized nomenclature for radiographic projections used in veterinary medicine. *Vet Radiol Ultrasound* 26:2-9, 1985.

CHAPTER 6

The Thorax

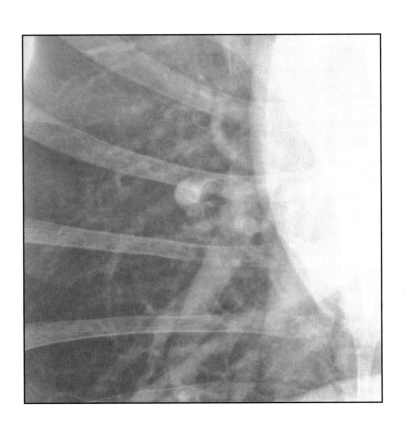

The thorax is well suited to radiographic imaging due to the inherent subject contrast afforded by the air-filled lungs. The craniocaudal field-of-view should extend from cranial to the manubrium to at least one to two vertebral body lengths caudal to the most dorsocaudal limit of the diaphragm (Figure 6-1). Dorsoventrally, the entire limits of the thoracic cavity should be covered by the primary x-ray beam.

For lateral views, the goal should be to have the caudodorsal rib heads superimposed (Figure 6-2). This is usually accomplished by elevating the sternum with a small piece of nonradiopaque padding, for example, a

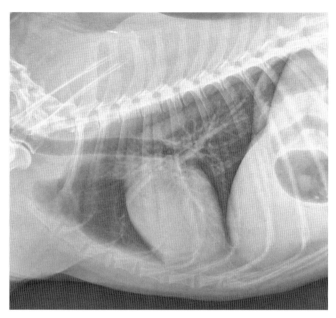

Figure 6-1. Left lateral radiograph of a canine thorax. The edges of the illustration represent minimum beam collimation.

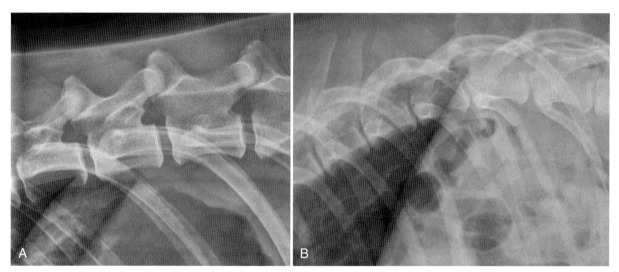

Figure 6-2. A, Lateral radiograph of a 4-year-old Rottweiler. The dog was positioned optimally for thoracic radiography. The most dorsal aspects of the caudal ribs are nearly perfectly superimposed, indicating that the sagittal plane of the patient is perpendicular to the central x-ray beam. Note that each entire rib is not directly superimposed on its counterpart. This is common and is not a prerequisite for adequate patient positioning. B, Lateral radiograph of a 6-year-old Doberman Pinscher. This dog was not positioned optimally for thoracic radiography. Note the superimposition of lung on the thoracic spine. One rib head is more dorsal than its counterpart. Most of the time, the most dorsal rib is the one closest to the x-ray table as a result of failure to elevate the sternum; that is, the sternum is closer to the table top than the spine. This is readily corrected by placing a nonradiopaque pad under the dependent axilla or sternum.

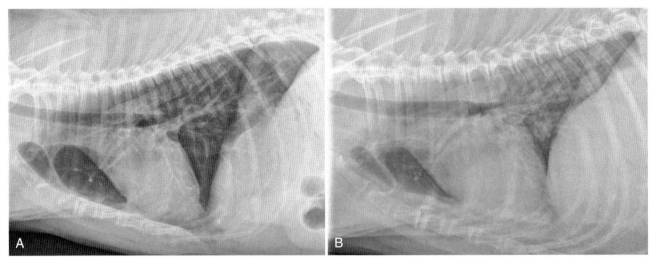

Figure 6-3. A, Left lateral radiographs of a 12-year-old Siberian Husky, acquired at peak inspiration with the patient unsedated. **B** was acquired at expiration with the patient sedated. The heart appears larger in the sedated image mainly due to less thoracic expansion; however, increased cardiac chamber filling associated with sedation-induced bradycardia also occurs. In addition, there is an increase in lung opacity in **B**. Decreased aeration and positional atelectasis will reduce pulmonary lesion conspicuity.

foam pad placed under the axilla so that the sternum is in the same horizontal plane as the spine. In brachycephalic breeds or animals with a dorsoventrally compressed thorax, the sternum may actually need to be rotated toward the table rather than elevated to achieve superimposition of the rib heads. The forelimbs should be pulled cranially and the head and the neck extended slightly. The central axis of the x-ray beam (*crosshairs*) should be centered just caudal to the caudal aspect of the scapula in dogs and approximately 1 inch caudal to the caudal aspect of the scapula in cats.

The radiographic exposure should be made during peak inspiration to ensure optimal lung aeration (discussed in detail later). If possible, thoracic radiographs should be acquired without patient sedation, because sedation decreases lung aeration, leading to increased pulmonary opacity. This increased opacity reduces the conspicuity of pulmonary lesions and can be misinterpreted as an abnormal lung pattern. The thorax is a complex moving structure, so a radiograph acquired at peak inspiration can appear very different from a radiograph acquired later at peak expiration. A single radiograph is only a "snapshot in time" and might not reflect intrathoracic pathophysiology or disease accurately (Figure 6-3).

In a lateral thoracic radiograph with correct image blackness and contrast, the skeletal structures, such as the thoracic spinous processes, and the pulmonary parenchymal structures in the narrowest part of the thorax, cranial to the heart, can both be visualized (Figure 6-4). With film-screen systems, this is best achieved by using high kVp/low mAs exposure factors and a film-screen combination that has wide latitude (long scale of contrast). These imaging parameters result in extended grayscale and improved lesion conspicuity. The ability to see all structures within the image regardless of physical density and thickness is achieved much more easily with a digital imaging system.

The thoracic region has large differences in physical density and patient thickness, which challenges any imaging system. In addition, the thorax is undergoing continuous motion during radiographic exposure due to breathing and cardiac pulsations. A short exposure time is critical to avoiding motion artifact, which results in image blurring (Figure 6-5). Quality thoracic radiographs are best achieved by using an x-ray machine with a high mA output (so that exposure times can be minimized), an optimally configured digital imaging system, and a high kVp and low mAs technique.

A minimum of three views should compose the basic thoracic study: left and right lateral recumbent views and one orthogonal view, either the dorsoventral (DV) view or the ventrodorsal (VD) view. Significant disease can be missed unless these three views are always acquired.

The reason for routinely acquiring both lateral views relates to the normal physiologic pulmonary atelectasis that occurs in the dependent lung. The dependent lung in a lateral view is less aerated, which often results in disease in that lung being obscured by the adjacent increased lung opacity. Acquiring both recumbent lateral views obviates this problem (Figure 6-6).

Recumbency-associated atelectasis also occurs in VD and DV views, but not to the same extent as in lateral

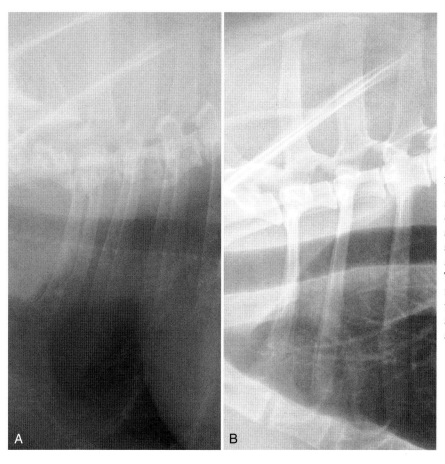

Figure 6-4. A, Lateral radiograph of the cranial aspect of the thorax of a large breed dog. With the digital imaging system used for A, it is not possible to manipulate image brightness and contrast so that the thoracic spinous processes and cranioventral lung could be displayed together optimally; this is similar to that achievable with film-screen systems. B, Lateral radiograph of the cranial aspect of the thorax of dog of similar size. The digital imaging system used to create this radiograph has superior processing capabilities, resulting in enhanced dynamic range, allowing for optimal visualization of bone and soft tissue structures simultaneously.

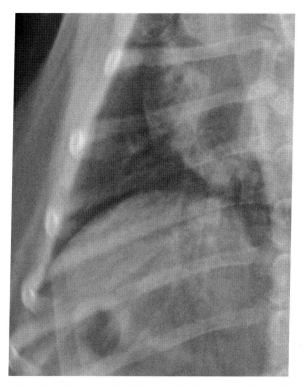

Figure 6-5. Ventrodorsal radiograph of a 10-year-old American Cocker Spaniel. The patient was panting during radiography, and motion artifact is present. The lateral thoracic wall has more blur than the spine. Motion artifact will compromise radiographic interpretation significantly.

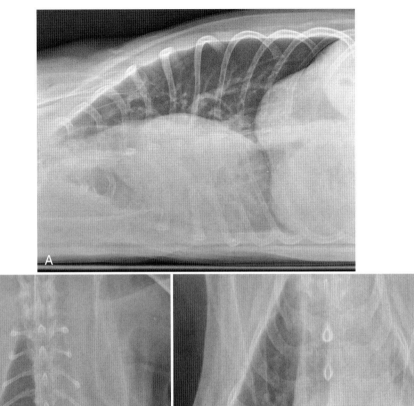
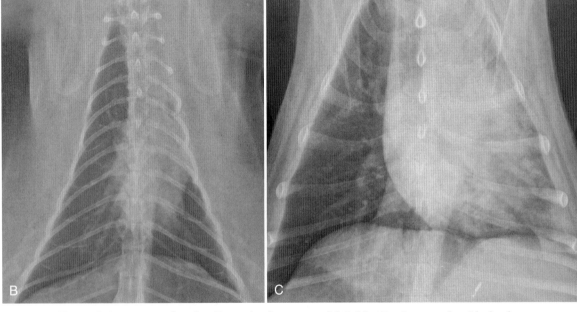

Figure 6-6. **A,** Ventrodorsal radiograph of a 12-year-old Golden Retriever made with the dog in left lateral recumbency and using a horizontally directed x-ray beam to illustrate the magnitude of atelectasis that occurs in the dependent lung. The patient had been lying on the left side for 15 minutes before radiography. The resulting atelectasis of the dependent left lung leads to a shift of the heart and mediastinal structures to the recumbent left side. The dependent lung has reduced aeration and increased opacity. Recumbency-associated atelectasis occurs in every patient placed in lateral recumbency for routine thoracic radiography; this normal physiologic phenomenon can result in misdiagnosis because disease in the dependent lung silhouettes with the atelectatic lung and reduces lesion conspicuity. To overcome this, both left and right lateral radiographs should be acquired in every routine thoracic radiographic examination. **B,** Ventrodorsal radiograph of a 7-year-old Domestic Medium-hair cat. There is increased opacity in the left lung, especially cranially, and a leftward mediastinal shift. Although the possibility of parenchymal disease cannot be ruled out, the radiographic appearance is most likely associated with recumbency-related atelectasis. Avoiding chemical restraint and ensuring the patient is not in prolonged lateral recumbency before radiography will minimize this. **C,** Ventrodorsal radiograph of a 12-year-old mixed breed dog. There is a marked leftward shift of mediastinal structures and increased opacity in the left hemithorax. This is due to recumbency-associated atelectasis and not pulmonary disease.

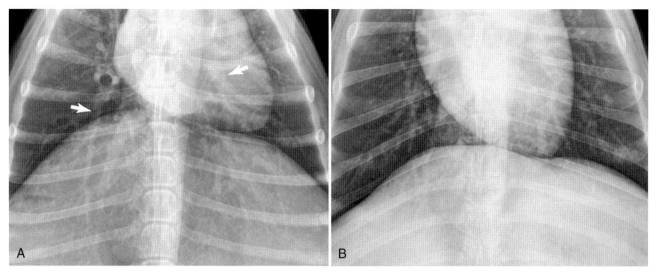

Figure 6-7. Dorsoventral (**A**) and ventrodorsal (**B**) radiographs of a 12-year-old mixed breed dog. In **A**, the dome of the diaphragm contacts the caudal border of the heart, and the apex of the heart is shifted to the left. This is relatively common in the dorsoventral view, especially in large dogs. There is much less cardiosternal contact in the ventrodorsal view (**B**), and the heart is not shifted. There is increased conspicuity of the dorsally located caudal lobar arteries on the dorsoventral view *(white arrows)*. A dorsally located pulmonary mass has increased conspicuity in the dorsoventral view compared with the ventrodorsal view.

views. The absolute least amount of recumbency-associated atelectasis occurs with the subject in sternal recumbency, that is, the DV view.

Although typically only either a VD view or a DV view is acquired, there are many situations in which obtaining both VD and DV views provides additional diagnostic information. Typically, dorsally located pulmonary lesions are more conspicuous in the DV view; conversely, ventrally located lesions, including disease in the accessory lobe, are more conspicuous in the VD view (Figure 6-7).

As with the abdomen, there are important anatomic differences between various views; understanding these variations enhances interpretation accuracy. Fairly consistent differences in appearance exist between left and right lateral views (and between DV and VD views) that, in most situations, readily distinguish them apart from each other. Figure 6-8 shows an overview of the major organs visible with thoracic radiography.

LEFT LATERAL VIEW

For a left lateral thoracic radiograph, the patient is in left recumbency and the x-ray beam enters the right side of the thorax. Based on correct radiographic nomenclature, this projection should be termed a *right-left view*[1] but this is usually abbreviated to just *left lateral view* (Figure 6-9). In the left lateral view, the diaphragmatic crura diverge dorsally, with the left crus typically being more cranial than the right crus. The left crus has no unique distinguishing features, but the right crus is identified by its confluence with the caudal vena cava, which passes through the caval hiatus in the right side of the diaphragm. The effect of recumbency on the relative position of the left and right diaphragmatic crus is fairly constant, but these relationships are not observed in every dog, or in cats. Sometimes the right crus is more cranial in the left lateral view. The more cranial location of the left crus typically seen in left lateral radiographs is due both to pressure from abdominal organs pushing the left side of the diaphragm cranially and to decreased lung aeration in the dependent left lung (Figure 6-10).

With regard to the heart, there is generally less heart-to-sternal contact in the left lateral view compared with the right lateral view.

Pulmonary vessels are arranged as a pulmonary artery–pulmonary vein pair, with an interposed bronchus. In lateral views, pulmonary arteries are dorsal to the bronchus and pulmonary veins are ventral. In VD or DV radiographs, pulmonary arteries are lateral to the bronchus, and pulmonary veins are medial. Careful examination of

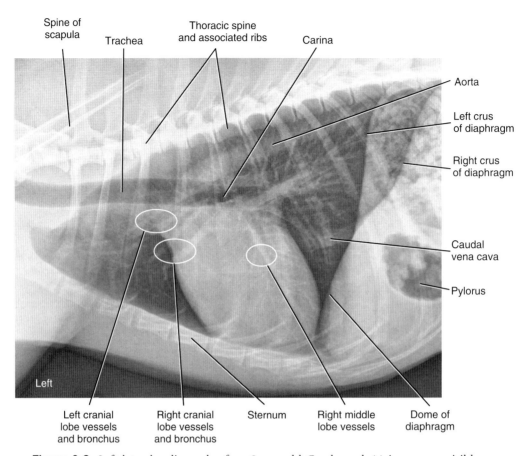

Figure 6-8. Left lateral radiograph of an 8-year-old Greyhound. Major organs visible on thoracic radiograph are identified.

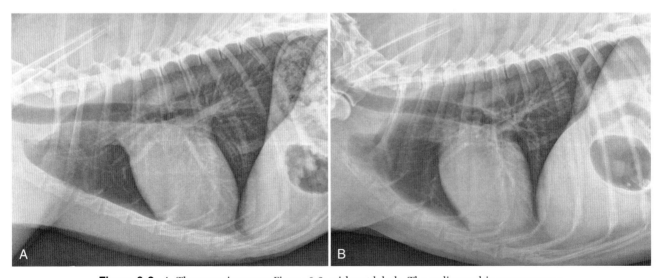

Figure 6-9. **A,** The same image as Figure 6-8, without labels. The radiographic exposure was made at optimal inspiration and the patient was positioned adequately. **B,** Left lateral radiograph of a 9-year-old mixed breed dog. The radiographic exposure was made at optimal inspiration, and the patient was positioned adequately.

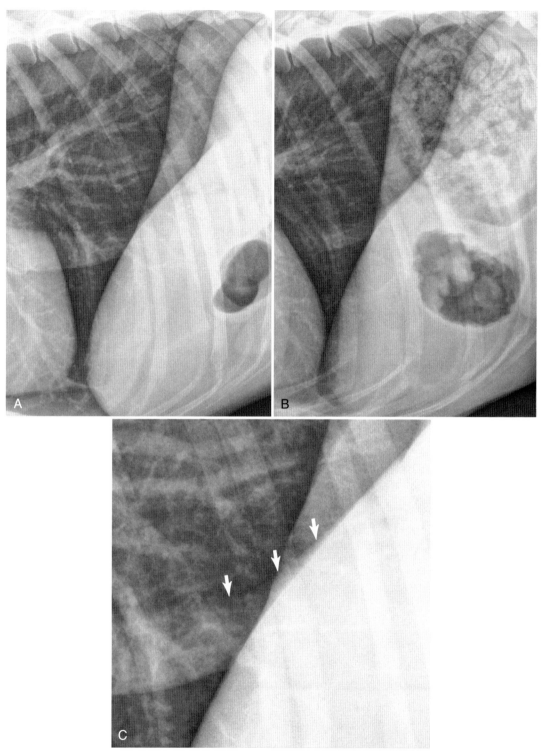

Figure 6-10. A, Left lateral radiograph of a 7-year-old Greyhound with an empty stomach. The left crus is cranial to the right crus; the caudal vena cava can be seen entering the right crus. A small amount of gas is present within the pylorus. B, Left lateral radiograph of an 8-year-old Greyhound. The fundus is moderately distended with ingesta and is readily visible caudal to the left crus. The position of the gastric fundus relative to the left crus can also be used to assist in differentiating the left crus from the right crus. In both A and B, the left crus is cranial to the right crus due to a combination of decreased aeration in the recumbent left lung and pressure from abdominal organs. The pylorus is gas-filled because gas rises to the right side. C, The same image as A, magnified over the caudal vena cava. The caudal vena cava (dorsal margin delineated by *white arrows*) extends beyond the left crus and enters the more caudal right crus.

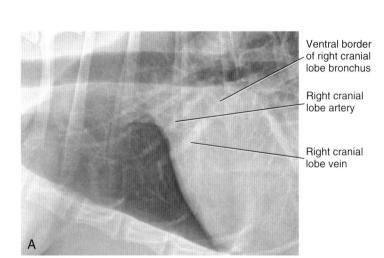
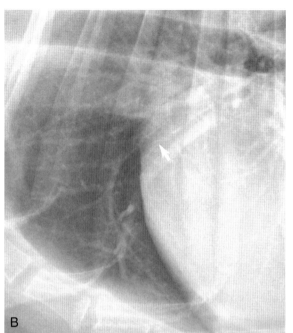

Figure 6-11. **A**, Left lateral radiograph of an 8-year-old Greyhound, centered on the cranial lobe pulmonary vessels. The left and right cranial lobe pulmonary vessels are almost always easier to discern in the left lateral view than in the right lateral view. The right cranial lobe pulmonary vessels and bronchus are more ventral than the left cranial lobe pulmonary vessels and bronchus. The most ventral structure is the right cranial lobe pulmonary vein. The dorsal vessel is the right cranial lobe pulmonary artery. The radiolucent structure between the two pulmonary vessels is the right cranial lobe bronchus. Importantly, the bronchus does not always occupy the entire space between the pulmonary artery and pulmonary vein, and the distance between these pulmonary vessels should not be inferred to represent the size of the bronchus. Identification of a mineralized bronchial wall facilitates assessment of the bronchus diameter, but mineralized bronchial walls are not visible in every patient. The vessel/bronchus triad seen dorsal to the right cranial lobe pulmonary artery–pulmonary vein pair in this figure is associated with the cranial part of the left cranial lung lobe. **B**, Left lateral radiograph of a 9-year-old mixed breed dog. The right and left cranial lobe pulmonary vessels are easy to differentiate. The wall of the right cranial lobe bronchus is mineralized *(white arrow)*, and the bronchus clearly does not occupy the entire distance between the right cranial lobe pulmonary artery and the pulmonary vein.

pulmonary artery–pulmonary vein pairs should be routine because a disparity in size is often the first radiographic sign of serious cardiovascular disease. The right cranial lobe pulmonary artery–pulmonary vein pair is typically the easiest to identify. This pulmonary artery–pulmonary vein pair is most conspicuous in the left lateral view where they are ventral to the left cranial lobe pulmonary vessels (see Figures 6-8, 6-11 and 6-12). Pulmonary artery–pulmonary vein pairs should be evaluated in terms of their size, relative to each other, and their absolute size. The cranial lobe pulmonary vessels are usually equal in size to the width of the fourth rib at its narrowest point.[2]

Interposed between any pulmonary artery and vein is the corresponding bronchus. Importantly, a bronchus does not always occupy the entire space between a pulmonary artery and vein; therefore, the distance between the pulmonary artery and pulmonary vein should not be inferred as the size of the bronchus. Identification of a mineralized bronchial wall facilitates assessment of the absolute bronchus diameter, but mineralized bronchial walls are not visible in every patient (see Figures 6-11 and 6-13).

The right lung, being nondependent in a left lateral radiograph, is not affected by abdominal pressure and is better aerated than the dependent left lung. This optimizes visualization of normal pulmonary structures on the right side and the conspicuity of right-sided lesions. In a left lateral view, the vessels superimposed over the heart are typically the right middle lobar pulmonary artery and pulmonary vein. These vessels are commonly mistaken for coronary vessels. Pulmonary vessels are readily apparent radiographically because they are surrounded by air in the lung. The air increases vessel

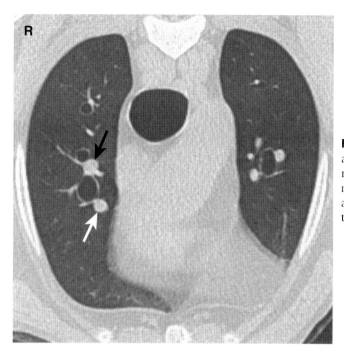

Figure 6-12. Transverse CT of a canine thorax at the level of the ascending aorta, optimized for lung. The patient was in sternal recumbency. The *black arrow* is the right cranial lobe artery. The *white arrow* is the right cranial lobe vein. The right cranial lobe artery/bronchus/vein triad is typically more ventral than the contralateral structures on the left.

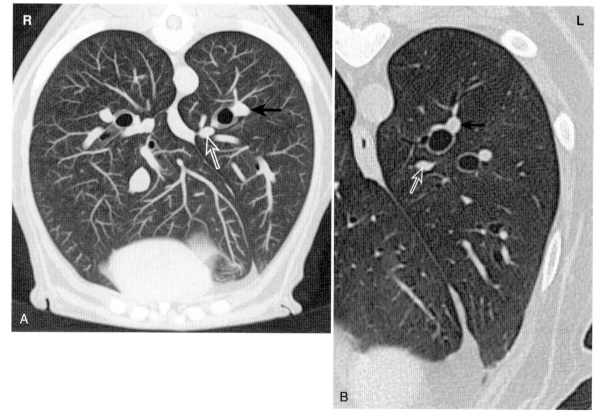

Figure 6-13. Transverse CT image of two canine patients imaged at a level between the caudal aspect of the heart and the diaphragm, optimized for lung. The patients are in sternal recumbency. In image **A**, the pulmonary vasculature is readily evident as the curvilinear arborizing soft tissue attenuating structures distributed throughout the lung. Image **B** is a thin slice image at approximately the same level. In both images, the structure depicted by the *solid black arrow* is the left caudal lobar artery. The structure delineated by the *hollow white arrow* is the left caudal lobar vein. The tubular structure between the two vessels is the left caudal lobar bronchus. The bronchus does not occupy the entire space between the artery and vein and is only evident in this study due to the enhanced contrast resolution of CT.

conspicuity because it has a lower physical density and is much more radiolucent. The better the lung aeration, the more conspicuous pulmonary vessels and other parenchymal structures become. Coronary vessels are not visible without the administration of contrast medium or unless they are mineralized. Normal coronary vessels are the same opacity as cardiac muscle and therefore silhouette with it. This results in border effacement and makes radiographic differentiation between heart and coronary vessel impossible (Figures 6-14 and 6-15, *A-B*).

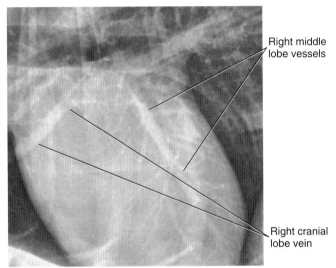

Figure 6-14. Left lateral radiograph of an 8-year-old Greyhound centered over the heart. The pulmonary vessels overlying the caudal third of the heart are associated with the right middle lung lobe. These do not represent coronary vessels. Visualization of pulmonary vessels is due to the presence of surrounding air. Coronary vessels are of the same opacity as myocardium; they are not visible radiographically unless injected with intravenous contrast medium or there is mineralization of the vessel wall.

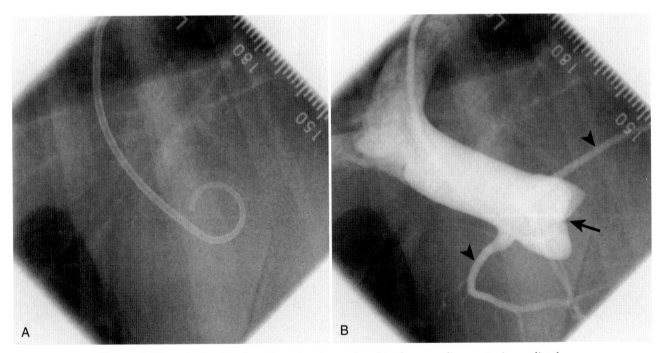

Figure 6-15. In **A**, the tip of a catheter has been placed in the ascending aorta, immediately above the aortic valve. Access to the aorta was via the femoral artery. The aortic root and coronary vessels are not visible. In **B**, radiopaque contrast medium has been injected into the ascending aorta. This is opacifying the ascending aorta and the coronary arteries *(black arrow heads)*. The vessels now have a differing attenuation (metallic) due to the presence of the contrast medium and are now readily visible. The normal aortic valve *(black arrow)* is preventing retrograde flow of contrast medium into the left ventricular outflow tract.

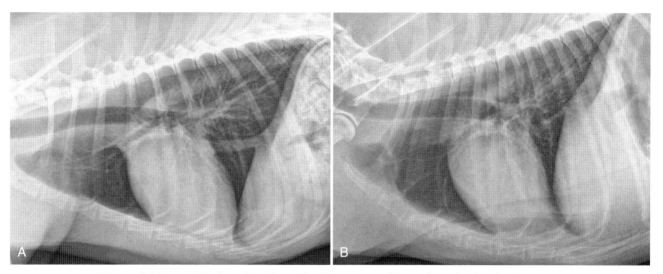

Figure 6-16. **A,** Right lateral radiograph of an 8-year-old Greyhound. The diaphragmatic crura are parallel with the right crus more cranial than the left crus. Ingesta is present within the fundus, caudal to the more caudal left crus of the diaphragm. The caudal vena cava can be seen entering the right crus. The left and right cranial lobe pulmonary vessels are superimposed and more difficult to discern accurately. This commonly occurs in the right lateral view. **B,** Right lateral radiograph of a 9-year-old mixed breed dog. The caudal vena cava can be seen entering the more cranial right crus. Cranial lobe pulmonary vessels from the left and right sides are superimposed, and accurate discrimination between arteries and veins is impossible.

RIGHT LATERAL VIEW

In the right lateral view, the patient lies in right recumbency and the x-ray beam enters the left side of the thorax. Similar to the left lateral view, based on correct radiographic nomenclature, this radiograph projection should be called a *left-right view*, but this is usually abbreviated to just a *right lateral view* (Figure 6-16). In the right lateral view, the diaphragmatic crura are usually parallel (Figure 6-17). The right crus is typically more cranial than the left; however, as mentioned previously, this relationship does not hold in every normal dog, or in cats. Opposite to the left lateral view, there is often more heart-to-sternal contact in the right lateral view. The right cranial lobe pulmonary artery and pulmonary vein are typically more dorsal than in the left lateral view and become superimposed on the left cranial pulmonary artery and

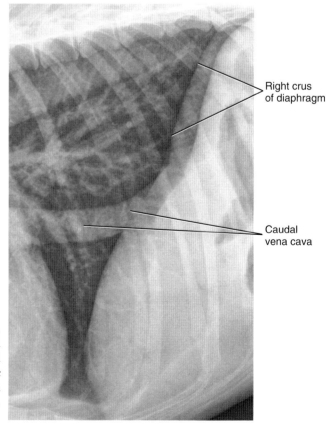

Figure 6-17. Right lateral radiograph of an 8-year-old Greyhound, centered on the diaphragm. The diaphragmatic crura are parallel to each other, with the right crus being more cranial. The caudal vena cava enters the right crus. The right crus is more cranial due to reduced aeration of the dependent right lung and pressure from abdominal organs. A small amount of gas is present in the stomach, including the pylorus.

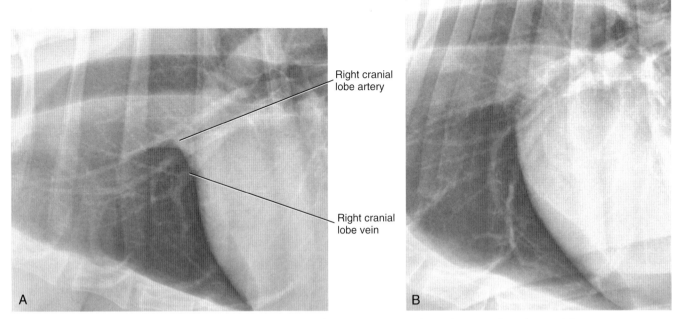

Figure 6-18. A, Right lateral radiograph of an 8-year-old Greyhound, centered on the cranial lobe pulmonary vessels. The left and right cranial lobe pulmonary vessels and the bronchi are superimposed, making accurate assessment of the absolute and relative size of the cranial lobe pulmonary vessels impossible. In the right lateral view, the cranial lobe pulmonary vessels are displaced dorsally compared with their location in the left lateral view, resulting in superimposition over the left-sided pulmonary vessels/bronchus. **B,** Right lateral radiograph of a 9-year-old mixed breed dog, centered on the cranioventral lung lobes. Superimposition of the left and right cranial lobe pulmonary vessels and bronchi makes accurate assessment of absolute and relative vessel size impossible. This superimposition of left and right cranial lobe pulmonary vessels occurs commonly in the right lateral view.

pulmonary vein (Figure 6-18). This results in confusion and inaccuracy with regard to assessing the relative and absolute size of the cranial lobe pulmonary vessels. The more dorsal positioning of the right cranial lobe pulmonary vessels on the right lateral view is due to the effect of compression of the dependent hemithorax. As the dependent hemithorax is compressed, the lung is forced dorsally. Not only do normal structures in the dependent lung become displaced dorsally, but also pulmonary lesions. This occurs in both left and right lateral views and is referred to as the *down pathology rises* phenomenon. The dorsal shift of a right-sided pulmonary mass in the right lateral radiographs illustrates the magnitude of anatomic shift (rise) that occurs in the dependent lung (Figure 6-19).

Vessels to the caudal part of the left cranial lung lobe and cranioventral aspect of the left caudal lung lobe are usually visible overlying the caudal aspect of the cardiac silhouette (Figure 6-20). These, too, are sometimes incorrectly identified as coronary vessels.

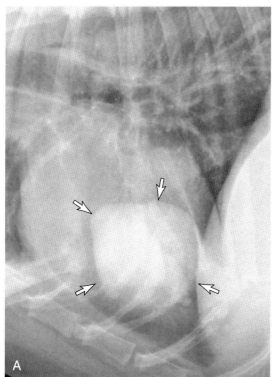

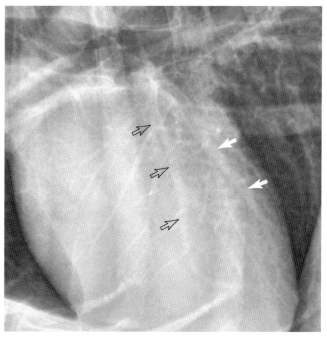

Figure 6-20. Right lateral radiograph of an 8-year-old Greyhound, centered on the midventral thorax. The *solid white arrows* indicate pulmonary vessels associated with the cranioventral aspect of the left caudal lung lobe. The *hollow black arrows* indicate pulmonary vessels associated with the caudal part of the left cranial lung lobe.

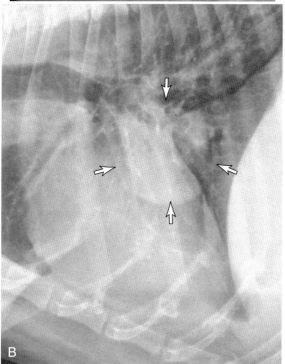

Figure 6-19. Left lateral (**A**) and right lateral (**B**) radiographs of an 11-year-old Boxer with a mass in the right middle lung lobe *(arrows)*. The mass is more dorsal on the right lateral view, illustrating the amount of anatomic shift that occurs due to recumbency. Both normal structures and pulmonary lesions are typically more dorsally located when in the dependent lung. This is due to reduced lung expansion in the dependent hemithorax. This phenomenon can be used to help assess the laterality of focal pulmonary lesions if they are not seen clearly in ventrodorsal or dorsoventral view.

DORSOVENTRAL VIEW

In the *dorsoventral view*, the patient is in sternal recumbency and the x-ray beam enters the patient dorsally and exits ventrally (Figure 6-21). The DV view can be difficult to position accurately, particularly in patients with hip osteoarthritis. Symmetrically positioning the forelimbs and ensuring that the head and neck are straight and lined up with the spinal axis increases positioning accuracy. The collimated field should be such that the entire thorax is imaged. The most common error occurs when the central beam is positioned too far caudally and a radiograph is obtained that includes too much of the abdomen and not enough of the cranial aspect of the thorax. In the DV view, the diaphragm has a domed appearance and extends more cranially than in the VD view, as a result of abdominal pressure. This results in physical contact between the diaphragm and the cardiac silhouette, often leading to the apex of the cardiac silhouette being shifted to the left. The extent of cardiac shift to the left is a function of body mass, with more cardiac

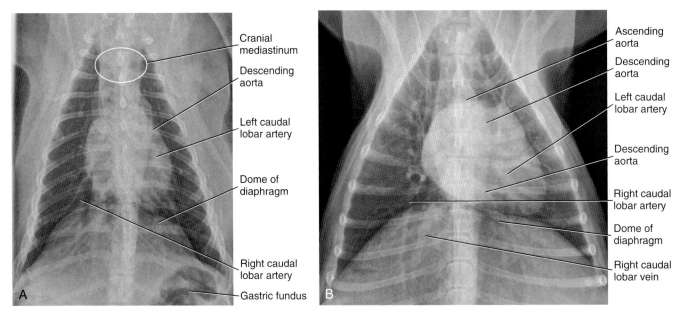

Figure 6-21. **A,** Dorsoventral radiograph of a 10-year-old Labrador Retriever. The caudal aspect of the heart is in contact with the diaphragm, and the diaphragm has a domed appearance. Gas is present in the fundus of the stomach because the fundus is more dorsally located when the dog is in sternal recumbency. The caudal lobe pulmonary vessels are readily visible. **B,** Dorsoventral radiograph of a 12-year-old mixed breed dog. The cranially located diaphragmatic dome is resulting in leftward displacement of the heart. The heart position is typically more variable in dorsoventral view.

displacement occurring in larger dogs.[3] The normal cardiac shifting to the left in the DV view is commonly misinterpreted as cardiomegaly. As in small dogs, cardiac shifting to the left in sternal recumbency is minimal in cats.[4]

The caudal lobar pulmonary vessels are more conspicuous in the DV view than in the VD view (Figures 6-22 and 6-23). In sternal recumbency, the caudal lobe pulmonary vessels are more perpendicular to the oncoming x-ray beam, which leads to minimal distortion. In addition, in sternal recumbency there is little atelectasis of the caudal lung lobes; this optimal aeration results in increased vessel conspicuity. As noted previously, the left and right caudal lobar pulmonary arteries are located lateral to the bronchus, and the left and right caudal lobar pulmonary veins are medial to the bronchus (see Figure 6-22). Often, the right caudal lobar pulmonary vein is superimposed over the caudal vena cava, making assessment of both the right caudal lobar pulmonary vein and the caudal vena cava difficult (see Figure 6-22). Caudal lobar pulmonary vessels should taper gradually toward the periphery of the thorax. Normal caudal lobe pulmonary arteries and veins should be approximately the same size in the dog and cat. With regard to absolute size, each vessel should be approximately the same size as the diameter of the ninth rib in VD or DV radiographs and therefore the summation opacity created by the overlap can be used as a gauge of vessel size (see Figure 6-22, C) Measurement data suggest that 1.2 times the size of the diameter of the ninth rib may be a more accurate upper limit of absolute size.[5] These size-estimation parameters are only a guide and must always be interpreted in concert with other assessments of cardiovascular morphology, especially heart size and shape.

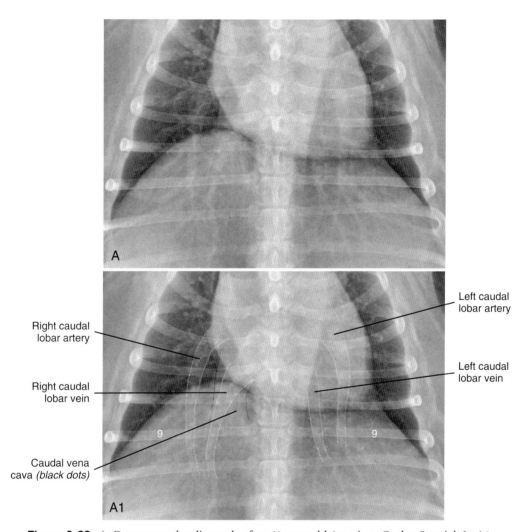

Figure 6-22. A, Dorsoventral radiograph of an 11-year-old American Cocker Spaniel. In **A1**, the caudal lobar vessels have been outlined *(white dots)*. Each caudal lobe artery is lateral to the respective vein. In this dog, the right caudal lobar vein can be readily identified, superimposed over the caudal vena cava *(black dots)*.

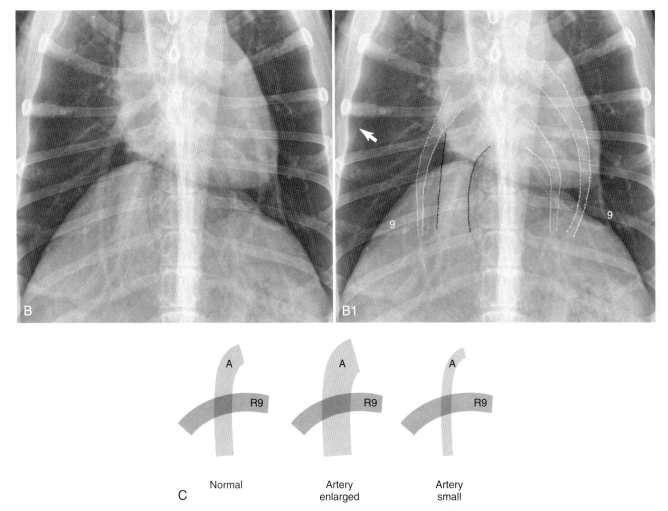

Figure 6-22, cont'd. B, Dorsoventral radiograph of a 15-year-old Labrador Retriever. In B1, the caudal lobar pulmonary arteries and left vein have been outlined *(white dots)*. The *dotted black lines* are the margins of the caudal vena cava. In this dog, the right caudal lobar vein is difficult to see as a result of superimposition with the caudal vena cava. A pleural fissure line between the caudal aspect of the right middle and cranial aspect of the right caudal lung lobes is present *(solid white arrow)*. C, Schematic of vessels as they cross the ninth rib. Each caudal lobar pulmonary artery and pulmonary vein should be approximately the same relative size; in the dog, the absolute size should be similar to the width (up to 1.2×) of the ninth rib at the point at which they intersect.

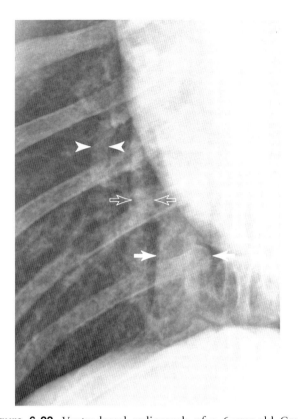

Figure 6-23. Ventrodorsal radiograph of a 6-year-old Great Dane. The caudal lobe vessels are often more difficult to identify in the ventrodorsal view. The *solid white arrows* are the caudal vena cava. The *solid white arrowheads* are probably the right caudal lobar artery. The *hollow white arrows* are presumed to be the right caudal lobar vein, in this dog, not superimposed over the caudal vena cava.

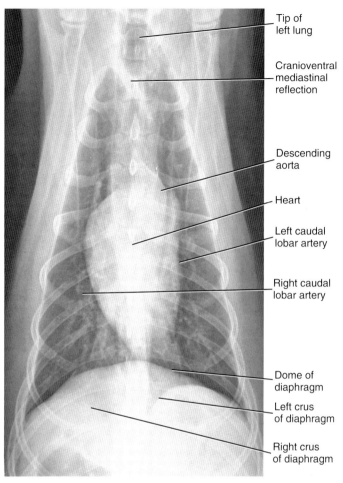

Figure 6-24. Ventrodorsal radiograph of a 9-year-old mixed breed dog. The cardiac silhouette has an elongated appearance, and there is an obvious space between the caudal aspect of the heart and the diaphragm. This is due to decreased abdominal pressure on the dome of the diaphragm when the patient is in dorsal recumbency. This allows better visualization of the accessory lung lobe region.

VENTRODORSAL VIEW

In the *ventrodorsal view*, the patient is in dorsal recumbency, and the x-ray beam enters the patient ventrally and exits dorsally. The VD view is often easier to position in patients with coxofemoral osteoarthritis because hindlimb extension is not required. Thin patients may benefit from padding under the spine to facilitate positioning. As with the DV view, it is important to position the forelimbs symmetrically and ensure that the neck is straight and the head is directed vertically. Asymmetrically positioning the forelimbs or bending the head and neck are common causes of thoracic malpositioning. The VD view should not be attempted in patients with obvious cardiopulmonary compromise because decompensation can result due to patient stress. In the VD view, the diaphragmatic crura have a convex appearance and the dome of the diaphragm is usually more caudal, because there is less abdominal pressure on the dome of the diaphragm when compared with a patient in sternal recumbency. This usually results in the presence of aerated lung between the caudal aspect of the heart and the diaphragm, thus allowing better visualization of the accessory lung lobe[3] (Figures 6-24 and 6-25). The caudal lobe pulmonary arteries and pulmonary veins are less conspicuous in the VD view than in the DV view because they are at a steep

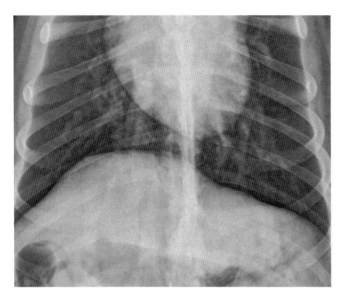

Figure 6-25. Ventrodorsal radiograph of a 6-year-old mixed breed dog. There is an obvious space between the cardiac silhouette and the diaphragm. A ventrodorsal view allows better evaluation of the accessory lung lobe region. Caudal lobe pulmonary vessels are more difficult to evaluate compared with the dorsoventral view because they are oriented at a steep angle with respect to the primary x-ray beam, and there is more surrounding atelectasis due to dependency.

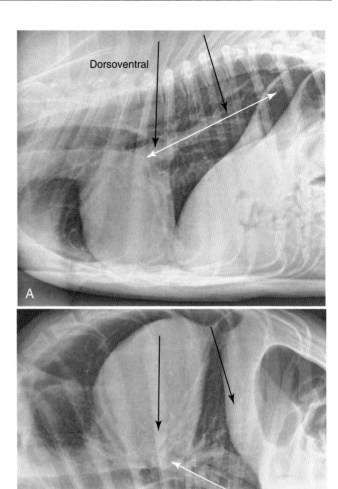

angle with respect to the primary x-ray beam and are therefore distorted. There also is more compression atelectasis of the caudal lobes with the subject in dorsal recumbency, which contributes to decreased vessel conspicuity in the caudal lung lobes (Figure 6-26).

The cardiac silhouette is usually more elongated and narrow in the VD view compared with the DV view. It is also less subject to displacement from diaphragmatic contact, leading to the heart typically being more centrally located within the thorax.[3]

The decision to use anesthesia or sedation for thoracic radiography depends on many factors, including local radiation safety regulations. When general anesthesia is used, modest positive pressure ventilation should be used during the radiographic procedure to achieve optimal lung aeration. Anesthesia should be induced with the patient in sternal recumbency, and the patient should remain in sternal recumbency until lateral radiographs are made. The time between anesthesia induction and radiography should be minimized.

Figure 6-26. Horizontal beam radiographs of a 3-year-old Doberman Pinscher. In **A**, the patient is in sternal recumbency, and in **B** the patient is in dorsal recumbency. When the patient is in sternal recumbency, the caudodorsal lungs are optimally aerated and the caudal lobar pulmonary artery and veins are more perpendicular to the diverging radiographic beam. These physiologic and positional factors result in increased conspicuity of the caudal lobar pulmonary vessels compared with the ventrodorsal view. With the patient in dorsal recumbency (**B**), as occurs in the ventrodorsal view, there is decreased aeration of the caudodorsal lung due to recumbency-related atelectasis, and the caudal lobar pulmonary vessels are not perpendicular to the primary beam. Consequently, caudal lobar pulmonary vessels are more difficult to see in the ventrodorsal view. The *white arrows* are the approximate direction of the caudal lobar pulmonary vessels. The *black arrows* are the approximate direction of the diverging radiographic beam.

Figure 6-27. A, Left lateral radiograph of a 9-year-old Beagle, at peak expiration. The diaphragm is in contact with the caudal aspect of the heart, and there is only a small triangle of aerated lung between the caudal aspect of the heart, the ventral aspect of the caudal vena cava, and the cranial aspect of the diaphragm *(white circle)*. This triangular region should be evaluated in every lateral radiograph as a way to assess the degree of lung aeration at the time of radiography. Reduced lung aeration results in a diffuse increase in lung opacity and reduced conspicuity of parenchymal disease. In addition, the heart appears larger in the incompletely expanded thoracic cavity, giving the false impression of cardiomegaly. Reduced lung aeration is exacerbated by sedation, obesity, and abdominal disease that results in cranial displacement of the diaphragm, such as significant hepatomegaly or voluminous peritoneal fluid. **B,** Left lateral radiograph of a 7-year-old Golden Retriever, at peak inspiration and with optimal lung aeration. The triangular region between the caudal vena cava, heart, and diaphragm is much larger due to optimal lung aeration. Optimal lung aeration increases the conspicuity of parenchymal disease and increases diagnostic accuracy.

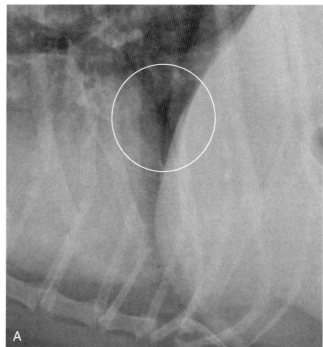

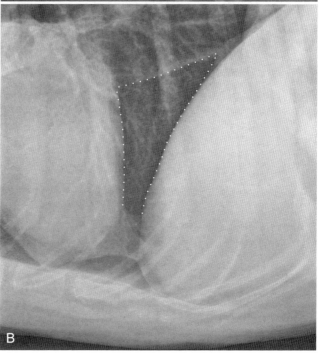

The interpretation of thoracic radiographs made after prolonged lateral recumbency or anesthesia is complicated by pulmonary atelectasis and a secondary mediastinal shift. The lack of support from the inadequately aerated lung results in a shift of mediastinal structures toward the side with atelectasis (see Figure 6-6). With respect to lateral views, the overall adequacy of lung aeration is best assessed by evaluating the size of the triangular region between the caudal aspect of the heart, the ventral aspect of the caudal vena cava, and the dome of the diaphragm. The larger the volume of this triangular region is, the better the overall pulmonary aeration (Figure 6-27).

THORACIC WALL

Variations in the development of the spine are commonly detected on thoracic radiographs; Chapter 3 discusses these findings.

With respect to the ribs and the costal arch, young animals have incompletely calcified costal cartilage, whereas mature animals invariably have more calcification of costal cartilage and costochondral junctions. The mineralization pattern of the costal cartilages is usually inconsequential, but this is often interpreted incorrectly as abnormal. The mineralized costal cartilages can also confound the assessment of the pulmonary parenchyma and lead to an incorrect assessment of pulmonary disease due to summation (Figure 6-28).

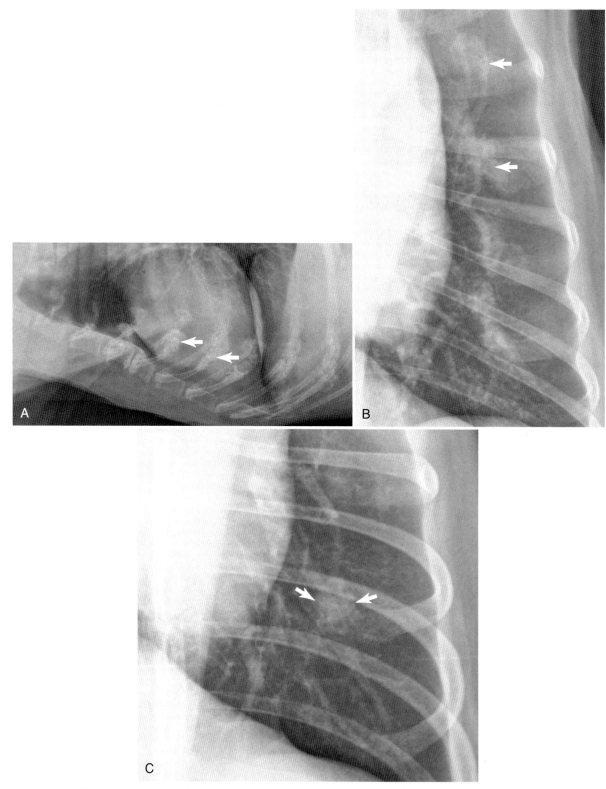

Figure 6-28. Lateral (**A**) and ventrodorsal (**B**) radiographs of a 9-year-old Golden Retriever. There is extensive new bone proliferation at the costochondral junctions (*white arrows* indicate only two of many costochondral junctions affected). This is a common occurrence, typically in older patients, and is of no clinical significance. It can, however, complicate interpretation of the ventral aspect of the thorax, and focal regions of new bone formation on the costal cartilages can be misinterpreted as a pulmonary nodule or nodules. Degenerative changes are present in the sternum of this dog. **C**, Ventrodorsal radiograph of a 12-year-old Labrador Retriever. Exuberant osseous proliferation at the seventh costochondral junction *(white arrows)* is giving the false impression of a pulmonary nodule.

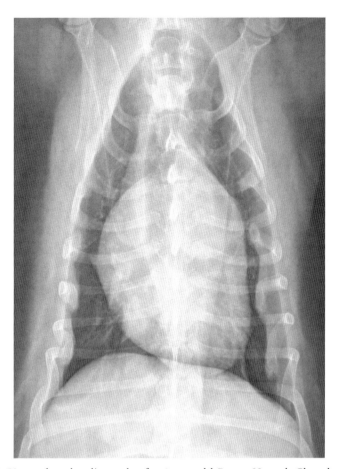

Figure 6-29. Ventrodorsal radiograph of a 6-year-old Basset Hound. Chondrodystrophoid breeds commonly have atypical curvature to the ribs at the costochondral junctions. The ribs deviate medially at the level of the costochondral junctions, which results in a soft tissue margin that can be misinterpreted as pleural effusion on ventrodorsal and dorsoventral radiographs. In this dog, this is most noticeable in the caudal aspect of the left hemithorax.

Chondrodystrophoid breeds, typically the Dachshund and the Basset Hound, have enlarged costochondral junctions and often an irregular contour to the costal arch. The costochondral junctions are often misinterpreted as pulmonary nodules in the VD or DV views. In addition, the configuration of the costochondral region leads to a peripheral region of increased opacity that can be misinterpreted as pleural fluid (Figure 6-29).

Skin folds can result in well-defined linear opacities that cross and extend beyond the borders of the thoracic cavity; these overlying opacities are quite obvious in some patients and can be distracting. In addition, especially in VD radiographs, the opacity lateral to skin folds is very radiolucent. When the skin fold is parallel to the lateral margin of the thorax, the reduced opacity lateral to the skin fold can be confused with pneumothorax (Figure 6-30). To make the correct interpretation, it is critical to note that the opacity of the skin fold extends beyond the limit of the thoracic cavity.[6]

The sternum comprises a series of eight bones that form the floor of the thoracic cavity. Each segment is joined by short blocks of cartilage, known as the

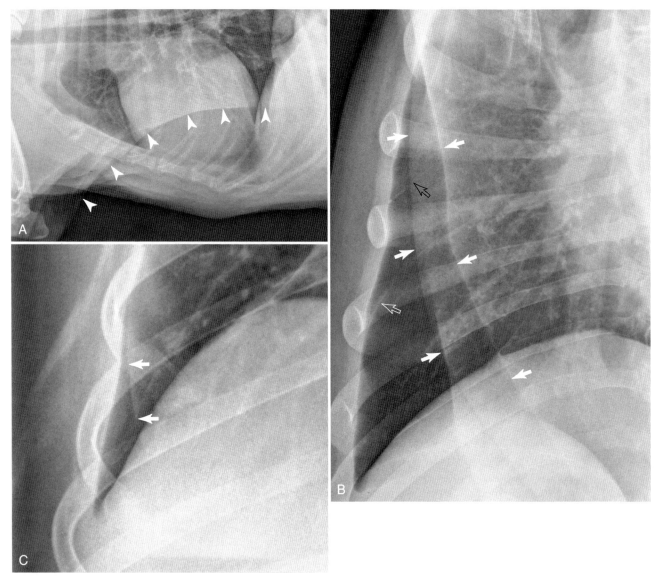

Figure 6-30. A, Lateral radiograph of an 11-year-old Golden Retriever. Prominent skin folds associated with the forelimbs are present. In the lateral view, skin-fold position and the relative opacity change observed is more severe when the limbs are not fully extended *(white arrowheads)*. **B,** Ventrodorsal radiograph of a 9-year-old Mastiff; the apparent opacity of the lung medial to the skin fold *(solid white arrows)* is increased, and lateral to the skin fold is reduced. The *hollow black arrow* is a pleural fissure between the right cranial and right middle lobes, and the *hollow white arrow* is a pleural fissure between the right middle and right caudal lobes. **C,** Ventrodorsal radiograph of a 10-year-old German Shepherd. The apparent reduction in lung opacity lateral to the skin fold can give the false impression of pneumothorax, particularly when the skin fold closely parallels the thoracic wall *(solid white arrows)*. A radiograph should have sufficient dynamic range to enable visualization of pulmonary markings throughout the entire thorax.

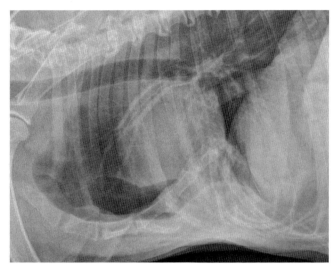

Figure 6-31. Right lateral radiograph of a 9-year-old German Shepherd. There is marked dorsal displacement of the caudal aspect of the sternum, resulting in mild displacement of the cardiac silhouette. There are only six sternal segments. This anomaly, known as *pectus excavatum*, is uncommon and is typically not associated with clinical signs. With a severe deformity, cardiac displacement can result in cardiopulmonary dysfunction.

intersternebral cartilages, which articulate bilaterally with the first nine rib pairs. Anomalies of the first external segment (manubrium) are uncommon, but variations in the morphology of the last sternebrae (xiphoid process) are common. Hypoplasia of the xiphoid process or absent caudal sternal segments is often associated with morphologic abnormalities of the caudal thorax and diaphragm, the most common being *pectus excavatum* (Figure 6-31). Although pectus excavatum is clearly not normal, it is often not associated with any clinically significant complications.

MEDIASTINUM

The mediastinum is the space between the right and left pleural sacs, bounded by mediastinal pleura. The mediastinum extends from the thoracic inlet to the diaphragm. It is fenestrated and typically cannot contain unilateral pleural disease.[7] It also communicates with the neck through the thoracic inlet and with the retroperitoneal space through the aortic hiatus.

The mediastinum is generally a midline structure; however, there are three mediastinal reflections where the mediastinum deviates laterally from midline:
- Cranioventral mediastinal reflection
- Caudoventral mediastinal reflection
- Plica vena cava reflection

The cranioventral mediastinal reflection is the border between the right cranial lung lobe and the cranial part of the left cranial lobe. The cranial part of the left cranial lobe is the most cranial of all pulmonary tissue in the dog and the cat. At the thoracic inlet, the cranial part of the left cranial lobe crosses the midline to the right. In lateral views, this portion of the left cranial lobe often appears as a sharply contained radiolucent region in the most cranioventral aspect of the thorax. Immediately caudal to the cranial part of the left cranial lobe, the right cranial lobe crosses the midline to the left. The mediastinal pleura between these lobes forms the cranioventral mediastinal reflection. This reflection is what causes the cranial part of the left cranial lobe to be so distinctly marginated in some lateral views (Figure 6-32). The internal thoracic

Figure 6-32. **A,** Right lateral radiograph of an 8-year-old Great Dane, centered on the cranioventral thorax. The soft tissue opacity outlined by the *solid white arrows* is the cranioventral mediastinal reflection. The lung, cranial to that outlined by the *hollow white arrows,* is the cranial most tip, or *cupula,* of the left cranial lung. The cupula of the left lung is actually located on the right side. **B,** Ventrodorsal radiograph of a 12-year-old Siberian Husky. The cranioventral mediastinal reflection is delineated by the *solid white arrows.* Cranially, the tip of the left cranial lung lobe extends to the right side, in the region of the circle. The right lung extends to the left side immediately cranial to the heart. **C,** Right lateral radiograph of a 7-year-old Golden Retriever. The cranioventral mediastinal reflection is not as conspicuous as in **A.** There is considerable variability in the appearance of this reflection, depending on the amount of mediastinal fat and its relationship with the primary x-ray beam.

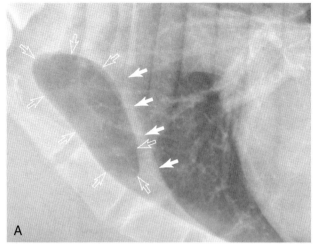

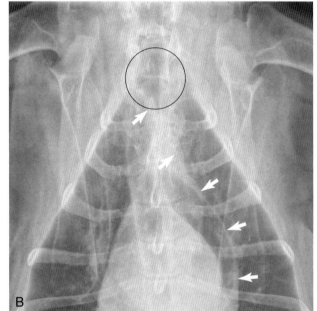

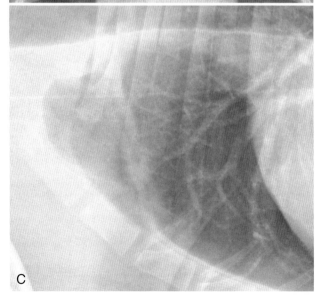

arteries and veins lie in the cranioventral mediastinal reflection.

The caudoventral mediastinal reflection is bounded by the accessory lobe on the right and the left caudal lobe on the left (Figures 6-33 and 6-34). The left aspect of the accessory lobe crosses the midline to the left, pushing the mediastinum to the left and forming the caudoventral mediastinal reflection. The thickness of the caudoventral mediastinal reflection varies considerably, mainly due to the amount of fat it contains.

The caudal vena cava extends from the caudal aspect of the cardiac silhouette to the right crus of the diaphragm and is enveloped within the caudal caval reflection, called the *plica vena cava.* This mediastinal reflection is not visible radiographically, but it is important to realize that there is a mediastinal reflection around the caudal vena cava (see Figures 6-1, 6-10, 6-17, 6-27 and 6-35). For example, mediastinal fluid can dissect into the plica vena cava and create the appearance of enlargement of the caudal vena cava. The size of the caudal vena cava is extremely variable, depending on the phase of cardiac and respiratory cycles. Normally, the caudal vena cava is less than 1½ times the width of the descending aorta.[8]

Mediastinal organs normally seen radiographically include the heart, trachea, caudal vena cava, aorta, and to varying degrees the esophagus, as discussed in the section on the esophagus in this chapter. The cranial vena cava and cranial mediastinal arteries are not typically visible because they are in contact with each other and with the adjacent esophagus, which reduces conspicuity. The dorsal aspects of the ascending aorta and the descending aorta are usually readily visible. The base of the ascending aorta is usually not visible because it silhouettes with adjacent mediastinal structures. In addition, both the cranial and caudal margins of the cardiac silhouette are visible (Figure 6-36). The proximal portion of the left subclavian artery can sometimes be seen as an arching

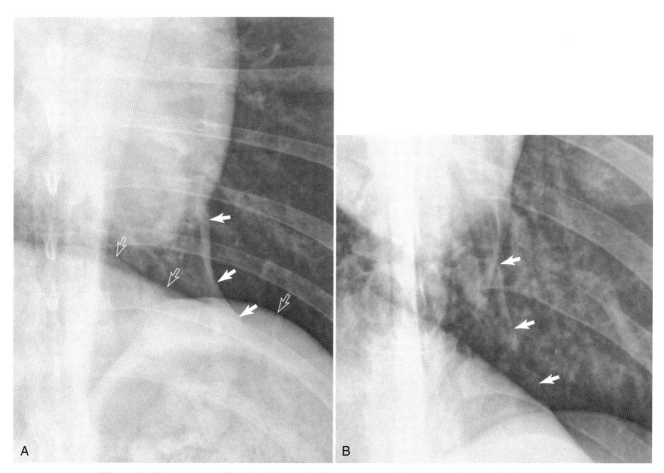

Figure 6-33. **A,** Ventrodorsal radiograph of an 11-year-old German Shepherd. The caudoventral mediastinal reflection, delineated by the *solid white arrows,* is the border between the medial margin of the left caudal lung lobe and the left lateral aspect of the accessory lung lobe, which is a lobe of the right lung. Conspicuity of the caudoventral mediastinal reflection depends on the amount of fat in the reflection, lung aeration, and patient conformation. The *hollow white arrows* are the cranial margin of the diaphragm. **B,** Ventrodorsal radiograph of a 6-year-old Great Dane. The caudoventral mediastinal reflection, outlined by the *white arrows,* is much thinner than in **A.** This is due to less fat and the plane of the reflection relative to the incident x-ray beam.

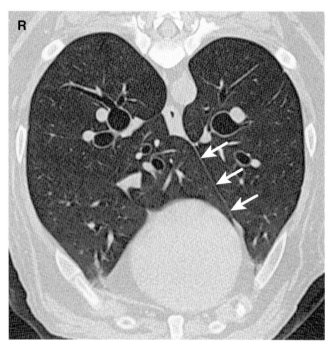

Figure 6-34. A transverse CT image of a canine thorax at the level of the caudal aspect of the cardiac silhouette, optimized for lung. The patient is in sternal recumbency. The caudoventral mediastinal reflection is denoted by the *solid white arrows* and is the physical boundary between the right and left sides of the thorax. The accessory lung lobe is on the right and the left caudal lobe on the left of the caudoventral mediastinal reflection.

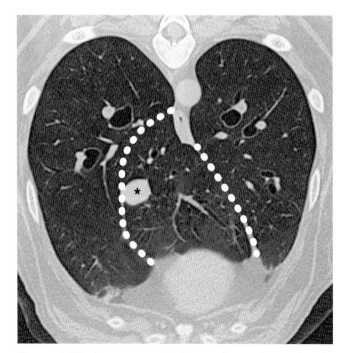

Figure 6-35. A transverse CT image of a canine patient in sternal recumbency, at the level of the apex of the cardiac silhouette, optimized for lung. The *dotted line* is the margin of the accessory lung lobe. The caudal vena cava is denoted by the *black asterisk*. It is within the *plica vena cava*, located between the right caudal and accessory lung lobes.

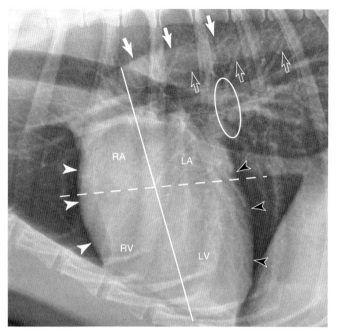

Figure 6-36. Right lateral radiograph of an 8-year-old Greyhound, centered on the heart. The *solid white arrows* indicate the dorsal margin of the descending aorta, and the *hollow white arrows* indicate the ventral border. The ellipse encompasses the caudal lobe arteries supplying caudal lung lobes and pulmonary veins from the caudal lobes draining into the left atrium. The *solid white arrowheads* delineate the cranial border of the heart, whereas the *black arrowheads* indicate the caudal border. The *solid white line* approximates the border between the left and right sides of the heart. The *dashed white line* is at the approximate level of the atrioventricular valves.

soft tissue opacity immediately cranial to the heart and dorsal to the trachea; this occurs most often in large, thin dogs without large amounts of mediastinal fat (Figure 6-37).

The width of the cranial mediastinum as seen in the VD or DV view is typically 1 to 2 times the width of a cranial thoracic vertebral body. This wide variation is due to obesity and patient conformation. Brachycephalic breeds, particularly bulldogs, typically have a wide mediastinum due to an excessive amount of mediastinal fat (Figures 6-38, *B*, and 6-39, *A-B*).

There are three mediastinal lymphocenters: sternal, cranial and caudal mediastinal, and tracheobronchial. These lymphocenters are not typically visible in a normal patient but become apparent radiographically when moderately enlarged. The sternal lymph nodes are located dorsal to the second sternebral segment. The cranial mediastinal lymph nodes are located in the cranial mediastinum immediately ventral to the intrathoracic portion of the trachea. The caudal mediastinal lymph nodes surround the caudal aspect of the trachea and heart base, and the tracheobronchial lymph nodes surround the carina and origin of the principal bronchi.

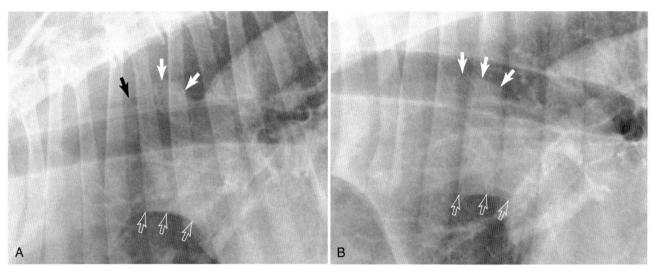

Figure 6-37. A, Left lateral radiograph of a 12-year-old Siberian Husky. The curving soft tissue structure outlined by the *solid white arrows* is the left subclavian artery. A small amount of gas is present in the esophagus *(solid black arrow)* immediately cranial to this artery. The ventral margin of the thicker dorsal portion of the cranial mediastinum that envelopes the esophagus, trachea, lymph nodes, and cranial mediastinal vessels is delineated by the *hollow white arrows*. The cranial vena cava, brachiocephalic trunk, and other cranial mediastinal vessels cannot be identified individually because they are in contact with each other and with the esophagus, causing border effacement. **B,** Right lateral radiograph of an 8-year-old Great Dane. The dorsal margin of the left subclavian artery, outlined by the *solid white arrows,* is superimposed over the trachea. The ventral margin of the thicker dorsal portion of the cranial mediastinum is delineated by the *hollow white arrows*.

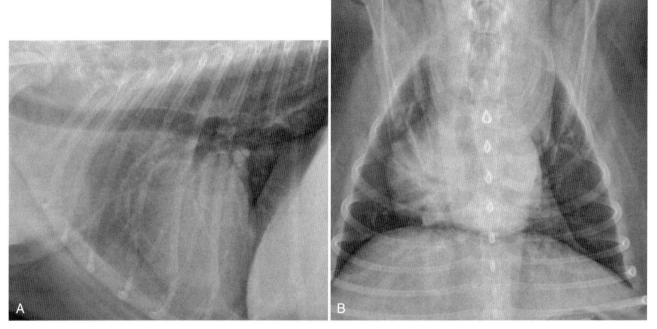

Figure 6-38. Lateral **(A)** and ventrodorsal **(B)** radiographs of an obese 8-year-old Miniature Schnauzer. A large amount of fat is present in the cranioventral mediastinum, in the mediastinum around the heart, and in the caudoventral mediastinal reflection. Although the cranial mediastinum appears wide in the dorsoventral view, mimicking a mass, there is no evidence of a cranial mediastinal mass on the lateral view.

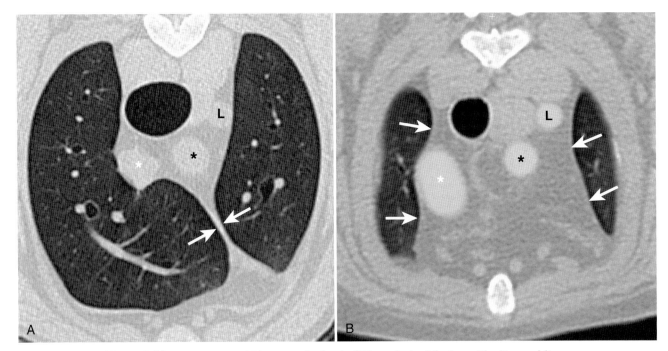

Figure 6-39. **A,** Transverse CT image at the level of T3, optimized for lung. The linear oblique soft tissue attenuating line delineated by the *white arrows* is the cranioventral mediastinum. It is thin because it contains only a small amount of fat. **B,** Transverse CT image at the same level in a Bulldog. A large amount of mediastinal fat is evident, and there is overall less lung aeration in the cranioventral thorax. In both images, the *white asterisk* is the cranial vena cava, the *black asterisk* the brachiocephalic trunk, and *L* is the left subclavian artery visible due to the presence of surrounding fat.

The thymus is often visible in young dogs (up to 6 months) as a triangle-shaped, soft tissue opacity within the ventral mediastinum, adjacent to the heart (Figure 6-40).

TRACHEA AND BRONCHI

There is extensive variability in the appearance of the intrathoracic trachea. The trachea is typically slightly to the right of midline on VD and VD views (Figure 6-41) and lies ventral to the esophagus. In most breeds, the intrathoracic trachea diverges from the spine from the level of the thoracic inlet caudally to the carina. In breeds with a dorsoventrally compressed thorax, for example, the Dachshund and Welsh Corgi, the intrathoracic trachea is typically more parallel to the spine but still diverges normally from the spine at the carina (Figure 6-42). With respect to tracheal size, typically, the mean ratio of tracheal diameter to thoracic inlet diameter has been reported to be 0.2 ± 0.03 in nonbrachycephalic breeds. In brachycephalic breeds, excluding Bulldogs, the ratio is 0.16 ± 0.03.[9] These figures must be interpreted with caution and should be used only as a guide.

The trachea bifurcates at the carina into the left and right principal bronchi. The naming convention of the

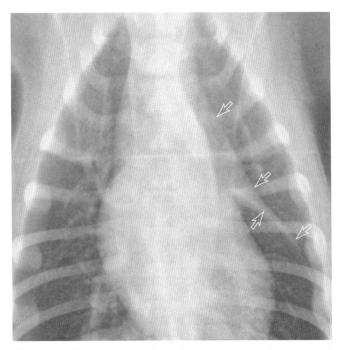

Figure 6-40. Ventrodorsal radiograph of a 2-month-old English Springer Spaniel. The thymus is delineated by the *hollow white arrows*. The thymus normally regresses by 6 months of age and should not be radiographically apparent after this time.

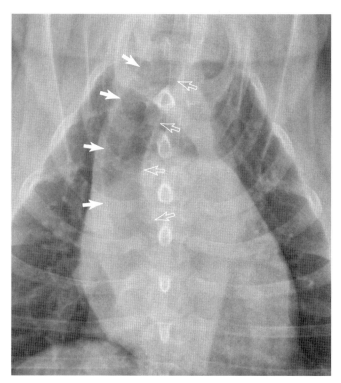

Figure 6-41. Dorsoventral view of a 7-year-old American Cocker Spaniel. The *solid white arrows* delineate the right margin of the trachea, and the *hollow white arrows* delineate the left margin of the trachea. The trachea is usually at midline or slightly to the right of midline. Sometimes, there is some rightward deviation of the intrathoracic trachea in normal patients. This also occurs with lymphomegaly and heart base masses.

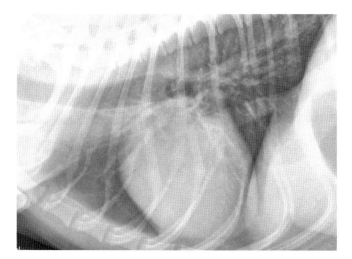

Figure 6-42. Lateral view of a 7-year-old Dachshund. There is less divergence of the trachea from the thoracic spine in lateral radiographs in dogs with a dorsoventrally compressed thorax, but there should still be obvious divergence between the trachea and spine at the carina.

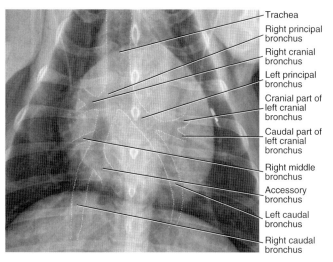

Figure 6-43. Dorsoventral radiograph of a 10-year-old Jack Russell Terrier. The bronchial tree has been outlined *(white dots)*. The right lung lobes comprise the right cranial, middle caudal, and accessory lobes. The left lung lobes comprise the cranial and caudal lobes. The left cranial lung lobe is further subdivided into cranial and caudal parts. A microchip is superimposed over the region of the left auricle.

bronchi is based on the origin of each lobar bronchus. The right lung is divided into cranial, middle, caudal, and accessory lobes, all with a bronchus arising from the right principal bronchus. Two lobes, the cranial and caudal, arise from the left principal bronchus. The left cranial lobe bronchus subdivides immediately after arising from the left principal bronchus to supply the cranial and caudal parts of the left cranial lung lobe (Figure 6-43). Overall, the volume of the right lung is greater than the volume of the left. Figure 6-44 outlines the approximate location of the lung lobes.

The appearance of the trachea in lateral radiographs can be affected by the position of the patient's head when radiographed. When the head is flexed during radiography, the intrathoracic portion of the trachea sometimes deviates dorsally and gives the false impression of a cranioventral mediastinal mass (Figure 6-45). In general, the lack of a soft tissue mass effect accounting for the tracheal elevation is sufficient evidence to conclude that the tracheal displacement is the result of neck position. If confirmation is necessary, the lateral radiograph can be repeated with the subject's neck extended.

In both the dog and the cat, the tracheal rings are incomplete dorsally, being connected by the tracheal muscle and anular ligament of the trachea. These structures, combined with the dorsal tracheal mucosa, are collectively known as the *dorsal tracheal membrane*. Sometimes the dorsal tracheal membrane prolapses

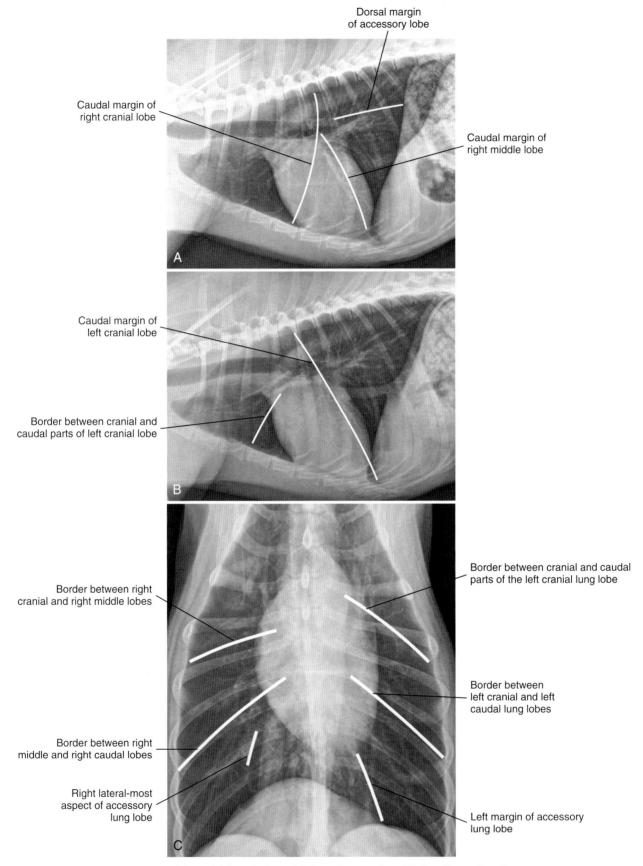

Figure 6-44. Left lateral (**A**), right lateral (**B**), and ventrodorsal (**C**) radiographs. The approximate location of lung lobe borders is indicated by the *solid white lines*. In some normal patients and commonly in disease, some of these borders are radiographically apparent and are referred to as *pleural fissures*.

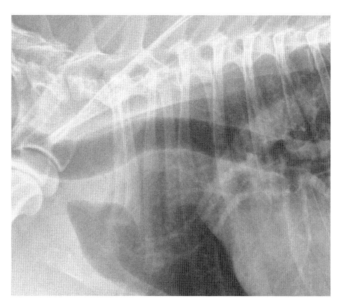

Figure 6-45. Right lateral radiograph of an 8-year-old English Pointer. There is focal dorsal deviation of the intrathoracic trachea. The mediastinum and other structures ventral to the trachea appear normal with no evidence of a space-occupying mass. Dorsal deviation of the trachea, as in this dog, occurs commonly with excessive head flexion during radiography. If there is any doubt whether there is a mediastinal mass causing tracheal elevation, the radiograph should be repeated with the neck extended.

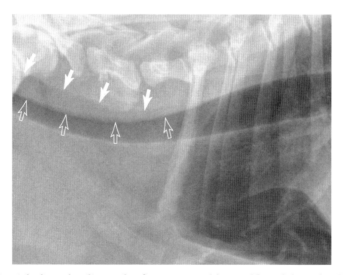

Figure 6-46. Right lateral radiograph of an 11-year-old mixed breed dog. The dorsal margin of the trachea is outlined by *solid white arrows*. The invaginated redundant trachea membrane is delineated by *hollow white arrows*. This should not be misdiagnosed as a collapsing trachea.

through the incomplete rings into the dorsal aspect of the tracheal lumen. The caudal cervical esophagus, located immediately dorsal to the trachea, can also invaginate into the dorsal tracheal lumen. These invaginations, commonly referred to as *redundancy or sagging of the dorsal tracheal membrane*, can result in an increase in opacity in the dorsal aspect of the trachea in the caudal cervical and thoracic inlet regions. Sagging of the dorsal tracheal membrane is usually not a clinically significant entity and should not be confused with tracheal collapse (Figure 6-46).

Bulldogs typically have a narrow trachea,[10] and measurement criteria have been defined to assess the adequacy of tracheal diameter in this breed. The trachea to thoracic inlet ratio in this breed has been quantified as 0.13 ± 0.38, indicating that the trachea is typically smaller but the variation in diameter is very large. The smallest ratio that was not associated with clinical signs was 0.09.[9]

ESOPHAGUS

The esophagus is dorsal to the trachea and principal bronchi as it courses from the cranial esophageal sphincter to the cardia. It is common for a small amount of gas to be present, normally within the cranial portion of the intrathoracic esophagus (Figure 6-47). Generalized esophageal distension with gas can occur in sedated or anesthetized patients (Figure 6-48).

The conspicuity of the esophagus caudal to the cardiac silhouette is variable. Often, the ventral margin of the caudal esophagus is apparent on the lateral views, most commonly when the patient is in left lateral recumbency (Figure 6-49). An ill-defined tubular increase in opacity through the region of the caudal esophagus is sometimes visible, particularly if the esophagus contains a small amount of fluid; this can occur with sedation-associated reflux but is also seen in conscious patients (Figure 6-50).

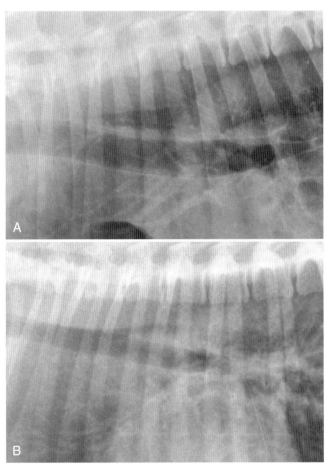

Figure 6-47. Lateral radiographs of a 9-year-old Akita (**A**) and an 11-year-old Nova Scotia Duck Tolling Retriever (**B**). A small amount of gas is present in the intrathoracic esophagus, cranial to the tracheal bifurcation. This is present commonly in both sedated and fully conscious patients. Patients under general anesthesia can have a large amount of gas in the esophagus secondary to central nervous system depression.

Chapter 6 ■ The Thorax 215

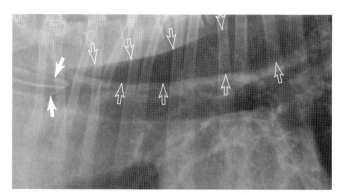

Figure 6-48. Left lateral radiograph of an 8-year-old Boston Terrier. The patient was under general anesthesia when radiographed, and an endotracheal tube can be seen *(solid white arrows)*. The cranial intrathoracic esophagus is dilated with gas *(hollow white arrows)*. The potential for gas distension of the esophagus increases with sedation and general anesthesia. Repeating radiographs when the patient is conscious help differentiate esophageal disease from the effects of chemical restraint.

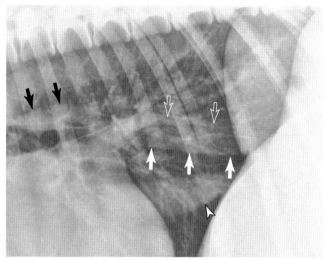

Figure 6-50. Lateral radiograph of a 9-year-old Akita. The caudal esophagus contains a small amount of fluid. The ventral border is delineated by the *solid white arrows* and the dorsal border by the *hollow white arrows*. The ventral border of the caudal vena cava is delineated by the *solid white arrowhead*. The caudal vena cava extends into the more caudal right diaphragmatic crus, typical of a left lateral view. The *solid black arrows* delineate the dorsal margin of the left caudal lobar pulmonary artery. The left pulmonary artery, located dorsal to the carina, is most commonly seen in the left lateral radiograph. This is probably because it is displaced slightly dorsally in the left lateral view (down anatomy rises), resulting in less superimposition.

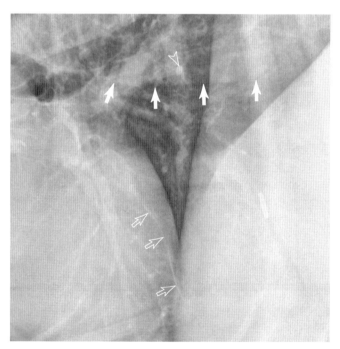

Figure 6-49. Left lateral radiograph of a 9-year-old Great Dane, centered on the caudal thorax. The ventral border of the esophagus is readily visible *(solid white arrows)*. The *hollow white arrows* outline a curvilinear soft tissue border, which is a fissure between the right middle and right caudal lung lobes. Visualization of this fissure is relatively common and does not automatically indicate the presence of pleural disease. A surgical staple superimposed over the liver is subsequent to cholecystectomy. An end-on vessel summating with normal structures is present over the esophagus *(hollow white arrowhead)*.

Brachycephalic breeds often have redundancy of the intrathoracic esophagus, which can manifest as a gas-filled tubular opacity in the cranial mediastinum, ventral to the trachea. This should not be confused with pathologic esophageal dilation secondary to an esophageal stricture or extrinsic compression associated with a vascular ring anomaly (Figure 6-51), although in some dogs esophageal redundancy may be associated with regurgitation.

HEART

The cardiac silhouette comprises the heart and pericardial fat. The appearance of the cardiac silhouette depends highly on radiographic positioning. A small degree of patient obliquity can result in significant changes to the appearance of the cardiac silhouette, potentially resulting in an incorrect assessment of chamber enlargement. In the dog, there is a wide range of normal heart size and shape, both within breeds and between breeds. Long, thin dogs have a heart that appears long and thin. Short,

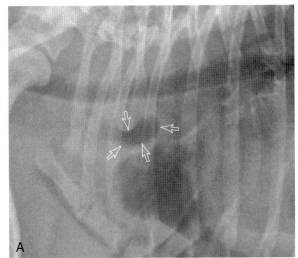
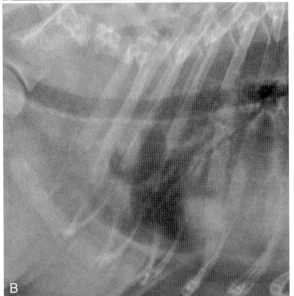
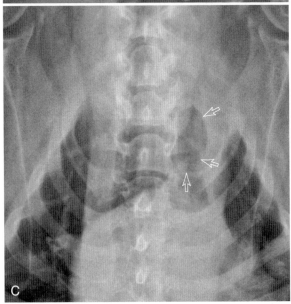

Figure 6-51. A, Lateral view of a 5-year-old Pug. A curving tubular gas opacity ventral to the trachea at the level of the second and third ribs *(hollow white arrows)* is due to redundancy of the cranial intrathoracic esophagus. Lateral view **(B)** and ventrodorsal view **(C)** of a 1-year-old Bulldog. The *hollow white arrows* outline esophageal redundancy. Esophageal redundancy in brachycephalic breeds should not be confused with pathologic esophageal dilation associated with vascular ring anomalies, although if severe, esophageal redundancy can result in clinical signs.

squat, or muscular dogs have a heart that appears round and relatively big. Cardiomegaly is often misdiagnosed in these muscular compact dogs (Figure 6-52). Sedation, an expiratory radiograph, or obesity can give a false impression of cardiomegaly because the thorax appears smaller overall with decreased lung expansion. In addition, it is probable that the cardiac silhouette is relatively larger in heavily sedated patients because bradycardia may result in increased cardiac chamber filling.

For these reasons, subjective assessment of the size of the heart on radiographs is inaccurate and can be misleading. To account for breed variation, a cardiac measurement technique, termed *vertebral heart score (VHS)*, has been developed.[11] In this method, the sum of the length and the width of the cardiac silhouette is normalized against the length of thoracic vertebrae. Unfortunately, the range of normal VHS values is quite wide. This method may be of limited value in assessing the size of the heart in individual patients; it is probably most useful for longitudinal comparisons in the same patient.

An approximation of cardiac chamber location, as seen on lateral views, is outlined in Figure 6-36. In lateral views, the right side of the heart is located cranially in the cardiac silhouette, and the left side is located caudally. The atria are dorsal and the ventricles ventral. Typically, the heart spans no more than 3½ rib spaces on the lateral view. On DV and VD views, a clock-face analogy is sometimes used to describe the various regions of the heart, with the 12 o'clock position being cranial and the 6 o'clock position caudal (Figure 6-53). The clock-face analogy is only a general guide; variations in patient conformation can affect the appearance and position of the heart within the thorax. Importantly, the thoracic vertebrae are not suitable as landmarks for dividing the heart into left versus right chambers because the interatrial and interventricular septa are oriented at an angle relative to the long axis of the thoracic spine.

Overall, the canine heart usually occupies no more than two-thirds of the width of the thorax on a DV or VD view and typically appears more narrow and elongated on a VD view than on a DV view. The left atrium is in the caudodorsocentral aspect of the cardiac silhouette, straddled by the principal bronchi (Figure 6-54).

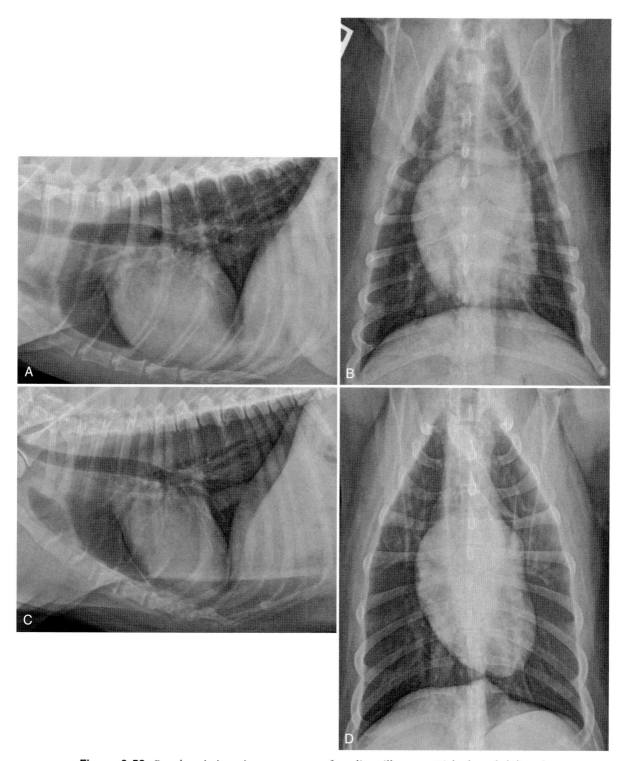

Figure 6-52. Breed variations in appearance of cardiac silhouette. Right lateral (**A**) and ventrodorsal (**B**) radiographs of an 8-year-old Welsh Corgi. The trachea is relatively parallel to the cranial aspect of the thoracic spine, and overall there is an impression of cardiomegaly. This is common in chondrodystrophoid breeds and should not be confused with cardiac disease. The impression of cardiomegaly is often exacerbated by decreased lung aeration and by the use of chemical restraint. Right lateral (**C**) and ventrodorsal (**D**) radiographs of a 9-year-old Rottweiler. The sharply delineated radiopaque margin at the cranial aspect of the heart on the left lateral view is due to fat within the mediastinum. The apparent bulge in the region of the pulmonary trunk is a normal anatomic variation, and, in this case, did not reflect pulmonary hypertension or turbulence.

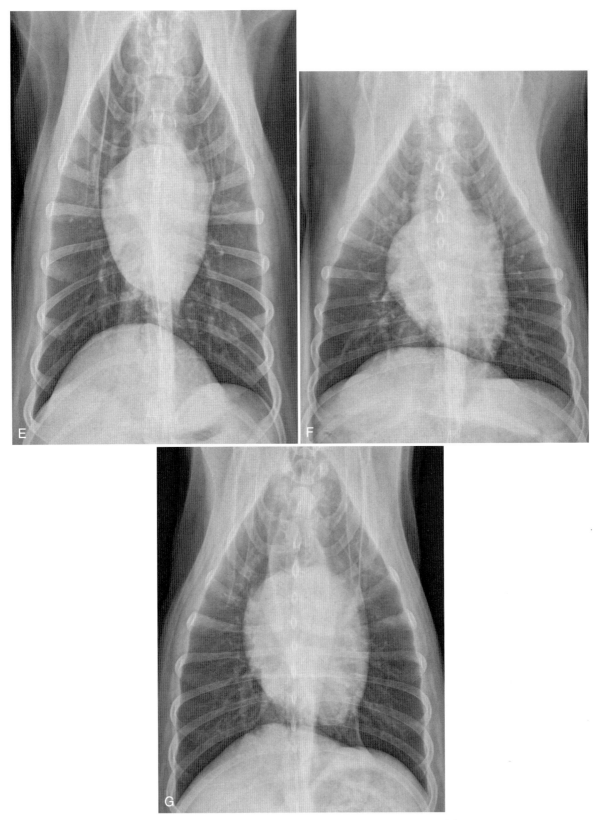

Figure 6-52, cont'd. E, Ventrodorsal radiograph of a 7-year-old Golden Retriever. F, Ventrodorsal radiograph of a 7-year-old American Bulldog. G, Ventrodorsal radiograph of an 11-year-old German Shepherd. E, F, and G all demonstrate typical breed-associated variation in appearance of the cardiac silhouette.

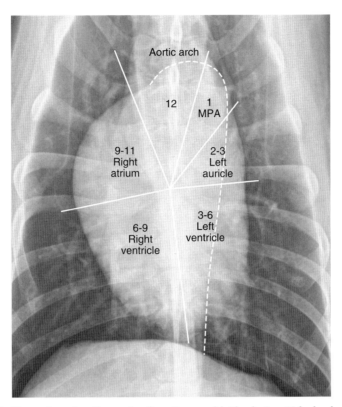

Figure 6-53. Ventrodorsal radiograph of an 8-year-old Rhodesian Ridgeback. A clock-face analogy is often used as a rule of thumb for locating various cardiac chambers and major vessels in the ventrodorsal view. It should be noted that the pulmonary trunk is on the left side and the thoracic spine is not an accurate reference point for dividing the heart into left versus right sides. The aortic arch is typically in the 12 o'clock position; the pulmonary trunk, otherwise known as the main pulmonary artery (MPA) at the 1 o'clock position; the left auricle at the 2 to 3 o'clock position; and the left ventricle at the 3 to 6 o'clock position. The right ventricle is at the 6 to 9 o'clock position and the right atrium at the 9 to 11 o'clock position.

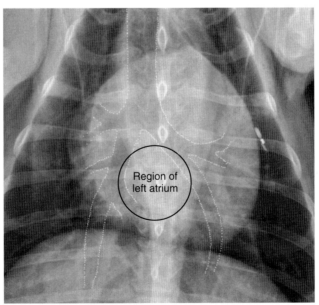

Figure 6-54. Dorsoventral radiograph of a 10-year-old Jack Russell Terrier. The left atrium is in the dorsocaudocentral aspect of the cardiac silhouette and is ventral to the principal bronchi. Enlargement of the left atrium can result in bronchial elevation and compression and bronchial splaying.

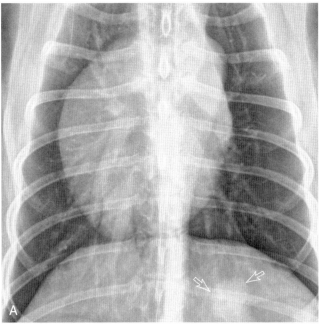

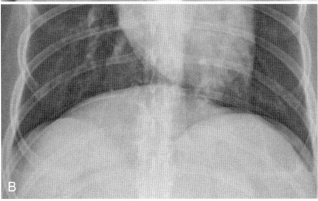

Figure 6-55. Dorsoventral view (**A**) and ventrodorsal view (**B**) of a 10-year-old Golden Retriever. Cardiovascular structures were considered normal in this patient but the cardiac apex is shifted to the right in **A**. This cardiac shift to the right in dorsoventral views occurs uncommonly but is more frequently seen in chondrodystrophoid breeds and is not necessarily an indicator of any abnormality. A cytologically confirmed neoplastic nodule is present in the dorsal aspect of the left caudal lung lobe *(hollow white arrows)*. This is not as conspicuous in the ventrodorsal view (**B**) due to recumbency-associated atelectasis.

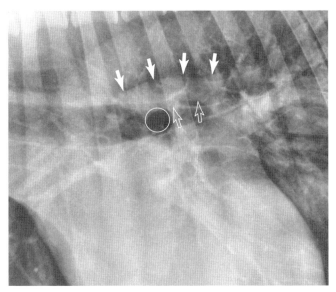

Figure 6-56. Left lateral radiograph of a 9-year-old Akita. The left caudal lobar artery (dorsal margin delineated by *solid white arrows*, ventral margin delineated by *hollow white arrows*) is located dorsal to the carina *(hollow white circle)* and is usually more easily seen on the left lateral view. The right caudal lobar artery crosses to the right side ventral to the carina and then courses caudodorsally. It can be seen as the poorly defined soft tissue opacity immediately ventral to the carina.

As noted earlier, in the DV view the apex of the heart is often shifted to the left and typically contacts the dome of the diaphragm (see Figure 6-7). In some chondrodystrophoid breeds, the apex of the heart can actually be shifted to the right in the DV view rather than to the left (Figure 6-55).

The left caudal lobar pulmonary artery is dorsal to the carina and is usually radiographically apparent, most commonly on the left lateral view. The right caudal lobar pulmonary artery courses ventral to the trachea from its main pulmonary artery origin on the left before turning and coursing caudodorsally. It is an apparent end-on as a circular opacity ventral to the carina that is often misinterpreted as a pulmonary mass (Figures 6-56, 6-57, *A-B* and 6-58).

The normal feline heart is proportionally smaller than the normal canine heart, and is also more consistent in shape (Figure 6-59). The caudal border of the feline heart should be characterized by a contiguous convex margin, and any significant change in contour suggests the presence of cardiac disease.

In older normal cats, the heart often lies at a more oblique angle, with the apex/base axis being more parallel with the sternum, and the aortic arch is often more angular in appearance. The angular appearance of the aortic arch can give the false impression of a pulmonary mass in the DV or VD view (Figure 6-60). A geriatric change in the shape of the aorta occurs much less commonly in dogs, but this does occur and should not be misinterpreted as aortic dilation (Figure 6-61).

Obese patients accumulate fat in the mediastinum and the pericardial sac. This can result in an enlarged cardiac silhouette and lead to a misdiagnosis of cardiomegaly (Figure 6-62).

Text continued on page 226

Chapter 6 ■ The Thorax 221

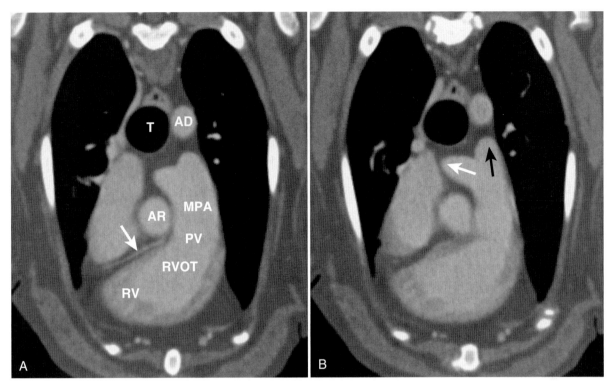

Figure 6-57. Transverse CT images at the level of the right ventricular outflow tract, optimized for mediastinum. Intravenous contrast medium has been administered. In **A**, *RV* is the right ventricle, *RVOT* is the right ventricular outflow tract, *PV* is the region of the pulmonic valve, *MPA* is the main pulmonary artery, *AR* is the aortic root, *AD* is the descending aorta and *T* is the trachea. The *white arrow* is the right coronary artery. Image **B** is 5 mm caudal to **A**. The right pulmonary artery is directed to the right, ventral to the trachea *(white arrow)*. The left pulmonary artery is directed dorsally, to the left of the trachea *(black arrow)*.

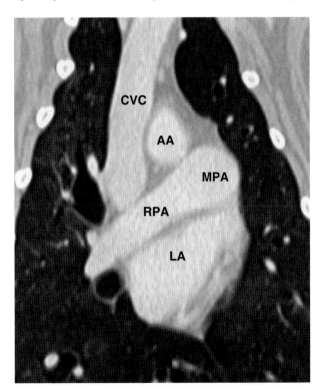

Figure 6-58. Dorsal plane CT image of the dorsal aspect of the heart, immediately ventral to the carina. *CVC* is cranial vena cava, *AA* is ascending aorta, *MPA* is the main pulmonary artery, *RPA* is the right pulmonary artery, *LA* is left atrium. This is the basis of the clock-face analogy.

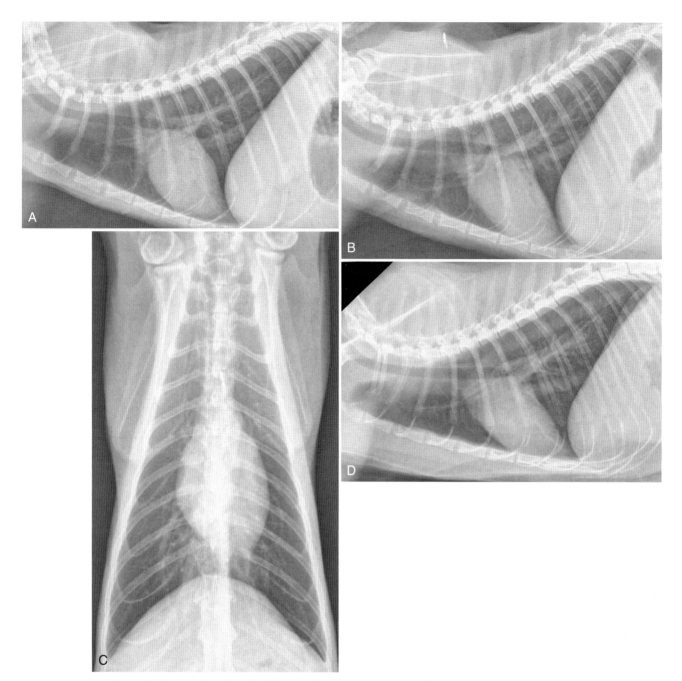

Figure 6-59. Left lateral (A), right lateral (B), and ventrodorsal (C) radiographs of a 9-year-old mixed breed cat. Left lateral (D), right lateral (E), and ventrodorsal (F) views of a 7-year-old Domestic Shorthair cat. Lateral view (G) of a 3-year-old Domestic Longhair cat and ventrodorsal view (H) of a 7-year-old Rex cat. The normal feline cardiac silhouette is relatively smaller than in the dog. The cardiac silhouette in images **D, E,** and **F** would be considered at the uppermost limits of normal for size. The cranial to caudal dimension of the normal feline heart is typically less than two and a half rib spaces. The separation of the diaphragmatic crura on the lateral views as a function of left versus right recumbency is less pronounced in cats than in dogs.

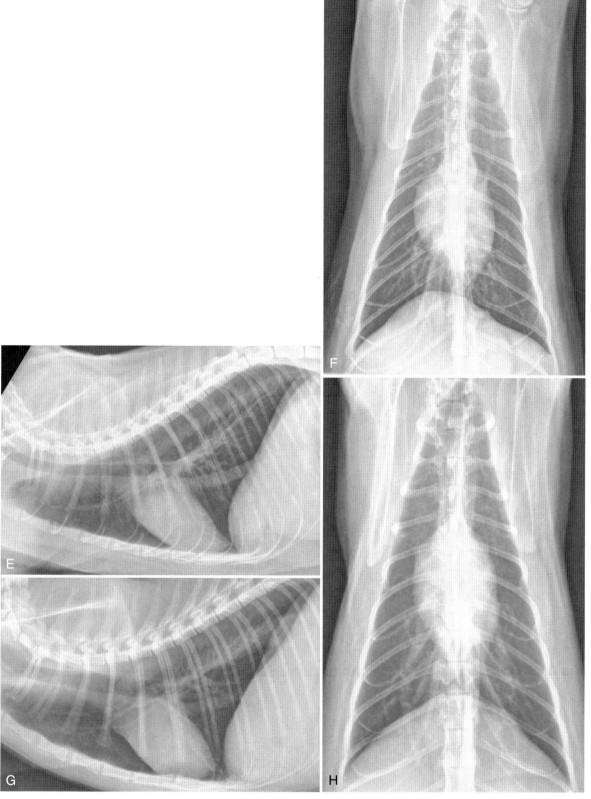

Figure 6-59, cont'd.

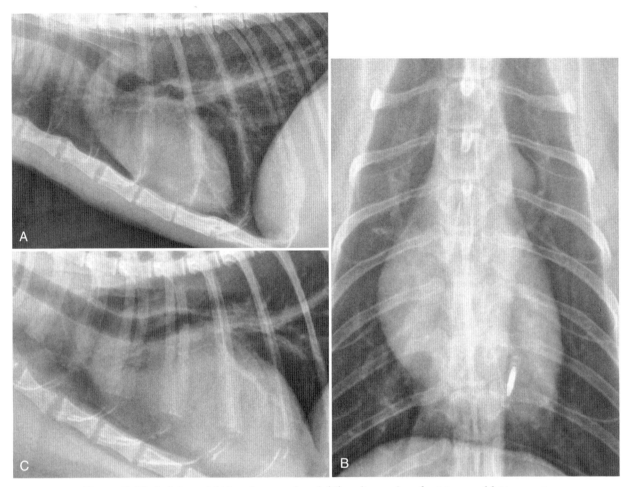

Figure 6-60. Left lateral (**A**) and ventrodorsal (**B**) radiographs of a 12-year-old Manx cat. The heart is more obliquely positioned in the thorax, and the ascending aorta is relatively angular. This is seen commonly in older cats. The end-on projection of the angular aorta on ventrodorsal view can appear as a circular structure immediately left of midline; this is sometimes confused with a pulmonary mass. A microchip is superimposed over the caudal aspect of the heart. **C**, Lateral radiograph of a 12-year-old Domestic Shorthair cat. The cardiac silhouette is nearly parallel with the sternum, and the ascending aorta has an angular appearance as it leaves the cardiac silhouette. The trachea at the heart base is abnormally deviated ventrally, which is a manifestation of altered heart position, exacerbated by a small amount of gas in the esophagus.

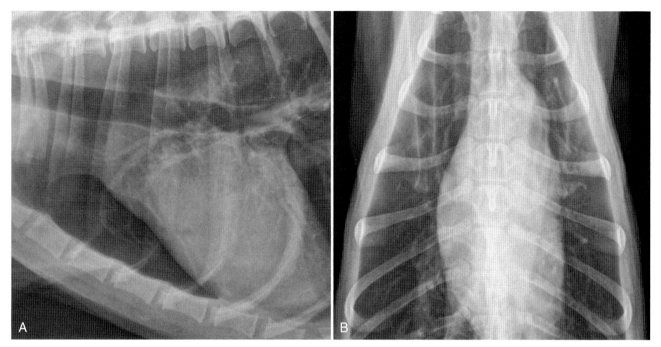

Figure 6-61. Right lateral (**A**) and ventrodorsal (**B**) radiographs of a 9-year-old Siberian Husky. The cardiac silhouette is more oblique than is typical, and there is redundancy of the aorta, which results in a bulge at the cranial aspect of the cardiac silhouette and gives an elongated appearance to the cardiac silhouette in the ventrodorsal view. Similar morphologic changes of the heart can occur with disease of the aortic valve. Clinical findings coupled with an echocardiogram can be required to assess the significance of aortic arch shape changes in dogs versus cats, where they are more common.

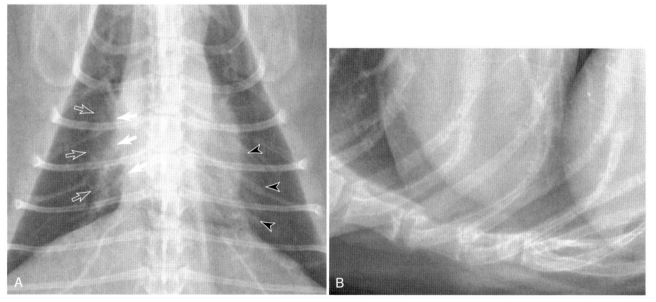

Figure 6-62. Ventrodorsal radiograph of a 10-year-old Domestic Shorthair cat (**A**) and lateral radiograph of an 8-year-old Doberman Pinscher (**B**). In **A**, the cardiac silhouette is enlarged; on close examination, however, the peripheral border is poorly defined, and there is an indistinct transition of radiographic opacity from the center of the cardiac silhouette peripherally. This is due to the presence of fat in the mediastinum surrounding the heart. The *solid white arrows* delineate the right margin of the cardiac silhouette. The *hollow white arrows* indicate the peripheral margin of the cardiac silhouette. The opacity between the two is pericardial fat. There is also a large amount of fat in the cranial mediastinum and caudoventral mediastinal reflection *(black arrowheads)*. In **B**, a fat opacity is present between the ventral aspect of the heart, the diaphragm, and the sternum. The fat opacity extends along the cranial border of the heart and is due to fat in the ventral aspect of the pericardial sac.

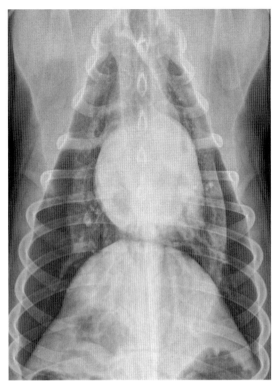

Figure 6-63. Dorsoventral view of a 3-year-old Doberman Pinscher. The heart is positioned more vertically in deep-chested dogs, giving it a more rounded appearance in the dorsoventral view.

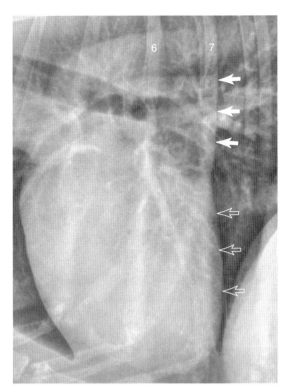

Figure 6-64. Left lateral radiograph of an 8-year-old mixed breed dog. The caudal border of the seventh rib *(solid white arrows)* is contiguous with the caudal border of the heart *(hollow white arrows)*, giving the false impression of left atrium enlargement.

In dogs with a deep thorax, the heart is often more vertical in location, which results in a more rounded overall appearance in DV and VD views (Figure 6-63).

A summation opacity created by a rib overlapping the caudal margin of the heart can give the false impression of left atrial enlargement, because the eye is drawn to the caudal border of the rib as it extends dorsally (Figure 6-64).

Infrequently, dystrophic mineralization of the aortic root occurs, apparent as a sharp linear mineral opacity in the region of the aortic outflow tract. This mineralization is usually conspicuous only on the lateral view and is considered a clinically insignificant incidental finding[12] (Figure 6-65).

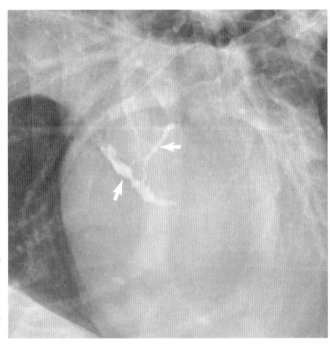

Figure 6-65. Left lateral radiograph of an 11-year-old Greyhound. Linear mineralization, not associated with the pulmonary vasculature, is superimposed over the region of the aortic bulb *(solid white arrows)*. This is likely idiopathic aortic bulb/root mineralization. This is not usually associated with an endocrinopathy, and patients are asymptomatic. The mineralization is typically in the fourth intercostal space in the craniodorsal quadrant of the cardiac silhouette.

LUNG

The lung comprises the following: connective tissue, the bronchial tree and terminal airspaces, and afferent and efferent vessels. Radiographically, it is normal to see pulmonary vessels, bronchi, and some interstitial markings. The interstitial markings should be visible to the lung periphery (Figure 6-66). Failure to see lung extending to the thoracic wall suggests the presence of either pneumothorax or radiographic overexposure. The skin-fold effect, described previously, is also important to recognize because it can also cause the impression of failure of the lung to extend to the thoracic wall.

Determining when the conspicuity of interstitial markings exceeds normal limits is imprecise, and even experienced radiologists cannot do this consistently. It is most important to be familiar with the patient variables and the technical factors that result in an artifactual increase in lung opacity. One of the most common mistakes is failure to realize that overall pulmonary opacity will always be greater in lateral radiographs than in DV or VD radiographs because of the greater amount of atelectasis that occurs in lateral recumbency. It is very common for the normally increased opacity occurring in lateral views to be misinterpreted as pulmonary disease, particularly in sedated patients (see Figure 6-3).

Bronchi appear as peripherally tapering hollow tubes with thin walls, whereas pulmonary vessels appear as soft tissue tubular structures that attenuate toward the periphery. Vessels and airways are seen end-on and side-on throughout the lung parenchyma. End-on pulmonary vessels appear as small nodules and end-on bronchi appear as small ring shadows. Bronchial and interstitial markings can develop increased conspicuity as the patient matures. Unfortunately, with many diseases, there is also a gradual transition from normal to diffuse pulmonary disease, which also complicates assessment of the lung.

The most common reason for poor visualization of the normal interstitial markings extending to the periphery of the thorax is overexposure with film-screen systems and inappropriate image processing in digital systems. The most common reasons for a generalized increase in unstructured lung opacity are decreased aeration or underexposure or poor development when using a film-screen system. Decreased aeration is often due to atelectasis occurring in lateral recumbency but also occurs when the image is acquired at peak expiration (see Figure 6-3).

The amount of tissue in the thoracic wall also has a dramatic effect on the overall opacity of the lung and, therefore, the conspicuity of interstitial markings. In subjects with a high body mass index, there is a tendency to misdiagnose the resulting generalized increased pulmonary opacity caused by the superimposed soft tissue as interstitial lung disease. Subjects that have undergone forelimb amputation provide a good example of this phenomenon. The forelimb amputation results in a decrease in lung opacity in the ipsilateral hemithorax due to loss of superimposed soft tissue resulting from the amputation itself and subsequent disuse atrophy of remaining soft tissue in the shoulder region on the side of the amputation. This reduction in the amount of superimposed soft tissue leads to a discrepancy in pulmonary opacity between sides. The more opaque lung on the side opposite the amputation can be misinterpreted as diseased (Figure 6-67).

Overlapping of normal intrathoracic structures with the creation of summation opacities can give the false impression of parenchymal disease, especially misdiagnosis of pulmonary nodules. This is particularly common when a pulmonary vessel is superimposed on a rib, creating a very obvious summation opacity that is misinterpreted as a pulmonary nodule (Figure 6-68). In addition, extrathoracic structures (particularly nipples, dermal masses, or even ectoparasites) can result in artifacts that can be misinterpreted as parenchymal nodules (Figure 6-69). It is imperative that these findings are fully assessed with an orthogonal view, a detailed clinical examination, and, in some instances, the application of a radiopaque marker on a body wall structure thought to be the source of a possible pulmonary nodule (Figure 6-70).

Visualization of an end-on pulmonary vessel is also often misinterpreted as a lung nodule. Fortunately, the distinction between an end-on vessel and a pulmonary nodule is usually easy to make. First, end-on pulmonary vessels often have a diameter smaller than the diameter required for visualization of a solitary pulmonary nodule. End-on pulmonary vessels with a diameter in the 2- to 3-mm range can be seen radiographically. A solitary soft tissue nodule with a diameter less than approximately 5 mm, however, usually is not visible because it does not absorb an adequate number of x-rays. End-on pulmonary vessels can be seen at such a small diameter because they are cylindrical and have depth when viewed end-on. This depth results in sufficient x-ray absorption for visualization. Second, end-on pulmonary vessels usually have a *tail*, which is an adjacent linear opacity created by a portion of the vessel that has turned and is being struck side-on by the x-ray beam rather than end-on (Figure 6-71). Third, due to the normal association between pulmonary vessels and airways, end-on pulmonary vessels are usually directly adjacent to an end-on bronchus (see Figure 6-66, *C-D*).

The term *fissure* is used to describe a radiopaque linear region between adjacent lung lobes. Fissures are usually not apparent in the normal patient, but they do become apparent when pleural fluid tracks between lung lobes. Occasionally, thin radiopaque lines reflecting the interface between two adjacent lung lobes can be seen in

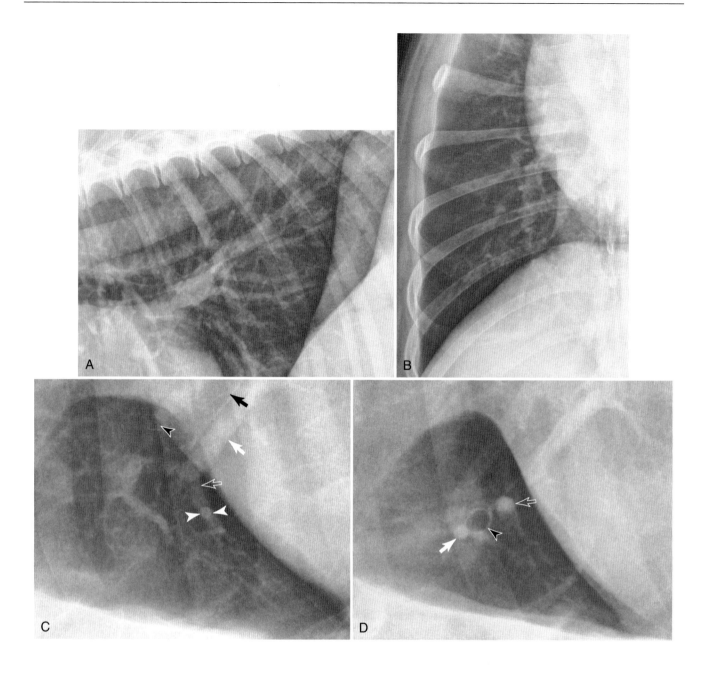

normal subjects (Figure 6-72). Increased conspicuity of this border in the absence of pleural fluid can be an age-related change due to mild pleural fibrosis.

Occasionally, a linear margin of increased opacity is present between right cranial and right middle lung lobes, particularly in well-conditioned patients. This is fat within a redundant mediastinal reflection (Figure 6-73) and if present in isolation, should not be confused with a pleural fissure line that might otherwise indicate pleural fluid. For normal pleural fissures to be apparent radiographically, the x-ray beam must be exactly parallel to the lung border. This occurs commonly at the junction between the right middle and right caudal lung lobe, resulting in a curvilinear soft tissue opacity superimposed over the caudal margin of the heart in lateral radiographs (Figure 6-74).

Figure 6-66. Normal pulmonar parenchyma. Pulmonary vessels are branching tubular soft tissue opacities emanating from the hilus. They become progressively smaller toward the periphery. Bronchi are hollow, thin-walled tubular structures that are closely associated with pulmonary vessels. Bronchi are easier to identify closer to the hilus, where they are larger. Bronchi become progressively more mineralized with age, and this is apparent radiographically. **A,** Left lateral radiograph of a 7-year-old Greyhound. **B,** Ventrodorsal radiograph of an 11-year-old German Shepherd. **C,** Left lateral radiograph of a 12-year-old Siberian Husky, centered on the cranial lobe pulmonary vessels. The *solid white arrow* is the ventral border of the right cranial lobe pulmonary vein. The *hollow white arrow* is a branch of the right cranial lobe pulmonary vein. The *solid white arrowheads* are outlining a circular structure superimposed on the branch of the right cranial lobar pulmonary vein. This circular structure reflects an additional branch emanating from that vessel, which is projected end on. The apparent increased opacity of this vessel compared with other pulmonary vessels is because the end-on vessel has depth in the vertical plane resulting in increased x-ray attenuation; that is, the structure is actually cylindrical and not spherical, as a nodule would be. This is a common occurrence, and an end-on vessel should not be confused with a pulmonary nodule. The *solid black arrow* is the ventral wall of the right cranial lobe bronchus. The *black arrowhead* is the ventral border of the right cranial lobe pulmonary artery. **D,** Left lateral radiograph of a 10-year-old mixed breed dog, centered on the cranioventral thorax. Peripherally, the right cranial lobe pulmonary artery, bronchus, and right cranial lobe pulmonary vein have turned and become oriented end on with the x-ray beam. The *hollow white arrow* is the pulmonary vein, the *solid white arrow* is the pulmonary artery, and the *black arrowhead* is the bronchus.

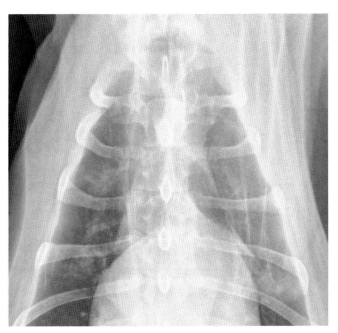

Figure 6-67. Ventrodorsal radiograph of a 9-year-old Labrador Retriever. The right forelimb has been amputated. There is a relative reduction in lung opacity in the cranioventral aspect of the right hemithorax due to a lack of superimposition of the forelimb and secondary muscle atrophy.

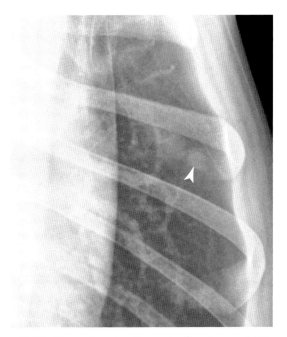

Figure 6-68. Ventrodorsal radiograph of an 8-year-old German Shepherd. There is a focus of increased opacity at the level of a costochondral junction *(solid white arrowhead)*. This has the appearance of a pulmonary nodule but is due to a focus of mineralization and remodeling at the costochondral junction. This is a common finding.

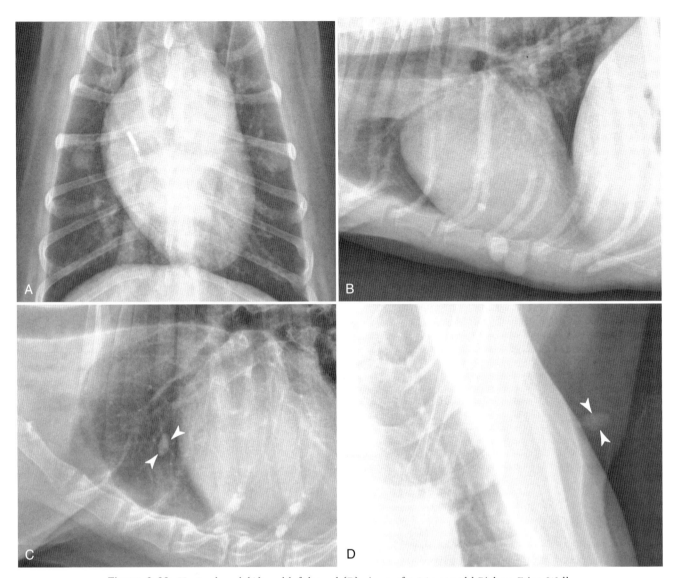

Figure 6-69. Ventrodorsal (**A**) and left lateral (**B**) views of a 14-year-old Bichon Frise. Well-defined soft tissue opacities superimposed over the left and right lung fields at the level of the midaspect of the heart are enlarged nipples. The nipples are readily apparent on the orthogonal left lateral view (**B**) as pendulous soft tissue opacities outside of the thoracic cavity. In the ventrodorsal view, a microchip is superimposed over the right hemithorax. Right lateral (**C**) and ventrodorsal (**D**) views of an 11-year-old Australian Heeler. In the lateral view, a well-delineated soft tissue opacity (outlined by the *solid white arrowheads*) is superimposed over the cranioventral lung lobes, immediately cranial to the heart and ventral to the cranial lobar pulmonary vessels. In **D**, the nodule is readily apparent on the left thoracic wall *(solid white arrowheads)*. Clinical inspection confirmed the presence of an engorged tick.

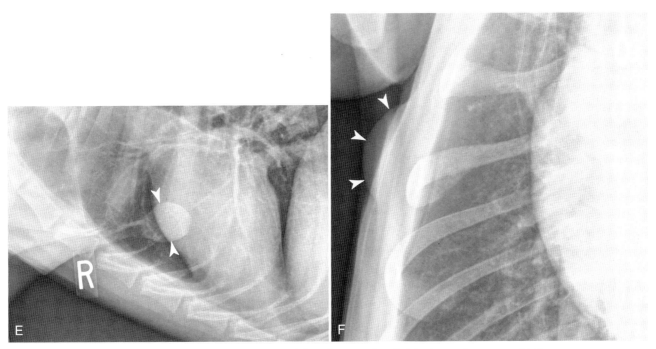

Figure 6-69, cont'd. Right lateral (**E**) and ventrodorsal (**F**) views of a 6-year-old Great Dane. A large soft tissue nodule is overlying the cranioventral aspect of the cardiac silhouette. The mass appears well-defined and, based on this view, is within the thoracic cavity *(solid white arrowheads)*. However, on the ventrodorsal view, the mass is readily apparent on the right thoracic wall *(solid white arrowheads)*.

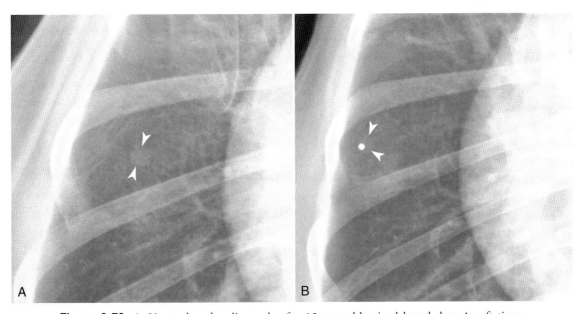

Figure 6-70. **A,** Ventrodorsal radiograph of a 13-year-old mixed breed dog. A soft tissue nodule is present in the right fifth intercostal space *(solid white arrowheads)*. An important differential is metastatic lung disease. This was not identified in the lateral views, and a small cutaneous nodule was identified on clinical examination. A small radiopaque bead was placed on the nodule, and the patient was re-radiographed (**B**). The focal increased opacity seen radiographically was confirmed to be a cutaneous nodule.

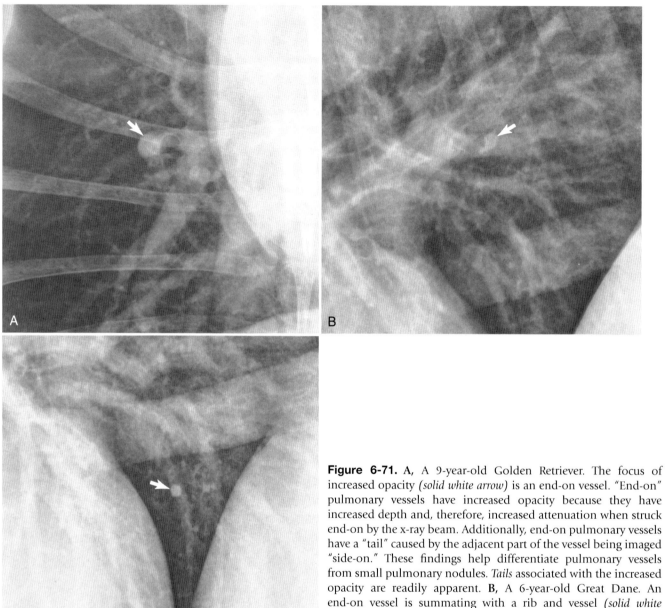

Figure 6-71. A, A 9-year-old Golden Retriever. The focus of increased opacity *(solid white arrow)* is an end-on vessel. "End-on" pulmonary vessels have increased opacity because they have increased depth and, therefore, increased attenuation when struck end-on by the x-ray beam. Additionally, end-on pulmonary vessels have a "tail" caused by the adjacent part of the vessel being imaged "side-on." These findings help differentiate pulmonary vessels from small pulmonary nodules. *Tails* associated with the increased opacity are readily apparent. **B,** A 6-year-old Great Dane. An end-on vessel is summating with a rib and vessel *(solid white arrow)*. The superimposed rib is resulting in increased opacity and making the vessel more conspicuous. **C,** The same patient as in B; an end-on vessel *(solid white arrow)* has increased conspicuity and a "tail" is readily evident.

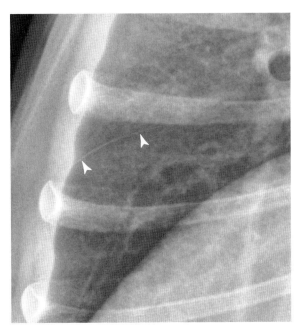

Figure 6-72. Dorsoventral radiograph of a 12-year-old mixed breed dog. A thin radiopaque line is seen in the right caudolateral aspect of the thorax *(solid white arrowheads)*. This is a fissure line and reflects the boundary between the caudal aspect of the right middle lung lobe and cranial aspect of the right caudal lung lobe. It is common to see thin pleural fissure lines in normal patients. Visualization of the margin is probably a manifestation of mild pleural thickening or opportunistic alignment of the x-ray beam with the plane of the fissure. Separation between lobes occurs with the presence of pleural fluid and is an important differential.

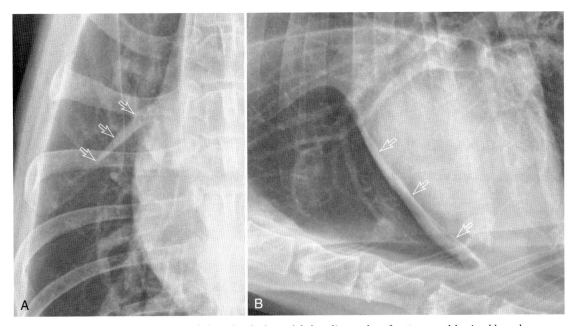

Figure 6-73. Ventrodorsal (**A**) and right lateral (**B**) radiographs of a 10-year-old mixed breed dog. The curving margin of increased opacity outlined by the *hollow white arrows* is fat between the right cranial and middle lung lobes, within a redundant mediastinal fold. In the right lateral view, the fat is silhouetting with the cranial border of the heart. This should not be confused with pleural fluid or collapse of the right middle lung lobe.

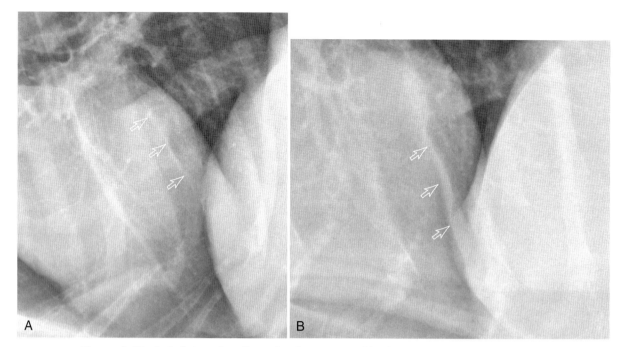

Figure 6-74. A, Left lateral radiograph of a 12-year-old Siberian Husky. B, A left lateral radiograph of a 10-year-old Labrador Retriever. The *hollow white arrows* outline the fissure between the caudal aspect of the right middle lung lobe and the cranial aspect of the right caudal lung lobe. This is commonly observed in the left lateral view, particularly when there is poor lung aeration. This fissure should not be confused with pleural fluid. It is radiographically apparent because the x-ray beam strikes the caudal border of the lung lobe end on. Skin folds arising from the axilla are apparent along the ventral aspect of the image.

Pulmonary osseous metaplasia, also termed *pulmonary heterotopic bone* or *pulmonary osteomas,* is a benign aging change encountered commonly in normal patients. Type I pneumocytes can produce osteoid; in some dogs this manifests as small punctate mineral opacities scattered throughout the lung parenchyma. This anomaly can occur in any breed but is most commonly seen in collies. These benign mineralized foci must not be confused with metastasis. Mineralized foci are typically more opaque than a vessel of the same size and more radiographically apparent than a soft tissue nodule of similar size (Figure 6-75). Because of their mineralization, pulmonary osteomas can be detected at a very small size, on the order of 1 or 2 mm. A soft tissue nodule in the lung, on the other hand, must reach a diameter of approximately 5 mm to be radiographically detectable.

Dolichocephalic breeds have a taller and narrower thoracic cavity compared with mesaticephalic and brachycephalic breeds. This can result in less cardiosternal contact. When the lungs are well aerated, there may actually be lung between the ventral aspect of the heart and the sternum in lateral views. The radiolucent lung and the dorsal heart positioning are often misinterpreted as evidence of pneumothorax (Figure 6-76).

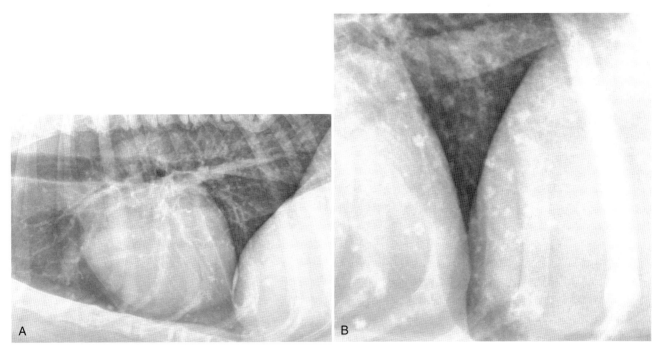

Figure 6-75. Left lateral (**A**) and magnified (**B**) views of an 11-year-old Great Pyrenees with pulmonary osseous metaplasia, which creates multiple punctate opacities throughout all lung lobes. The opacities are small and more radiopaque than pulmonary vessels of a similar size because they are mineralized. Pulmonary osseous metaplasia is common in older patients, especially in collie-type breeds. A sharply marginated linear soft tissue opacity superimposed over the caudal aspect of the heart most likely reflects the border between the caudal aspect of the right middle lung lobe and the cranial aspect of the right caudal lung lobe. It is radiographically apparent because the x-ray beam strikes the caudal border of the lung lobe end on. In **A**, spinal degenerative changes are present with endplate sclerosis, disc space collapse, and ventral spondylosis. The ventral border of the esophagus is also visible in **A**.

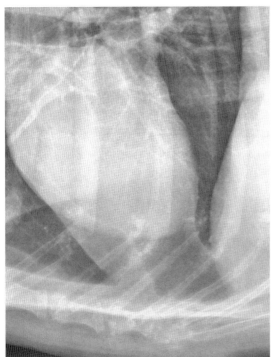

Figure 6-76. Left lateral radiograph of a 12-year-old Golden Retriever. The heart appears dorsally displaced from the sternum, and an important differential is pneumothorax. However, lung markings extend to the periphery of the thorax, and the opacity ventral to the heart is intermediate between gas and soft tissue (i.e., fat). This patient did not have pneumothorax, and the changes present are a manifestation of patient conformation.

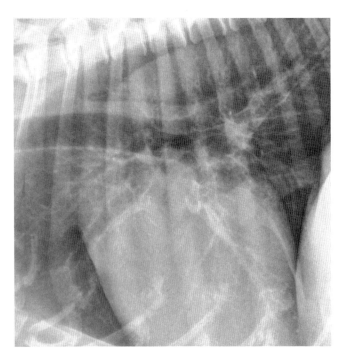

Figure 6-77. Lateral radiograph of a 12-year-old mixed breed dog. There are mildly increased bronchial markings, a common finding in older patients. The bronchi are the thin linear opacities adjacent to the pulmonary vasculature and are visible due to wall mineralization. The clinical significance of this must be interpreted in light of clinical signs. It is common to see such changes in older asymptomatic patients.

Older dogs typically have more conspicuous bronchial markings compared with younger patients. The relative significance of increased bronchial markings must be interpreted in light of clinical signs (Figure 6-77). Often, this increased bronchial visualization is due to mild wall mineralization that could be a consequence of chronic low-grade airway inflammation that is no longer causing clinical signs.

Cats typically have fewer bronchial markings than dogs. The presence of bronchial markings in a cat to the extent that would be considered normal in a dog is usually indicative of airway disease (Figure 6-78).

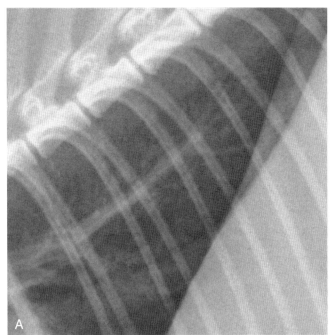

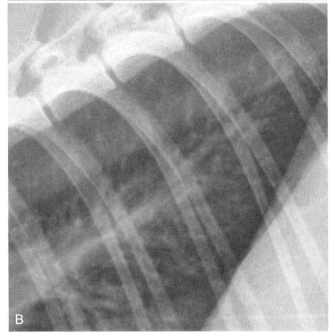

Figure 6-78. Lateral views of normal 1-year-old (**A**) and normal 7-year-old (**B**) Domestic Shorthair cat. Cats have fewer bronchial and interstitial markings compared with dogs. An increase in bronchial markings is more typically associated with respiratory disease in cats. Overall, younger cats (**A**) usually have fewer bronchial markings than older cats (**B**) for the same reason as in dogs, but the changes are much less common or pronounced.

DIAPHRAGM

The diaphragm is the musculotendinous division between the thoracic and abdominal cavities. Radiographically, it comprises a ventral dome and two dorsally located crura (see Figures 6-10, *A-B*; 6-16, *A-B*; and 6-17). As described earlier, the position of the diaphragm is affected by recumbency and phase of respiration. In a DV view, the dome of the diaphragm is typically more cranial and rounded, which results in increased cardiac contact with subsequent cardiac displacement as already discussed (see Figure 6-21, *A-B*). The crura of the diaphragm are more readily apparent in a VD view compared with a DV view (see Figure 6-24).

The diaphragm has three normal openings, each called a *hiatus*. Dorsally, the aortic hiatus provides for passage of the aorta and para-aortic vessels into the retroperitoneal space; the esophageal hiatus allows passage of the esophagus and vagal nerves, and the caval hiatus allows passage of the caudal vena cava from the abdomen into the right atrium. The crura of the diaphragm extend dorsocaudally to attach to the midlumbar vertebral bodies. The position of the diaphragm is dramatically affected by pulmonary aeration or hyperinflation. Typically, the caudodorsal lung tip extends to the level of T11 to T12, but this depends on patient conformation and lung aeration. The caudodorsal lung border in cats is typically more ventral when compared with dogs (Figure 6-79), a feature sometimes accentuated in outdoor cats compared with indoor sessile cats. This is possibly associated with increased hypaxial muscle mass in physically active cats and should not be confused with pleural effusion.

Mild asymmetry of the diaphragm is common as a normal variant and is most apparent on the VD view (Figure 6-80). It is impossible to determine the significance of such asymmetry without dynamic imaging, because diaphragmatic palsy, although less common, can also lead to this appearance in thoracic radiographs.

A sliding hiatal hernia is occasionally visible as an ill-defined mass effect in the dorsocentral aspect of the thorax, at the level of the esophageal hiatus. Hiatal hernias, which are often clinically insignificant, are more often seen in the left lateral view, most likely due to mechanical pressure on the stomach from compression in the dependent hemiabdomen. A hiatal hernia can be confused with a pulmonary or mediastinal mass. Typically, however, the hiatal hernia is not present on every view of a three-view thoracic study (Figure 6-81). In some instances, it may be necessary to use more specialized imaging, such as a barium gastrogram or computed tomography (CT) imaging, to make the correct interpretation.

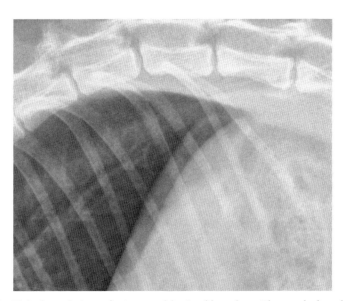

Figure 6-79. Right lateral view of a 9-year-old mixed breed cat. The caudodorsal lung border diverges from the thoracic spine. This is typical in cats and is associated with the relatively increased mass of the hypaxial musculature in the cat versus the dog. This should not be confused with pleural effusion. The dorsocaudal tip of the lung lobes is typically at the level of T12 or T13-L1, depending on patient conformation and degree of lung aeration.

Figure 6-80. Ventrodorsal radiograph of an 8-year-old Rhodesian Ridgeback. There is asymmetry of the diaphragm, with the right diaphragmatic crus being more cranial than the left. Mild diaphragmatic asymmetry is common in normal patients. An important differential is diaphragmatic paralysis or decreased aeration of the ipsilateral lung. Abdominal masses might also cause cranial displacement of the diaphragm as a result of increased intra-abdominal pressure.

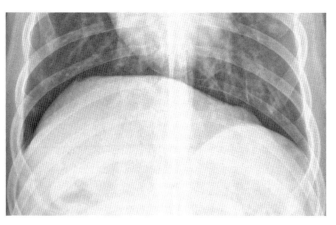

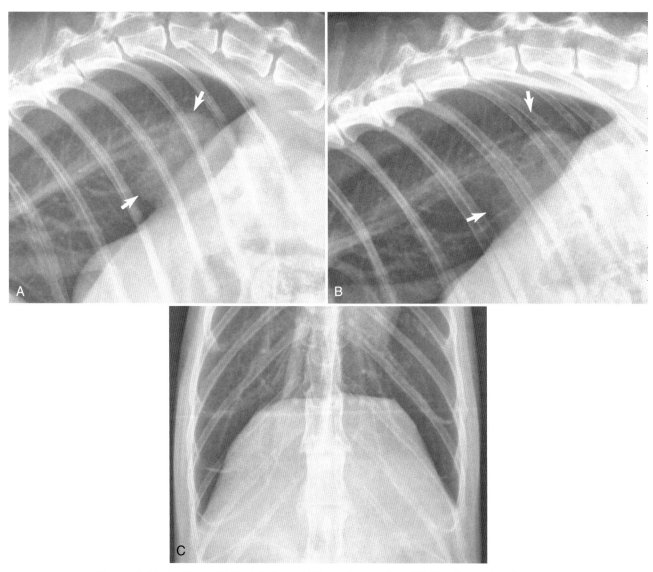

Figure 6-81. Left lateral (**A**), right lateral (**B**), and ventrodorsal (**C**) radiographs of a 6-year-old Siamese cat. There is an ill-defined soft tissue mass effect within the caudodorsal thorax superimposed over the cranial border of the diaphragm at approximately the level of the esophageal hiatus (*solid white arrows* in **A** and **B**). The mass effect is not apparent in the ventrodorsal view. This was confirmed to be an asymptomatic sliding hiatal hernia. Such a finding could easily be confused with a pulmonary or mediastinal mass.

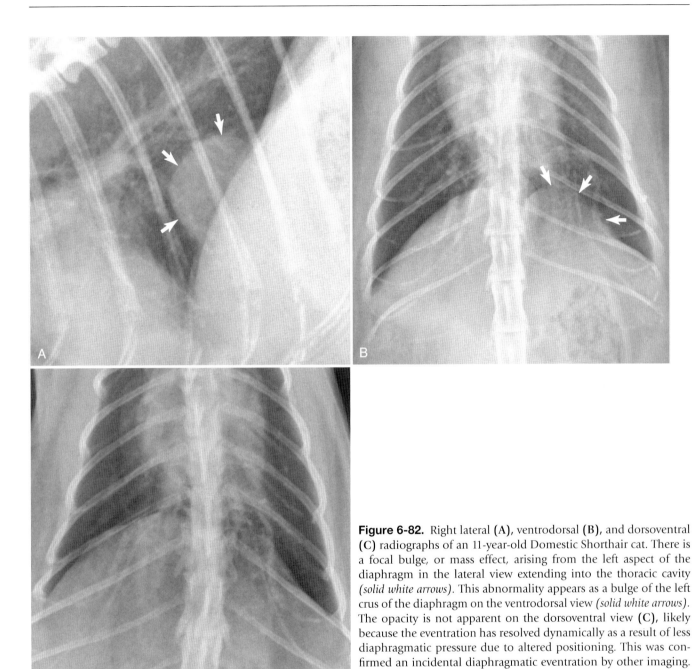

Figure 6-82. Right lateral (A), ventrodorsal (B), and dorsoventral (C) radiographs of an 11-year-old Domestic Shorthair cat. There is a focal bulge, or mass effect, arising from the left aspect of the diaphragm in the lateral view extending into the thoracic cavity *(solid white arrows)*. This abnormality appears as a bulge of the left crus of the diaphragm on the ventrodorsal view *(solid white arrows)*. The opacity is not apparent on the dorsoventral view (C), likely because the eventration has resolved dynamically as a result of less diaphragmatic pressure due to altered positioning. This was confirmed an incidental diaphragmatic eventration by other imaging. Important differentials are a caudally located pulmonary mass or a diaphragmatic mass.

Diaphragmatic eventration is characterized by regional failure of the muscular or tendinous portion of the diaphragm, without complete disruption. This can be congenital, acquired (as with trauma, such as an incomplete tear), or associated with regional neurogenic dysfunction. The diaphragmatic defect causes the affected portion of the diaphragm to be located more cranially than normal, and its appearance often changes with patient position. The anomaly can be very focal and can give the impression of a caudal thoracic or diaphragmatic mass (Figure 6-82). As with hiatal hernia, more specialized imaging (such as ultrasonography, CT, or positive-contrast peritoneography) might be necessary to make the correct interpretation.

Occasionally, the caudoventral aspect of the thorax is characterized by increased opacity ventral to the caudal vena cava, the dorsal margin of which is defined by a relatively sharp linear border that is nearly parallel to the ventral border of the caudal vena cava. This is not a pleural fissure line and is consistent with a mesothelial remnant associated with failure of closure of the embryologic division between the thoracic and abdominal cavities. Depending on the patency of this remnant, it can provide a route for abdominal organs to migrate into the

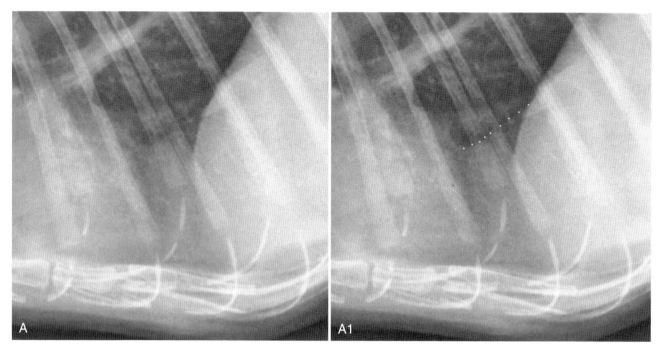

Figure 6-83. Left lateral radiograph of a 9-year-old Domestic Longhair cat. In **A**, there is increased opacity caudal to the heart and cranial to the diaphragm. The dorsal margin of this opacity has been highlighted in **A1** with a *white dotted line*. This reflects the dorsal margin of a mesothelial remnant. There may be a reduction in the number of sternebrae, especially absence of the xiphoid, in combination with this anomaly but that was not present in this cat.

peritoneal sac, as in peritoneopericardial diaphragmatic hernia (Figure 6-83). A reduction in number of sternebrae, especially the xiphoid, can accompany this anomaly.

REFERENCES

1. Berry C, Thrall D: Introduction to radiographic interpretation. In Thrall D, editor: *Textbook of veterinary diagnostic radiology*, ed 5, Philadelphia, 2007, Saunders.
2. Losonsky J, Thrall D, Lewis R: Thoracic radiographic abnormalities in 200 dogs with spontaneous heartworm disease. *Vet Radiol Ultrasound* 24:120, 1983.
3. Ruehl W, Jr, Thrall D: The effect of dorsal versus ventral recumbency on the radiographic appearance of the canine thorax. *Vet Radiol Ultrasound* 22:10–16, 1981.
4. Carlisle C, Thrall D: A comparison of normal feline thoracic radiographs made in dorsal versus ventral recumbency. *Vet Radiol Ultrasound* 23:3–9, 1982.
5. Oui H, Oh J, Keh S, et al: Measurements of the pulmonary vasculature on thoracic radiographs in healthy dogs compared to dogs with mitral regurgitation. *Vet Radiol Ultrasound* 56:251–256, 2015.
6. Thrall D: Radiology Corner: Misidentification of a skin fold as pneumothorax. *Vet Radiol Ultrasound* 34:242–243, 1993.
7. Schummer A, Nickel R, Sack W: *Viscera of domestic animals*, ed 2, New York, 1985, Springer Verlag.
8. Lehmukhl L, Bonagura J, Biller D, et al: Radiographic evaluation of caudal vena cava size in dogs. *Vet Radiol Ultrasound* 38:94–100, 1997.
9. Harvey C, Fink E: Tracheal diameter: Analysis of radiographic measurements in brachycephalic and nonbrachycephalic dogs. *J Am Anim Hosp Assoc* 18:570–576, 1982.
10. Suter P, Colgrove D, Ewing G: Congenital hypoplasia of the canine trachea. *J Am Anim Hosp Assoc* 8:120–127, 1972.
11. Buchanan J, Bucheler H: Vertebral scale system to measure canine heart size in radiographs. *J Am Vet Med Assoc* 206:194–199, 1995.
12. Douglass J, Berry C, Thrall D, et al: Radiographic features of aortic bulb/valve mineralization in 20 dogs. *Vet Radiol Ultrasound* 44:20–27, 2003.

CHAPTER 7

The Abdomen

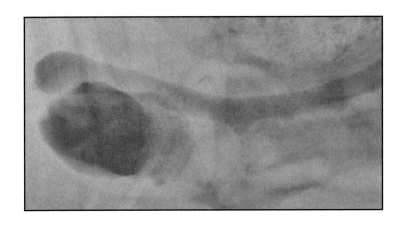

Standard radiographic views of the abdomen include lateral and ventrodorsal (VD) views. Dorsoventral (DV) views of the abdomen are characterized by distortion of the caudal aspect of the abdominal cavity created by pressure from the position of the pelvic limbs and as a result are rarely made. In general, abdominal radiographs should be made without fasting or administering an enema. This way the image represents the natural status of the gastrointestinal tract, which can be important for making the diagnosis. All figures in this chapter are of patients for which no special preparation was undertaken prior to radiography.

Gas is usually present in the some portion of the gastrointestinal tract, and the distribution of this gas is very important in formulating a correct interpretation. Obtaining both right lateral and left lateral radiographs in addition to the VD view is recommended for all abdominal radiographic examinations because the distribution of gas in opposing lateral views will be different and can provide important diagnostic information. In other words, gas is acting like an endogenous *contrast medium*.* Left versus right recumbency also influences the radiographic appearance of some other organs, notably the spleen and kidneys. The effect of left versus right lateral recumbency on the normal radiographic appearance of the abdomen will be illustrated for the stomach, spleen, and kidneys.

Pulling the pelvic limbs caudally for lateral and VD views should be avoided. In the VD view, pulling the pelvic limbs caudally leads to skin folds, which create linear opacities that will interfere with assessment of the caudal aspect of the abdomen. It is preferable to flex the pelvic limbs for the VD radiograph so that the caudal aspect of the abdominal wall is not taut but is relaxed with greater lateral expansion and less crowding (Figure 7-1). In lateral view, the femurs should be approximately perpendicular to the lumbar spine. Pulling the femurs caudally causes crowding of the caudal abdomen and pushing them cranially leads to them being superimposed on the abdomen.

A brief review of abdominal anatomy is needed to understand the terminology used in this atlas.[1] The abdomen is the part of the trunk extending from the diaphragm to the pelvis, and it contains the abdominal cavity, which is contiguous with the pelvic cavity. The abdominal and pelvic cavities are lined by the parietal peritoneum. Organs within the abdominal cavity are covered by reflections of the parietal peritoneum, termed the visceral peritoneum, and these organs are termed *intraperitoneal*. It is important to recognize that the term intraperitoneal does not imply within the peritoneal cavity, only that the organ is covered by visceral peritoneum.

The space between the parietal and visceral peritoneal layers is the peritoneal cavity. Normally, there is nothing within the peritoneal cavity except a small amount of fluid to serve as a lubricant. Organs located near the wall of the abdominal cavity, such as a kidney, which are only partially covered by peritoneum are termed *retroperitoneal*.

Radiographic visualization of abdominal organs requires the presence of normal surrounding fat to provide contrast. Fat provides enhanced visualization of the edge of abdominal organs, because of its lower physical density and lower effective atomic number, making it slightly more radiolucent than soft tissue (Figure 7-2). The radiographically visible margin of abdominal organs created by the adjacent fat is termed the *serosal margin*. Normal fat deposits in the abdominal cavity include fat in the mesentery, omentum, retroperitoneal space, and in the falciform ligament. Fat in the mesentery and omentum is *intraperitoneal* while fat in the retroperitoneal space and falciform ligament is *extraperitoneal*. A misconception is that the falciform ligament is intraperitoneal. However, the falciform ligament is formed by the two layers of the ventral aspect of the mesogastrum, making it extraperitoneal.[2]

The radiographic opacity of all fat within the abdominal cavity should be similar. However, a difference between the opacity of intraperitoneal fat, as in the omentum or mesentery, and extraperitoneal fat, as in the retroperitoneal space or falciform ligament, can be a hint as to the existence of intraperitoneal versus extraperitoneal disease, such as fluid or inflammation.[3]

A relative lack of fat in the abdominal cavity, as seen in emaciated animals, reduces the conspicuity of the serosal margin of abdominal organs because without interposed fat the adjacent tissues have the same or nearly the same radiographic opacity, leading to border effacement (Figure 7-3). In addition to serosal margin conspicuity being reduced by lack of fat, it is also reduced in very young patients (Figure 7-4). It has been theorized that intra-abdominal fat in young patients may be more hydrated than fat in adults, thus increasing its opacity and decreasing its ability to provide radiographic contrast.[4]

Cats, particularly those in a state of positive caloric balance, can accumulate massive amounts of fat in the abdominal cavity, especially in the retroperitoneal space and in the falciform ligament (Figures 7-5 and 7-6). Fat in the falciform ligament can create a large mass effect in the cranioventral aspect of the abdominal cavity, and is sometimes misinterpreted as peritoneal effusion by

---

*A contrast medium is a substance that is used to enhance the contrast between structures in images. Most contrast media are exogenous, being administered parenterally or intravenously. Gas in the bowel can be considered an endogenous contrast medium, and its change in location as a function of body position will alter the contrast of different portions of the gastrointestinal tract. This can be extremely helpful from a diagnostic standpoint.

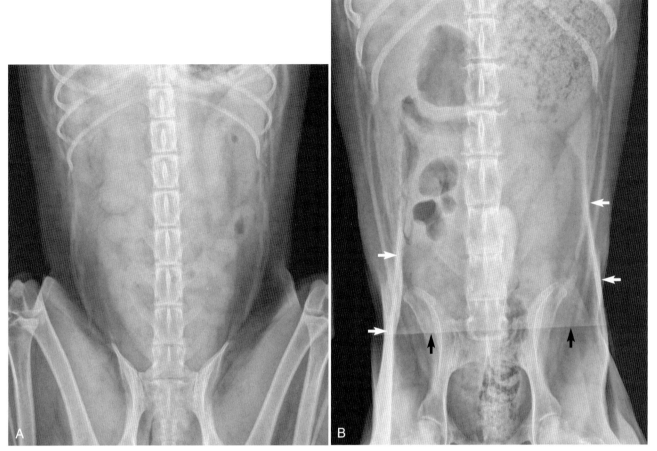

Figure 7-1. Ventrodorsal radiographs from two dogs. In **A**, the pelvic limbs are flexed, allowing relaxation of the caudal abdominal muscles and greater expansion of the caudal aspect of the abdomen. In **B**, the pelvic limbs are pulled caudally, creating skin folds *(solid white arrows)* that interfere with interpretation, and the caudal aspect of the abdominal cavity is narrower and more crowded. Also in **B**, the edge of the positioning trough has created a linear opacity *(solid black arrows)* that will also interfere with interpretation. The edge of positioning devices should not be included in the primary x-ray beam.

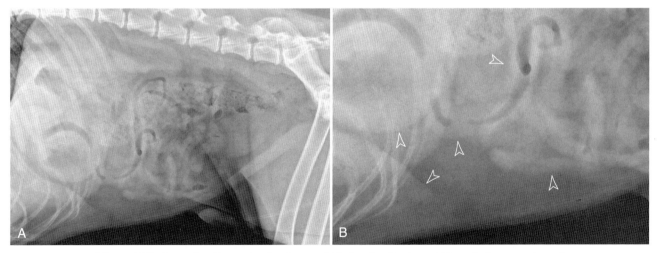

Figure 7-2. Lateral radiograph of a 6-year-old Labrador Retriever (**A**), and close-up view of the cranioventral (**B**) aspect of the abdomen. Various organs are visible. Fat provides contrast for visualization of the outer margin, also called the serosal margin, of organs *(hollow white arrowheads* in **B**).

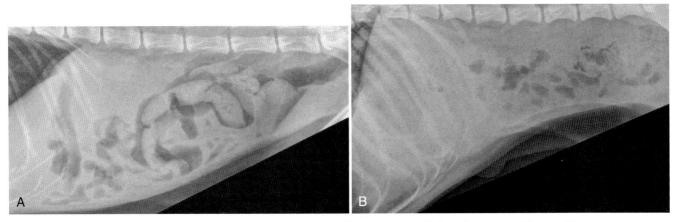

Figure 7-3. Lateral radiographs of a 16-year-old domestic cat (**A**) and a 2-year-old Basenji (**B**). The conspicuity of the serosal margin of abdominal organs in each patient is poor due to a thin body condition leading to a lack of intraperitoneal and extraperitoneal fat to provide contrast. Lack of abdominal fat compromises radiographic assessment of the abdomen.

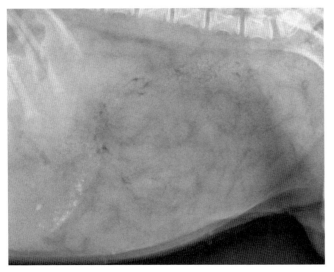

Figure 7-4. Lateral radiograph of an 11-week-old Golden Retriever. Conspicuity of serosal margins is diminished due to a relative lack of abdominal fat; this is a common finding in young animals. Margin visualization is, however, better than in patients that are in thin body condition (compare with Figure 7-3).

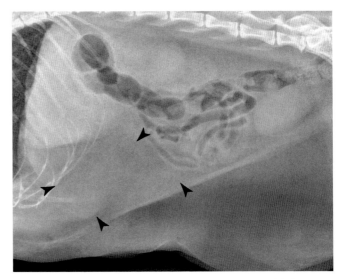

Figure 7-5. Right lateral radiograph of an 8-year-old domestic cat. There is a large collection of fat in the cranioventral aspect of the abdomen *(solid black arrowheads)* that is causing dorsal displacement of the liver and caudal displacement of the small bowel. This fat, which is in the falciform ligament, is sometimes misinterpreted by inexperienced observers as peritoneal fluid. This cannot be fluid, however, due to its lower radiographic opacity than adjacent soft tissue organs. Fluid would cause border effacement of liver and bowel. This cat also has abundant fat in the retroperitoneal space surrounding the kidneys and in the inguinal region.

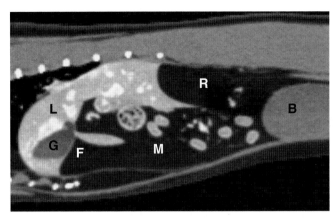

Figure 7-6. Sagittal computed tomography (CT) image of a feline abdomen, acquired approximately 2.3 cm to the right of the midsagittal plane. There is a massive amount of abdominal fat present in the falciform ligament (*F*), mesentery (*M*) and retroperitoneal space (*R*). *L*, liver; *G*, gallbladder; *B*, urinary bladder. Hepatic vessels appear hyperattenuating *(white)* as this patient received intravenous iodinated contrast medium immediately prior to image acquisition.

inexperienced interpreters. However, basic principles of radiographic opacity dictate this cannot be true or there would be border effacement of the liver margin and loss of serosal margin detail (see Figure 7-5).

While fat enhances visualization of the periphery of abdominal organs, the opacity of the interior of abdominal organs relates to their constituency. Most abdominal organs are of soft tissue opacity, but the lumen of the gastrointestinal tract commonly contains material that is either more radiopaque than soft tissue, such as bone fragments, or less radiopaque than soft tissue, such as gas.

LIVER

The liver is located in the cranial aspect of the abdominal cavity, between the diaphragm and stomach. Most of the visible liver mass is to the right of midline (Figure 7-7). The liver should normally be of homogeneous soft tissue opacity.

There is individual variation in the size of the normal liver in the dog and the cat, making the distinction between a mildly enlarged or mildly small liver from a normal liver highly subjective and relatively inaccurate. The position of the stomach is one parameter that is commonly used as an indicator of liver size. It is generally accepted that if the liver is normal, a line connecting the gastric fundus with the gastric pylorus in a lateral radiograph, termed the *gastric axis,* will range between being perpendicular to the spine or angled caudally such that it is parallel with the intercostal spaces (see Figure 7-7, *A1*). If the gastric axis extends outside of this range, cranially or caudally, then the possibility that the liver is decreased or increased in size, respectively, should be considered. It is important to note that the accepted normal range of the gastric axis in lateral radiographs is only a guideline, and slight deviation in the position of the gastric axis outside of the accepted normal range can be encountered when there is no pathologic liver abnormality.

Another commonly used criterion for liver size in lateral radiographs is extension of the liver beyond the *costal arch,* which is formed by the border of the last few costal cartilages. However, the costal arch is not a precise demarcation between a normal liver and an enlarged liver, and it cannot be used as a definitive landmark to assess liver size. It is important to recognize that a normal liver can extend beyond the costal arch (Figure 7-8). Unfortunately, the cutoff point between a normal liver and an enlarged liver based on extension beyond the costal arch is vague.

Another situation in which a normal liver might extend beyond the costal arch is a distended stomach placing pressure on the liver and causing its ventral aspect to slide caudally beyond the costal arch (Figure 7-9).

A change in shape to the margin of the portion of the liver that extends caudal to the costal arch from a gradual taper (see Figures 7-7, *A*, and 7-8, *A*) to a more rounded configuration is another finding often associated with liver enlargement.

In a deep-chested dog (such as an Afghan Hound or Collie), the gastric axis is normally more perpendicular. In contrast, the gastric axis in a shallow-chested dog (such as a Pug or a Bulldog) is typically angled more caudally.

The gastric axis is not used as a measure of liver size in the VD view. Rather, the position of the pyloric part of the stomach is an alternate landmark. With liver enlargement, the pylorus is displaced caudally and medially in the VD view. In the lateral view, liver enlargement causes dorsal and caudal displacement of the pylorus.

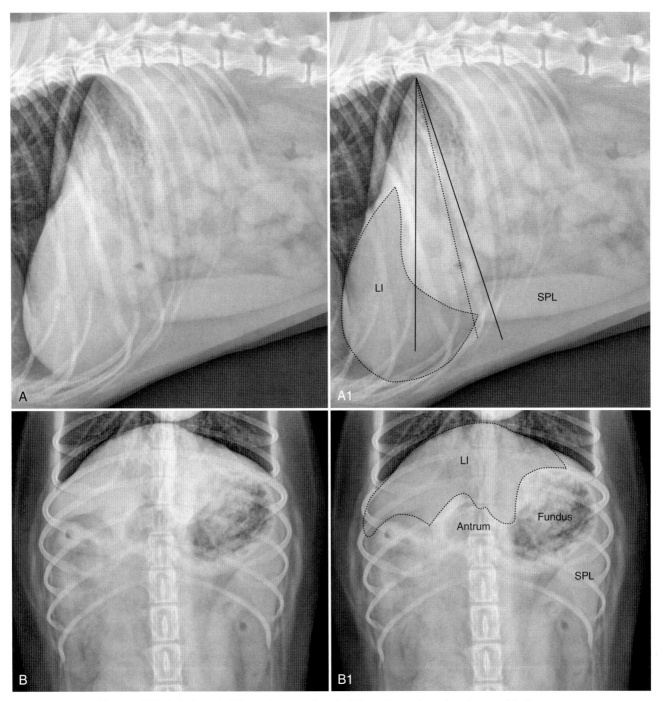

Figure 7-7. Left lateral (**A**) and ventrodorsal (**B**) radiographs of a 5-year-old German Shepherd. The liver is the homogeneous soft tissue opacity between the diaphragm and the stomach. In **B**, note that most of the visible liver mass is to the right of midline. **A1** and **B1**, corresponding labeled radiographs. In **A1**, the *solid lines* represent the accepted normal range of the gastric axis (the gastric axis is an imaginary line connecting the fundus and pylorus). The *dotted line* represents the approximate location of the gastric axis in this dog. The visible portions of the liver are *shaded gray* in **A1** and **B1**; this shading does not encompass the entire liver volume because there are portions of the liver that are superimposed on stomach that cannot be seen distinctly. *Antrum*, gastric antrum; *fundus*, gastric fundus; *LI*, liver; *SPL*, spleen.

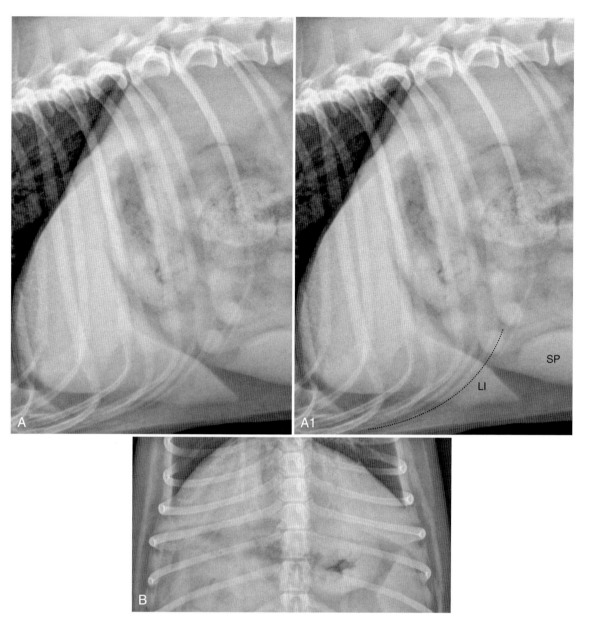

Figure 7-8. Lateral (**A**) and ventrodorsal (**B**) radiographs of a 4-year-old Miniature Schnauzer and corresponding labeled radiograph (**A1**). In **A1,** the costal arch, defined by the caudal limit of the costal cartilages, has been marked with a *dotted line*. The liver in this patient clearly extends beyond the costal arch but is normal. There were no clinical or laboratory signs of liver dysfunction, and sonographic examination of the liver was normal. Extension of the liver beyond the costal arch cannot be taken as indisputable evidence of liver enlargement. *LI,* Liver; *SP,* spleen.

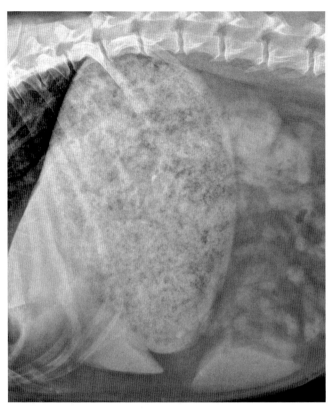

Figure 7-9. Lateral radiograph of a 9-year-old Rottweiler that ingested an excessive amount of dog food. The stomach is markedly distended with food. The liver extends beyond the costal arch in the cranioventral aspect of the abdomen, which likely is due to displacement caused by pressure from the enlarged stomach and not liver enlargement.

The criteria for evaluating liver size in cats are the same as in dogs. However, it is important to recognize that the gastric pylorus in cats is normally located more medially than in dogs, residing nearly on the midline. Thus, a change in the position of the pylorus is rarely used to assess liver size in VD abdominal radiographs in cats (Figure 7-10).

The gallbladder is situated between the quadrate lobe medially and the right medial lobe laterally, in the cranioventral aspect of the liver (Figure 7-11). In most patients, the normal gallbladder is not seen radiographically. Occasionally, especially in cats, a round opacity will be seen superimposed on the ventral aspect of the liver or extending slightly beyond the margin of the ventral aspect of the liver. This round opacity can be created by a normal, but modestly distended, gallbladder (Figure 7-12). This can be seen when there is no associated evidence of liver or gallbladder disease. However, visualization of this opacity may not always be normal and, if clinically indicated, the liver should be assessed using other methods, such as ultrasound examination.

SPLEEN

The spleen, a lymphatic organ, is located in the left hemiabdomen. The spleen should be of homogeneous soft tissue opacity. The proximal aspect of the spleen, commonly referred to as the *head* of the spleen, is located in close approximation to the gastric fundus. It is held loosely in this position by the gastrosplenic ligament and the short gastric arteries and veins that communicate with the splenic vessels.[5] These short gastric arteries and veins can be less well-developed in the cat.[6] The distal extremity of the spleen, commonly called the *tail* of the spleen, is more mobile and variable in position than the proximal extremity.

The size of a normal spleen is highly variable, especially in the dog. In the dog, the spleen is usually of adequate size to be identified in both lateral and VD abdominal radiographs. In general, the spleen is smaller in the cat than in the dog. In the cat, a normal spleen usually can be identified in the VD but not the lateral radiograph. Occasionally, no portion of the normal feline

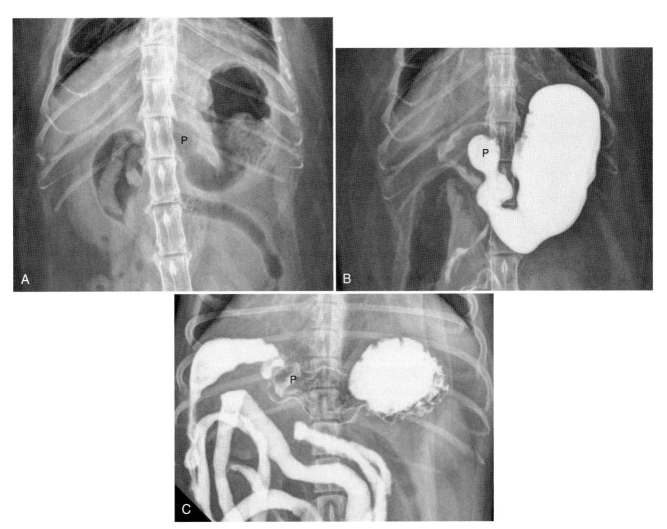

Figure 7-10. Ventrodorsal radiographs of a 12-year-old domestic cat without (**A**) and with (**B**) barium in the stomach, and of a 7-year-old Miniature Schnauzer with barium in the stomach (**C**). In the cat (**A**), note the midline location of the gastric pylorus *(P)*. In **A**, the patient was slightly rotated when the radiograph was made, giving the pylorus a mildly exaggerated midline position. The pylorus is easier to identify when there is barium in the stomach (**B**). The undulations in the pyloric antrum seen in **A** and **B** are due to normal peristalsis. In the dog (**C**), note the more rightward position of the gastric pylorus *(P)* versus the position in the cat.

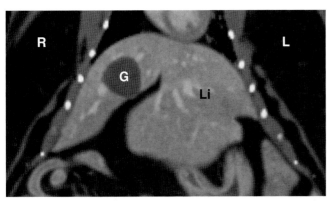

Figure 7-11. Dorsal plane CT image through the level of the gallbladder in a cat. Note the right-sided location of the gallbladder *(G)* in the liver *(Li)*. The hepatic vasculature is hyperattenuating as this patient received iodinated contrast medium prior to image acquisition. The overall attenuation of the liver is also increased due to the contrast medium in the sinusoids. There is no difference in the opacity of the gallbladder versus liver in radiographs. It is hypoattenuating to normal liver in CT images due to the inherent contrast resolution of CT compared to radiographs, and even more so in this patient due to the use of contrast medium. *L,* Left; *R,* right.

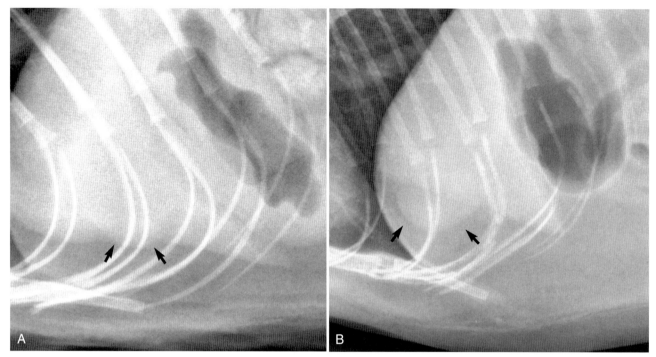

Figure 7-12. Lateral radiograph of a 1½-year-old Persian (**A**) and a 4-year-old domestic cat (**B**). In each radiograph, there is a round opacity extending beyond the ventral margin of the liver that is consistent with enlargement of the gallbladder *(solid black arrows)*. This can be found without evidence of biliary disease and can be a normal variant in some subjects.

spleen can be seen radiographically. If the spleen is seen in lateral *and* VD abdominal radiographs of a cat, the possibility of enlargement should be considered.

As noted, the size of the normal spleen varies widely, according to its functional status and whether chemical restraint has been used for radiography. Determining the clinical significance of a spleen that appears radiographically enlarged is not usually possible without other assessments, such as ultrasonography and/or cytologic sampling.

When the spleen is visualized radiographically, it is important to realize that the entire spleen is not seen in any single radiographic view. When the x-ray beam is oriented along or parallel to (*end-on* projection) a length of the spleen, it typically has a triangular shape. If the body of the spleen is struck perpendicularly by the primary x-ray beam (*side-on* projection), there usually is insufficient absorption of the x-rays to see that portion of the spleen (Figure 7-13). The extent of the spleen that can be seen radiographically depends on the thickness of the spleen, its orientation to the primary x-ray beam, and the contrast resolution of the imaging system.

In lateral abdominal radiographs of the dog, the distal extremity of the spleen is the portion seen most commonly (Figure 7-14, *A-B*). In lateral recumbency, the distal extremity lying adjacent to the ventral abdominal

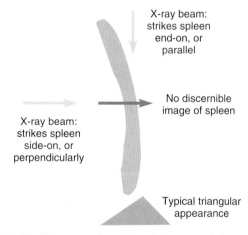

Figure 7-13. Diagram of a spleen being struck by a primary x-ray beam oriented either side-on (perpendicular) or end-on (parallel). When the long axis of the spleen is struck side-on, there is inadequate absorption of x-rays for the spleen to be seen. When the long axis is struck end-on, the number of x-rays absorbed is increased, creating an opacity sufficient to be seen. What is seen radiographically when the spleen is struck end-on is only a fraction of the total spleen volume.

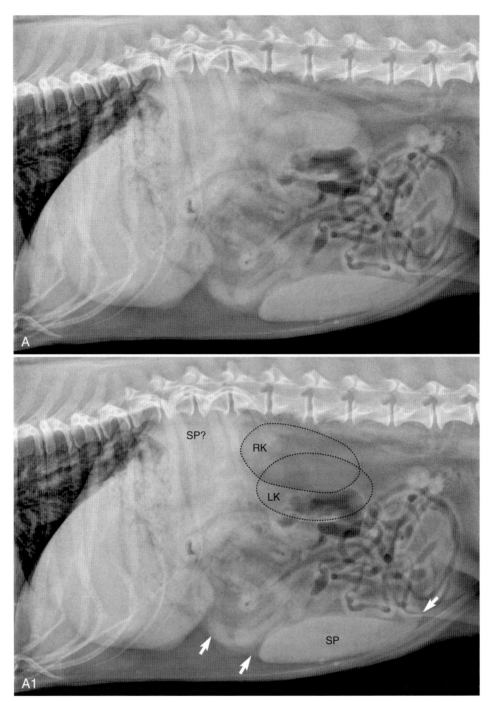

Figure 7-14. Left lateral (**A**) and right lateral (**B**) radiographs and corresponding labeled left lateral (**A1**) and right lateral (**B1**) radiographs of a normal 9-year-old Dachshund. The distal extremity of the spleen has its typical triangular shape in the ventral aspect of the abdomen. The spleen is relatively large in this dog, likely due to the chemical restraint used to facilitate radiographic positioning. Conclusions regarding splenic disease cannot be made from a spleen with this appearance. As noted in the text, the triangular structure is only a fraction of the total splenic volume. In both the left and right lateral views, a portion of the spleen extending from the triangular region can be seen (*solid white arrows* in **A1** and **B1**). Identifying this opacity as spleen is facilitated by it being contiguous with the familiar triangular splenic opacity; in isolation this opacity would likely not be diagnosed as spleen. In **A**, the two focal opacities ventral to the spleen, and in **B** the focal opacity superimposed on the spleen, are summation shadows due to superimposed nipples. *LK,* Left kidney; *RK,* right kidney; *SP,* spleen; *SP?,* opacity possibly due to proximal extremity of spleen.

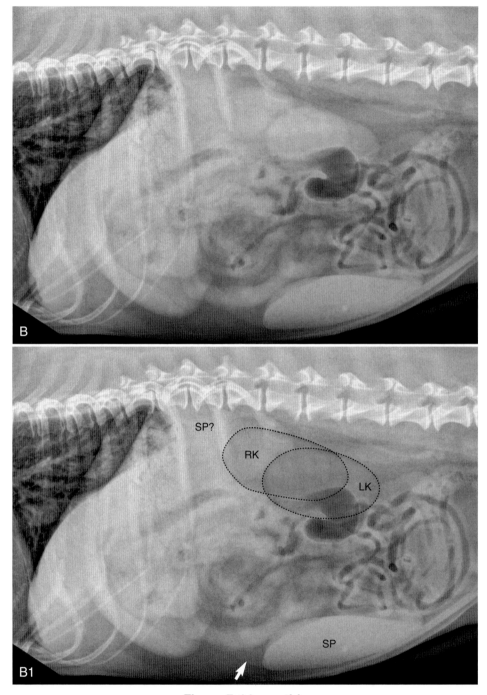

Figure 7-14, cont'd.

wall is oriented end-on to the primary x-ray beam. Occasionally, the portions of the spleen extending peripherally from the distal extremity can also be seen (see Figure 7-14, *A1* and *B1*, *white arrows*).

With a moderately sized normal spleen, the proximal portion might be oriented in a way that creates an ill-defined mass effect in lateral radiographs in the region between the gastric fundus and the kidneys, without being visualized as a distinct triangular structure (see Figure 7-14, *A1* and *B1*). Other imaging modalities would be needed to interpret this finding correctly, because a mass effect in this area in lateral radiographs could represent an abnormal mass.

In VD abdominal radiographs of the dog, it is the proximal extremity of the spleen that typically is seen (Figure 7-15). Again, this is due to the portion of the proximal extremity lying adjacent to the left abdominal wall being oriented end-on to the primary x-ray beam.

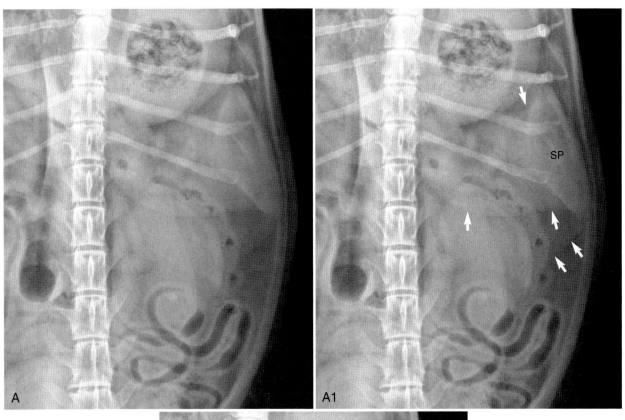
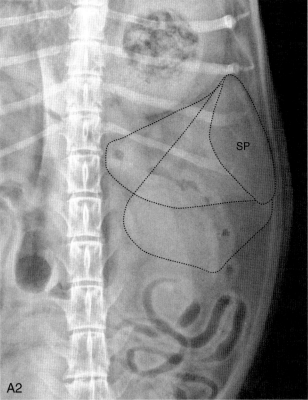

Figure 7-15. Ventrodorsal radiograph of a normal 9-year-old Dachshund (**A**), and corresponding labeled radiographs (**A1**, **A2**). The triangle lateral and caudal to the gastric fundus in (**A**) is the typical appearance of the proximal extremity of the spleen as seen in ventrodorsal views. In this dog, portions of the spleen extending from the region struck end-on by the primary x-ray beam can be seen (*solid white arrows* in **A1**, *dotted lines* in **A2**). These peripheral regions of the spleen cannot be seen in every dog. Their identification as spleen is facilitated by the fact that they are contiguous with the familiar triangular splenic opacity. *SP*, Spleen.

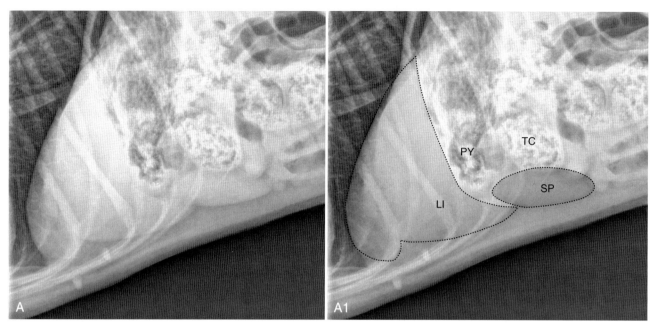

Figure 7-16. Lateral radiograph of a 2-year-old Belgian Sheepdog (**A**) and corresponding labeled radiograph (**A1**). In **A**, there is an oval opacity in the cranioventral aspect of the abdomen representing the distal extremity of the spleen. Facts supporting this opacity being spleen and not liver are (1) the cranial limit of this opacity does not extend cranial to the gastric pylorus, and (2) there is a fat plane between the liver and the oval opacity. The liver and spleen are shaded in **A1**. *LI*, Liver; *PY*, gastric pylorus; *SP*, spleen; *TC*, transverse colon.

Similar to lateral views, portions of the spleen extending peripherally from the proximal extremity can sometimes be seen in VD radiographs (see Figure 7-15).

In the dog, due to its mobility, the distal extremity of the spleen can be located in the cranioventral aspect of the abdomen and can create an oblong opacity that is often misdiagnosed as an enlarged liver lobe (Figure 7-16). The key to correctly distinguishing between spleen and liver in this circumstance relates to the location of the cranial aspect of the opacity in question. If the opacity is the distal extremity of the spleen, its cranial margin will not extend cranial to the gastric pylorus, and there will be a plane of fat opacity between it and the liver (see Figure 7-16), except in emaciated subjects. If the suspicious opacity is an enlarged liver lobe, the opacity can be traced cranial to the gastric pylorus, and there will not be a plane of fat opacity between it and the liver.

As noted previously, the distal extremity of the feline spleen is rarely seen in lateral radiographs. Most commonly, the normal feline spleen is seen in the VD view, lateral to the gastric fundus, with no aspect of the spleen being seen in either a left or right lateral view (Figure 7-17). Occasionally, in cats with abundant abdominal fat, the proximal extremity of the spleen can be seen in lateral radiographs just caudal to the stomach (Figure 7-18). This appearance can be confused with an adrenal mass. If there is any question of the identity of a radiographic opacity caudal to the stomach in the cat, more specific imaging modalities of the region should be used.

The great mobility of the spleen results in it having a different appearance in left versus right lateral radiographs of the abdomen, and, although the exact changes in splenic position that occur as a function of recumbency have not been completely characterized, the distal extremity seems to be more conspicuous in right lateral radiographs (Figure 7-19). The effect of spleen conspicuity as a function of recumbency is most likely due to compression of the dependent portion of the abdomen leading to a change in the location of the spleen.

An accessory spleen, also called a *splenunculus*, is a focus of ectopic splenic tissue supplied by the splenic

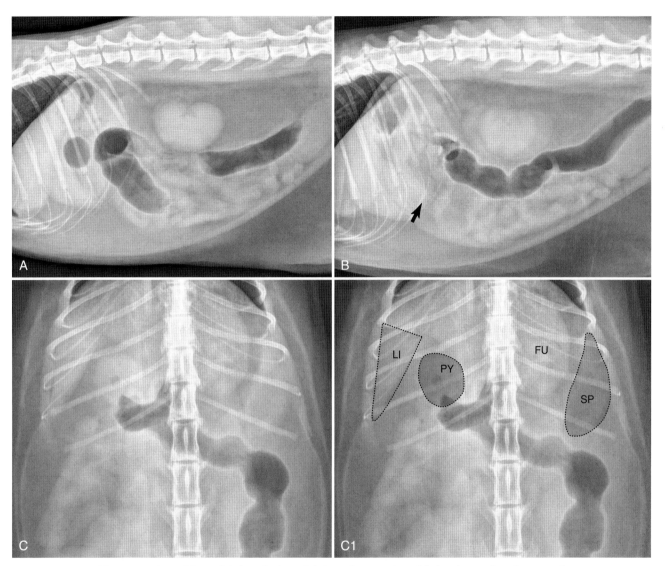

Figure 7-17. Left lateral (**A**), right lateral (**B**), and ventrodorsal (**C**) radiographs of a normal 7-year-old domestic cat, and corresponding labeled ventrodorsal radiograph (**C1**). Note that the distal extremity of the spleen is not visible in either lateral view; this is typical for the cat. In **B**, the ventral edge of the spleen may be visible *(arrow)*, but the expected triangular appearance is not seen. In **C** and **C1**, the spleen is lateral to the gastric fundus. *FU,* Gastric fundus; *LI,* liver; *PY,* gastric pylorus; *SP,* spleen.

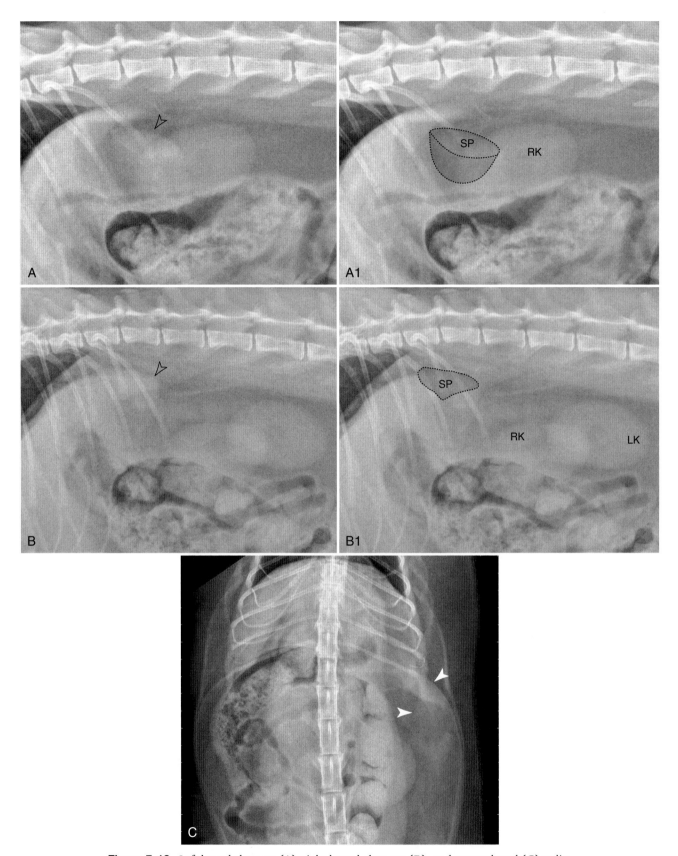

Figure 7-18. Left lateral close-up (**A**), right lateral close-up (**B**), and ventrodorsal (**C**) radiographs of a 5-year-old domestic cat, and corresponding labeled left lateral (**A1**) and right lateral (**B1**) radiographs. This cat is obese. The spleen is seen clearly in the ventrodorsal view, as expected *(solid white arrowheads)*. In each lateral view, however, the proximal extremity of the spleen can be seen as an approximately triangular opacity just caudal to the gastric fundus *(hollow black arrowhead)*. The left kidney is not seen in **A** or **A1** because it is superimposed on the descending colon. *LK*, Left kidney; *RK*, right kidney; *SP*, spleen.

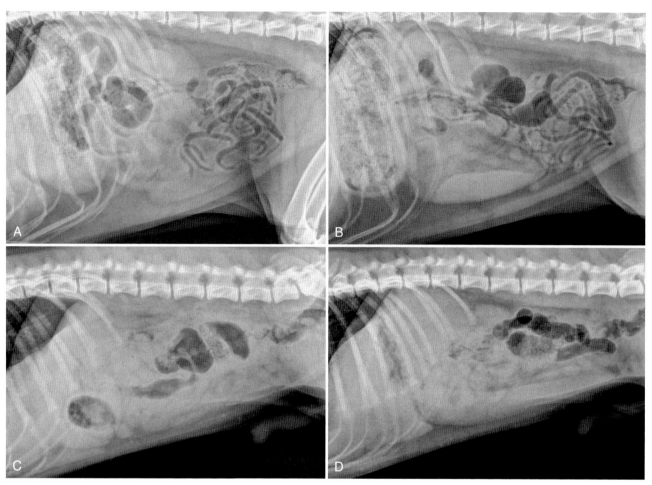

Figure 7-19. Left (**A**) and right (**B**) lateral radiographs of a 6-year-old Labrador Retriever, and left (**C**) and right (**D**) lateral radiographs of a 6-month-old German Pinscher. The distal extremity of the spleen is more conspicuous in these right lateral radiographs than in the left lateral radiographs.

artery (Figure 7-20).[7] Splenunculi can be located within the pancreas, whereas they will not be visible radiographically, but they can also be separate and isolated from the spleen, being either of congenital or traumatic origin secondary to autoimplantation.[8] If of adequate size and surrounded by adequate fat, isolated splenunduli can appear radiographically as a small mass adjacent to the spleen. It is important to be aware of the possible existence of accessory splenic tissue so as not to immediately misdiagnose small masses adjacent to the spleen as abnormal. Other imaging modalities will be needed to confirm that the abnormal mass is typical of ectopic splenic tissue.

PANCREAS

The normal pancreas is not visible radiographically in the dog. In the cat, the left lobe of the pancreas extends lateral to the gastric fundus and, in this location, is not in contact with any other soft tissue organ (Figure 7-21). If there is adequate fat surrounding the pancreas in this area, it can be identified in VD abdominal radiographs as a fusiform opacity medial to the midportion of the spleen (Figure 7-22).

KIDNEYS

The kidneys are paired retroperitoneal organs with one surface in contact with either hypaxial muscles or retroperitoneal fat, depending on the body habitus. The other renal surface is covered by peritoneum.[9,10] The parenchyma of the kidney should have uniform soft tissue opacity, except in some subjects, especially cats, the presence of fat in the renal hilus will cause a focal decreased opacity in this region (Figures 7-23 and 7-24).

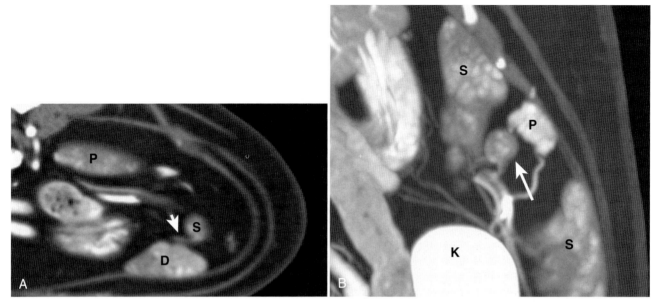

Figure 7-20. A, Transverse post-contrast CT image of the abdomen of a cat having a splenunculus *(S)*. The splenunculus was not connected to the spleen in any CT image slice. The linear structure from the ventral aspect of the splenunculus *(white arrow)* is a tributary of the splenic vein. *P*, Proximal extremity of spleen; *D*, distal extremity of spleen. B, Dorsal plane Maximum Intensity Projection CT image of the same cat as in (A). The splenunculus *(white arrow)* can be seen to have the same attenuation characteristics as the spleen *(S)*. *P*, Pancreas; *K*, left kidney. (A Maximum Intensity Projection, or MIP, image is a series, or summation, of 2-dimensional CT slice images to create a volumetric image. The thickness of the volume is controlled by the number of individual slices that are summed.)

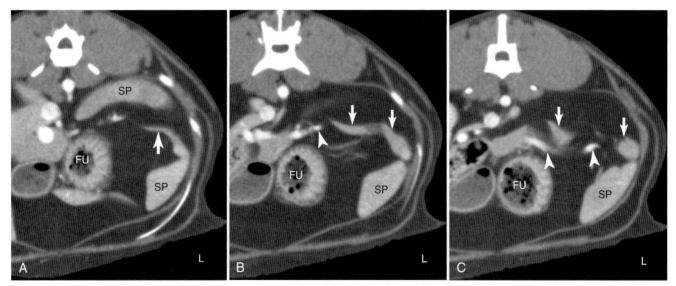

Figure 7-21. Contrast-enhanced transverse CT images of the cranial abdomen of a normal cat. The image in the *left panel* is most cranial, and the image in the *right panel* is most caudal; images are 3-mm thick and span a length of 1.5 cm. The pancreas *(white arrows)* can be seen between the gastric fundus *(FU)* and spleen *(SP)*. Note the abundant fat surrounding the pancreas, which increases its radiographic conspicuity. The contrast-enhanced splenic vein *(solid white arrowheads)* can also be seen.

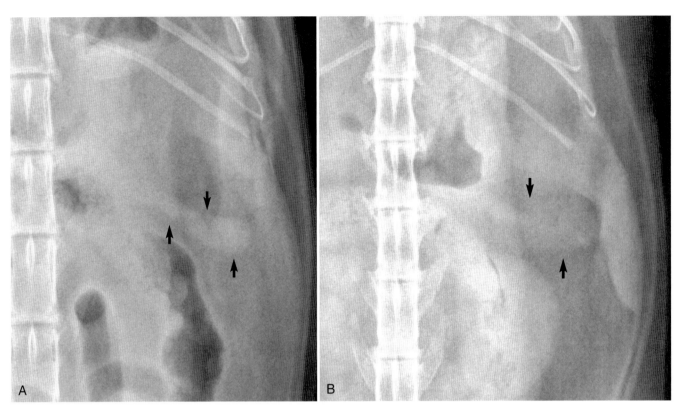

Figure 7-22. Ventrodorsal radiographs of a 1½-year-old domestic cat (**A**) and an 8-year-old domestic cat (**B**). In each cat, there is a fusiform opacity medial to the spleen that represents the left lobe of the pancreas *(solid black arrows)*. The pancreas is slightly less distinct in **B** than in **A**, likely due to its orientation with respect to the primary x-ray beam.

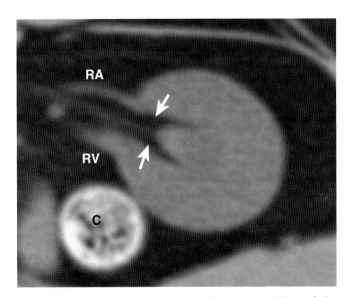

Figure 7-23. Transverse CT image through the hilus of the left kidney in a cat. Note the large amount of fat in the hilus *(white arrows)*. This will create a radiolucent region if the hilus is projected end-on. The thin linear structure between the arrows is the left ureter. *RA*, Renal artery; *RV*, renal vein; *C*, descending colon.

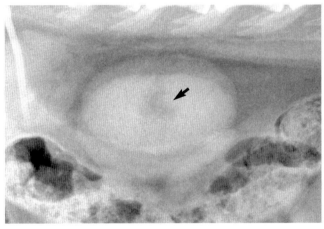

Figure 7-24. Right lateral radiograph of an 8-year-old domestic cat. The kidneys are superimposed. There is a radiolucent area near the renal hilus that is due to hilar fat *(solid black arrow)*.

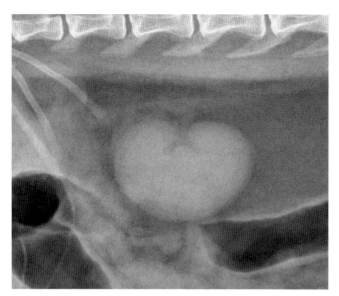

Figure 7-25. Left lateral radiograph of a 7-year-old domestic cat. The kidneys are nearly perfectly superimposed. Typically, there is less craniocaudal separation of the kidneys in cats than in dogs. This cat has a large amount of retroperitoneal fat that is displacing the kidneys ventrally; this is a common finding in obese cats.

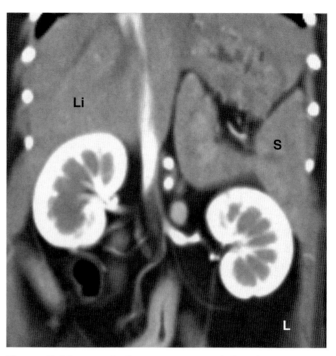

Figure 7-26. Dorsal plane CT image of the abdomen of a dog immediately after administration of intravenous contrast medium. Note how the right kidney is embedded in the renal fossa of the caudate lobe of the liver while the left kidney is surrounded by fat. This renders the right kidney, especially the cranial pole, less conspicuous in radiographs than the left kidney. *L*, Left; *Li*, liver; *S*, spleen.

The radiographic size of the kidneys has typically been assessed by computing the ratio of the length of the kidney to the length of the body of the second lumbar vertebra in VD radiographs. In the dog, normal kidneys typically measure between 2.5 and 3.5 times the length of the second lumbar vertebra,[11] whereas, in the cat, normal kidneys typically measure between 2.4 to 3 times the length of the second lumbar vertebra.[12] Some exceptions to these ratios have been identified in cats. For example, older cats without signs of renal disease can have smaller kidneys, 1.9 to 2.6 times the length of L2, but it is not known whether subclinical renal disease was present.[13] Also, intact cats may have larger kidneys (2.1 to 3.2 times the length of L2) than neutered cats (1.9 to 2.6 times the length of L2).[14] It is important to note that these ratios are only guidelines because the overall number of subjects is small, and the normal status may not have been rigorously documented. Renal function must not be inferred based on the radiographic assessment of renal size.

The left kidney is positioned slightly more caudal than the right kidney; this difference is more consistent in the dog. In the cat, the kidneys often lie at nearly the same craniocaudal level (see Figures 7-24 and 7-25). In the dog, the cranial pole of the right kidney is embedded in the renal fossa of the caudate lobe of the liver; thus either part of or the entire right kidney may not be visible radiographically (Figures 7-26 and 7-27). With enlargement of

Figure 7-27. Left lateral (**A**), right lateral (**B**), and ventrodorsal (**C**) radiographs of the abdomen of an 11-year-old Bluetick Coonhound, and corresponding labeled left lateral (**A1**), right lateral (**B1**), and ventrodorsal (**C1**) radiographs. In **A1**, **B1**, and **C1**, the left kidney is outlined by a *dashed line*, and the right kidney is outlined by a *solid line*. In the right lateral view (**B, B1**), the left kidney is more ventral than in the left lateral view (**A, A1**), and its entire margin can be seen. In the left lateral view (**A, A1**), the entire margin of the left kidney is not seen due to its cranial pole being superimposed on the right kidney. The right kidney is never seen in its entirety; this is normal because the cranial pole is embedded in the renal fossa of the caudate lobe of the liver, and thus the kidney border is effaced in this region. In the right lateral view (**B, B1**), the amorphous opacity between the right kidney, gastric fundus, and gastric pylorus (*asterisk* in **B1**) is most likely due to the proximal extremity of the spleen. *GF*, Gastric fundus; *GP*, gastric pylorus; *SP*, spleen.

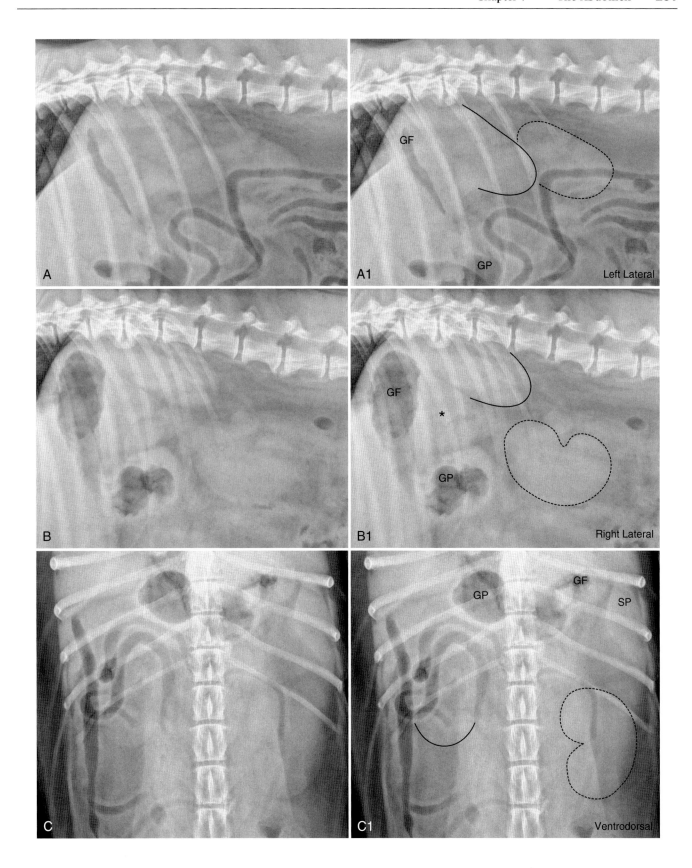

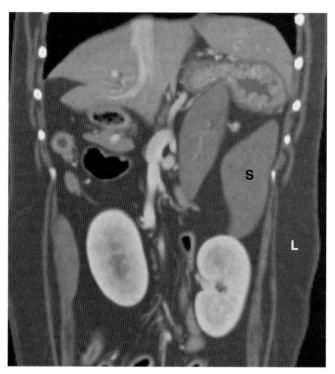

Figure 7-28. Dorsal plane CT image of a dog where the right kidney is more caudally positioned than normal. The cranial pole is not embedded in the renal fossa of the caudate lobe and is instead surrounded by fat, as is the left kidney. Both kidneys should be completely visualized in radiographs of this dog due to the fact that both are completely surrounded by fat. *L,* Left; *S,* spleen.

the right side of the liver, the right kidney may become displaced caudally. In that instance, the cranial pole still cannot be seen as its border remains embedded in the caudate lobe fossa. When the right kidney is situated more caudally than normal as a result of a variant in anatomy rather than hepatomegaly, the cranial pole should be clearly seen, assuming there is adequate retroperitoneal fat, because the kidney is no longer embedded in the caudate lobe of the liver (Figure 7-28).

The left kidney is loosely attached and may be positioned farther ventral in the abdomen than anticipated; this is typically more pronounced in cats than in dogs[8] (Figure 7-29, *B*), but a ventrally located left kidney can also be seen in dogs (see Figure 7-27, *B*).

In general, cats have more fat surrounding the kidneys than dogs, leading to very good renal conspicuity; thus the left kidney and much of the right kidney can usually be seen (see Figure 7-29).

Because the left kidney is relatively mobile, it is expected that the relative position of the kidneys will change in left versus right lateral abdominal radiographs. This is especially true in cats in which retroperitoneal fat has displaced the kidneys ventrally from the hypaxial muscles (Figure 7-30). The rationale behind discussing this concept is not to define a set pattern of kidney overlap in left versus right recumbency but rather to illustrate that the relative position of the kidneys can be quite different between left and right lateral views.

Figure 7-29. Left lateral **(A)**, right lateral **(B)**, and ventrodorsal **(C)** radiographs of an 8-year-old domestic cat, and corresponding labeled left lateral **(A1)**, right lateral **(B1)**, and ventrodorsal **(C1)** radiographs. In **A1, B1,** and **C1,** the left kidney is outlined by a *dashed line,* and the visible portions of the right kidney are outlined by a *solid line.* In lateral abdominal radiographs of this cat, the left kidney is more caudal than the right (see Figures 7-20 and 7-21, where the kidneys are nearly superimposed in lateral radiographs). Despite abundant peritoneal fat and relatively little fecal material in the colon, the entire right kidney is difficult to identify. A larger amount of fecal material than present here reduces visualization of the kidneys in the ventrodorsal view even further. *AC,* Ascending colon; *TC,* transverse colon.

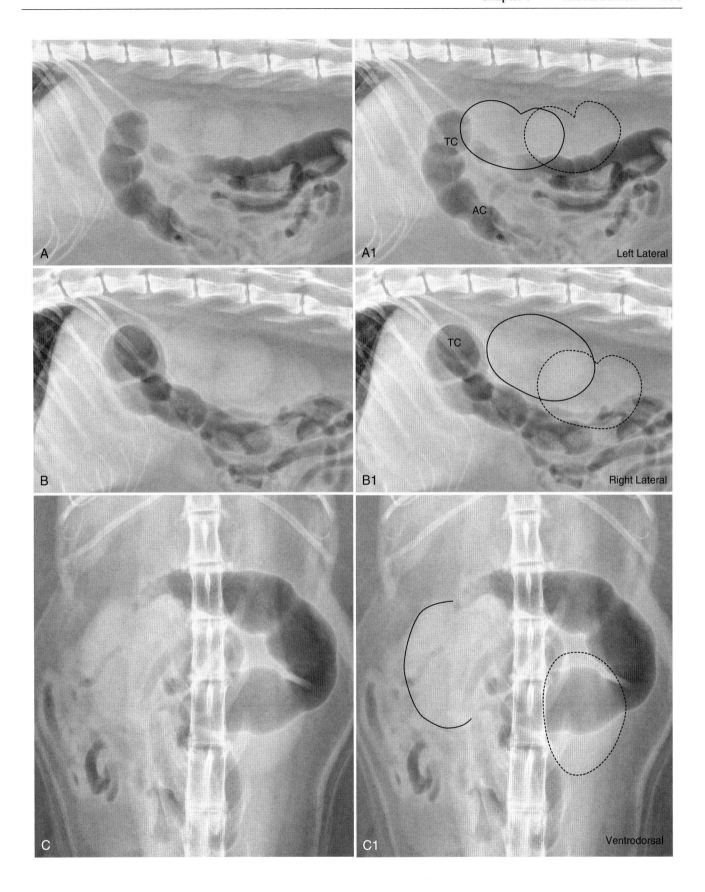

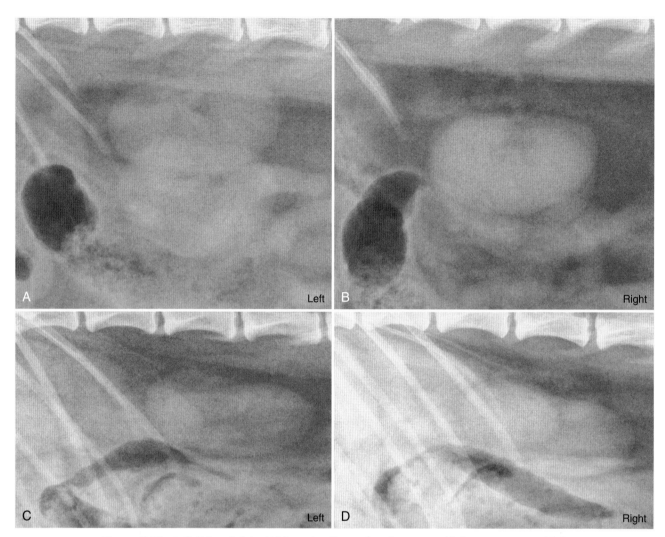

Figure 7-30. Left (**A**) and right (**B**) lateral radiographs of a 1-year-old domestic cat, and left (**C**) and right (**D**) lateral radiographs of a 5-year-old Miniature Poodle. In the cat, note the greater dorsoventral separation of the kidneys in the left lateral view. In the dog, note the greater overlap of the caudal pole of the right kidney and the cranial pole of the left kidney in the right lateral view. As noted in the text, the rationale behind these illustrations is not to define a set pattern of kidney overlap in left versus right recumbency but to illustrate the fact that the relative position of the kidneys can be quite different between left and right lateral views.

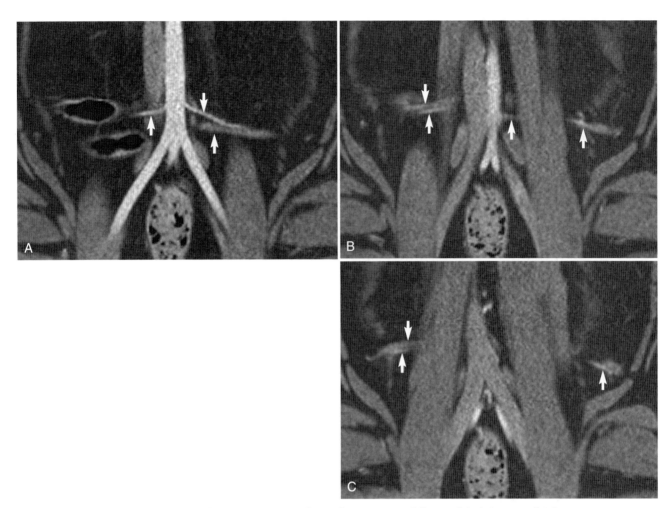

Figure 7-31. Dorsal plane, contrast-enhanced CT images of the caudal abdomen of a dog. The deep circumflex iliac arteries and veins *(solid white arrows)* are branches of the abdominal aorta and caudal vena cava, respectively.

URETERS

Normal ureters are not visible radiographically. The deep circumflex iliac arteries and veins are branches of the abdominal aorta and caudal vena cava, respectively, which extend laterally from the aorta (Figure 7-31). If surrounded by adequate fat, these vessels create focal retroperitoneal opacities in lateral radiographs that may be misinterpreted as ureteral calculi (Figure 7-32).

URINARY BLADDER

The urinary bladder is positioned at the caudal aspect of the abdominal cavity and normally should be of homogeneous soft tissue opacity. Occasionally, gas will be present in a normal urinary bladder following catheterization or cystocentesis (Figure 7-33). Some urinary calculi are referred to as being radiolucent. However, the term *radiolucent* is relative, and, with reference to urinary calculi, the term *radiolucent* is used to describe calculi that

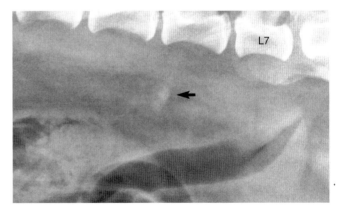

Figure 7-32. Lateral radiograph of a 4-year-old Welsh Corgi. The focal opacities in the caudal aspect of the retroperitoneal space *(solid black arrow)* are due to end-on projection of the deep circumflex iliac arteries and veins. These opacities should not be misinterpreted as ureteral calculi. Visualization of these vessels in lateral radiographs is more common in patients with abundant retroperitoneal fat. *L7,* Seventh lumbar vertebral body.

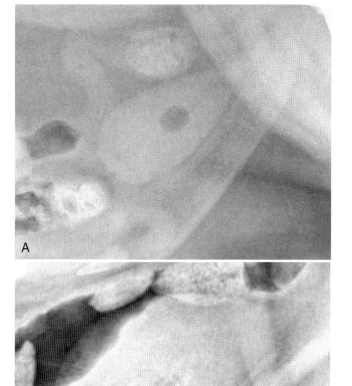

Figure 7-33. A, Lateral radiograph of an 8-year-old domestic cat. The large oval radiolucency in the center of the urinary bladder is an air bubble introduced following voiding hydropulsion. B, Lateral radiograph of a 4-year-old domestic cat. There are multiple small focal radiolucencies in the center of the urinary bladder that represent small air bubbles introduced during cystocentesis. The streaky appearance in B is an artifact due to ultrasound gel contamination of the hair.

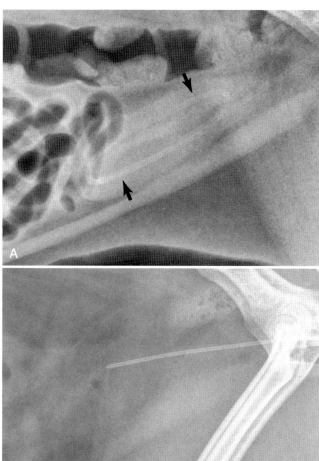

Figure 7-34. A, Lateral radiograph of a 16-year-old domestic cat. There is a tubular opacity in the urinary bladder *(black arrows)* that represents a rubber urinary catheter. The length of catheter in the urinary bladder is excessive. B, Lateral radiograph of a 4-year-old domestic cat. The linear opacity extending into the urinary bladder is typical for the appearance of a stiff plastic catheter. Multiple small air bubbles are also present in the urinary bladder in B.

are not visible in survey radiographs. This does not mean that the calculi are really radiolucent; it means they have the same radiographic opacity as soft tissue or urine. Thus, calculi thought of as being radiolucent will never appear as the air bubbles illustrated in Figure 7-33 because they are of soft tissue opacity and silhouette with surrounding urine.

Indwelling urinary catheters can also create opacities in the urinary bladder, assuming they are not of soft tissue opacity. In general, urinary catheters are more opaque than soft tissue (Figure 7-34).

The size of the urinary bladder varies considerably, depending on the volume of urine it contains. Housebroken animals can have a very large urinary bladder simply due to lack of urination (Figure 7-35). In cats, a distended urinary bladder becomes cranially displaced in the abdominal cavity, and, because of the large amount of abdominal fat commonly present, the urethra can usually be seen as a tubular structure extending caudally from the urinary bladder (see Figure 7-29, B). It should not be inferred that every distended urinary bladder is due to normal urine retention. Pathologic causes of

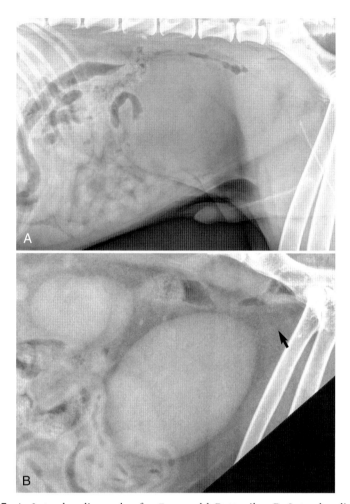

Figure 7-35. **A,** Lateral radiograph of a 7-year-old Rottweiler. **B,** Lateral radiograph of a 14-year-old domestic cat. These animals have urinary bladder distension related to lack of urination. The Rottweiler is receiving intravenous fluids; an intravenous catheter is visible in the caudal aspect of the image. In the cat (**B**), note the cranial position of the distended urinary bladder. Cranial positioning of the bladder is a common finding in cats with urinary bladder distension. The urethra is the linear opacity extending caudally from the urinary bladder neck *(solid black arrow)*. In **B**, the opacities superimposed on the dorsal aspect of the urinary bladder are the result of ultrasound gel staining the hair.

urinary bladder distension should always be investigated if a grossly distended bladder is identified.

PROSTATE GLAND

A normal prostate gland is not visible radiographically in either the dog or cat; this is because of its small size and its position in the pelvic canal where it silhouettes with adjacent soft tissues. However, benign prostatic hypertrophy is such a common development in middle-aged and old male dogs that many consider prostate gland enlargement from that cause to be a normal variant. In benign prostatic hypertrophy, the prostate gland typically is mildly enlarged and is positioned just cranial to the iliopubic eminence of the pelvis. In this location, the prostate gland can usually be easily identified in a lateral radiograph. A triangle of fat is usually present between the caudoventral aspect of the urinary bladder, the cranioventral aspect of the prostate gland, and the subjacent portion of the abdominal wall. This triangle is a reliable

Figure 7-36. A, Lateral radiograph of an 8-year-old male Labrador Retriever. The prostate gland is enlarged and appears as a mass caudal to the bladder. The colon contains a large amount of fecal material. The triangle of fat between the cranioventral aspect of the prostate gland, the caudoventral aspect of the urinary bladder, and the ventral abdominal wall is very conspicuous. B, Same dog as in A with the triangle of fat indicated by arrows. C, Lateral radiograph of a 10-year-old neutered male Siberian Husky. The mildly enlarged prostate gland *(black arrows)* has more of a fusiform shape than in the dog illustrated in A and B.

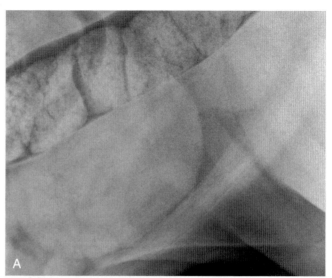

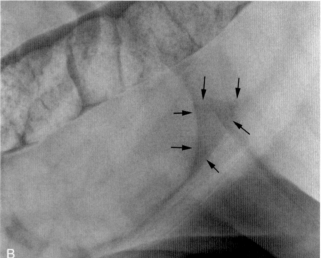

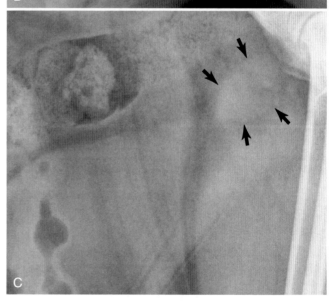

radiographic sign of mild prostate gland enlargement, such as with benign prostatic hypertrophy (Figure 7-36). As for distension of the urinary bladder, it should not be inferred that every mildly enlarged prostate gland is due to benign prostatic hyperplasia. Pathologic causes of prostate gland enlargement should always be considered if the prostate gland is visible in survey abdominal radiographs.

URETHRA

The normal urethra is not visible radiographically in either the dog or the cat. If urethral calculi are suspected in a male dog, a lateral radiograph made with the pelvic limbs pulled cranially should be acquired. This allows for an unobstructed view of the entire urethra. In normally positioned lateral abdominal radiographs, the pelvic limbs are usually superimposed on the urethra (Figure 7-37). When pulling the pelvic limbs cranially to assess the urethra in lateral radiographs, it is critical that the legs be pulled sufficiently forward so that no part of either pelvic limb remains superimposed on any portion of the urethra; otherwise, a superimposition artifact can lead to an incorrect assessment of a urethral calculus (Figure 7-38).

Another normal variant that can be confused with a urethral calculus is an opacity created when the os penis has more than one ossification center. These small ossification centers can be confused with a urethral calculus, an os penis fracture, or a foreign body (Figure 7-39).

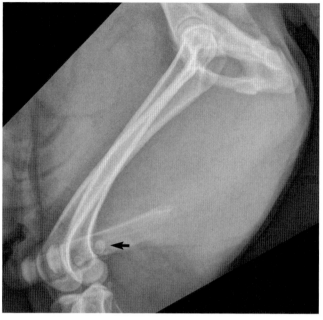

Figure 7-38. Lateral radiograph of the urethral region of a 10-year-old Standard Schnauzer made by pulling the pelvic limbs cranially, with the intent of achieving an unobstructed view of the urethral region to assess for urethral calculi. In this instance, the pelvic limbs were not pulled adequately cranially, and superimposition of one gastrocnemius fabella on the os penis creates an opacity *(solid black arrow)* that could be misinterpreted as a urethral calculus.

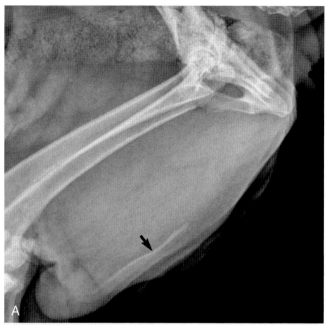

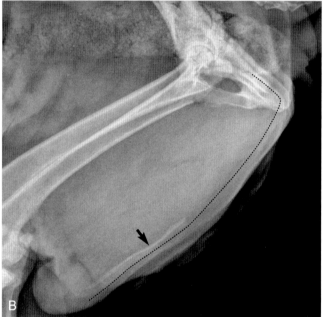

Figure 7-37. Lateral radiograph **(A)** and corresponding illustrated radiograph **(B)** of an 8-year-old Chow Chow suspected of having urinary calculi. The pelvic limbs have been pulled cranially to provide an unobstructed view of the urethra. The os penis is identified by the *solid black arrow.* The path of the normal urethra is shown by the *dotted line* in **B.** With this patient positioning, the *dotted line* represents the region that should be scrutinized carefully for radiopaque calculi.

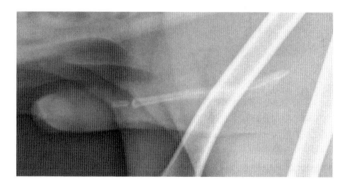

Figure 7-39. The os penis in this 7-year-old Soft Coated Wheaten Terrier has developed from two centers of ossification, a large one making up the majority of the structure and a small one located cranially. The small cranial ossification center could be confused with a foreign body, a fracture, or a urethral calculus, although this is an unusual location for a calculus. Sometimes, a small secondary center of ossification is located caudal to the main body of the bone, rather than cranially as illustrated here. In that instance, the appearance could be confused with a urethral calculus.

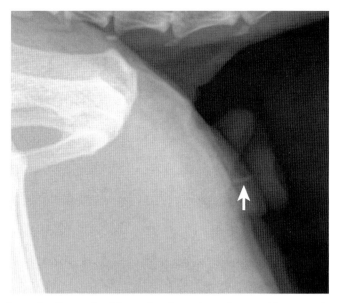

Figure 7-40. Lateral radiograph of the perineal area in an 8-year-old neutered male Domestic Shorthair cat. There is a faint linear opacity in the penis that is consistent with an os penis. This appearance could also be due to superficial debris or wet hair, or opaque urine sediment. This opacity was not proven to be an os penis, but this cat was urinating normally.

The os penis is very conspicuous in dogs, surrounding the penile urethra. Little mention is made of the presence of an os penis in a cat, but a small linear opacity is occasionally detected in the perineal area of male cats.[15] In a survey of 100 cats, an opacity consistent with a feline os penis was found in 19/50 (38%) cats imaged with computed radiography, and 8/50 (16%) cats imaged with an analog technology.[15] That the linear opacity was due to penile bone, however, was confirmed in only one cat. Subjectively, the reported incidence of a feline os penis seems much higher than typically observed clinically. Special attention has to be paid to the feline perineum for detection of an os penis as it is much less conspicuous than in dogs (Figure 7-40). A superimposition artifact, wet hair, opaque urine sediment or urethral mineralization could all be confused with a feline os penis. Therefore, careful evaluation of cats with a penile opacity is needed to distinguish between an os penis and these other conditions.

STOMACH

The stomach lies just caudal to the liver. The fundus is to the left of midline in the dorsal aspect of the abdomen; the pylorus is on the right in the ventral aspect of the abdomen; and the body spans the midline from left dorsal to right ventral, connecting the fundus and the pylorus. As noted earlier, the pylorus in the cat is located in a more midline position than in the dog (see Figure 7-10).

The stomach commonly contains both fluid and gas, and there is also often heterogeneous material in the stomach from a recent meal. Differentiating ingesta from foreign material is often not possible radiographically. The relative position of gas versus fluid in the stomach depends on the position of the patient during radiography (Figure 7-41). The redistribution of gas and fluid as a function of left versus right, and dorsal versus ventral, recumbency has a profound effect on the radiographic appearance of the stomach.

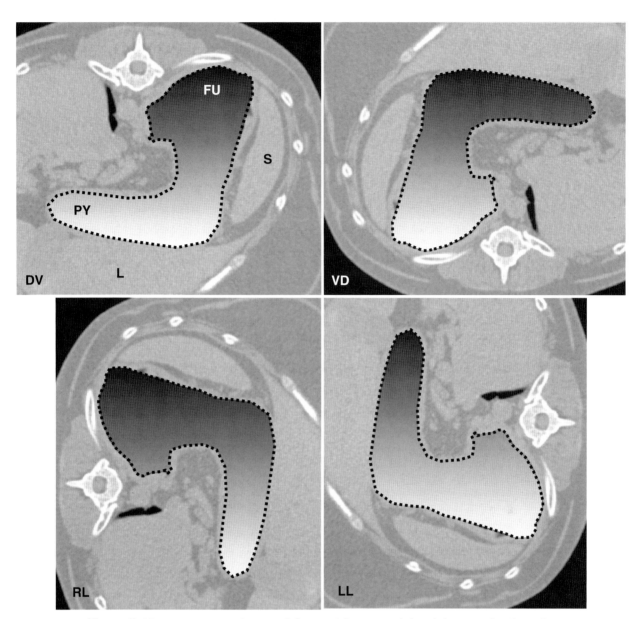

Figure 7-41. Transverse CT image of the cranial aspect of the abdomen of a dog. The stomach is outlined by a *dotted line*. The image has been rotated to illustrate the relative position of the gastric fundus *(FU)* and gastric pylorus *(PY)* in dorsoventral *(DV)*, ventrodorsal *(VD)*, right lateral *(RL)*, and left lateral *(LL)* radiographs. The lumen of the stomach has been shaded to illustrate the effect of gravity on redistribution of fluid *(white)* and gas *(black)* as a function of body position. This redistribution has a profound effect on the radiographic appearance of the stomach. The principle of this illustration is valid only if there is both gas and fluid in the stomach. The radiographic appearance of a stomach containing only gas or only fluid (or only ingesta) is less affected by body position. *L,* Liver; *S,* spleen.

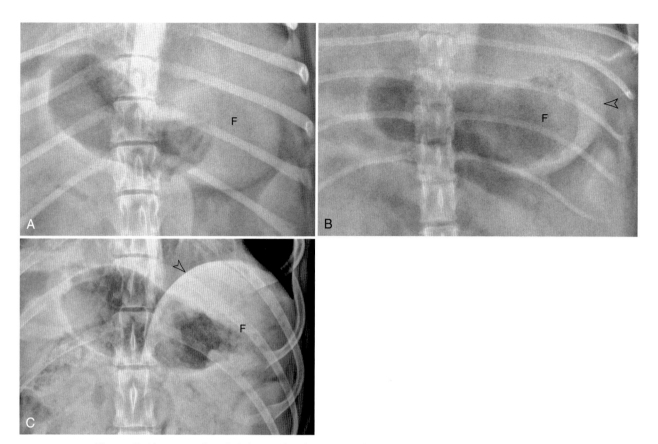

Figure 7-42. Ventrodorsal abdominal radiographs of a 6-month-old Labrador Retriever **(A)**, a 4-year-old Yorkshire Terrier **(B)**, and a 4-year-old Greyhound. Gas is present in the pylorus and body of the stomach in each of these dogs. In **A**, the fundus *(F)* is clearly seen and appears as a round opacity, being filled with homogeneous fluid. In **B**, the fundus has less fluid than in **A**; the fluid in the fundus is silhouetting with the gastric wall and creating a false impression of gastric wall thickening *(hollow black arrowhead)*. The thickness of the gastrointestinal wall cannot be accurately assessed radiographically in most patients due to luminal contents causing border effacement of the wall. In **C**, the material in the fundus is more heterogeneous and the serosal margin of the fundus is less clearly seen, creating more of a mass effect. Knowledge of the position of the fundus and that it usually contains fluid or food in the ventrodorsal view should reduce the possibility of interpreting the fundus as a soft tissue mass. In **C**, the curving line superimposed on the body of the stomach *(hollow black arrowhead)* is a margin of the diaphragm; this appearance is typical in ventrodorsal radiographs for a deep-chested dog such as this Greyhound.

In VD abdominal radiographs, fluid gravitates to the fundus and gas will be present in the pylorus and possibly the body, depending on the volume of gas (see Figures 7-41 and 7-42). The appearance of the fundus in a VD radiograph depends on the amount of adjacent fat to provide contrast and the character of the contents within the fundus (see Figure 7-42).

Even though dorsoventral (DV) abdominal radiographs are rarely made, it is important to understand the distribution of gas versus fluid in that position (Figure 7-41). In DV radiographs, fluid will gravitate to the pylorus and gas will rise to the fundus. The appearance of the body will depend on the relative amount of gas versus fluid and the absolute volume of both.

When the patient is in left recumbency, fluid gravitates to the fundus, and gas will collect in the pylorus (see Figures 7-41 and 7-43, *A-B*). For a clearly demarcated gas collection to be seen in the pylorus, the stomach must contain both gas and fluid. If the stomach is empty, or contains mostly food, the pylorus appears small and/or heterogeneous, respectively (see Figure 7-43, *C*).

When the patient is in right recumbency, fluid gravitates to the pylorus, and gas will collect in the fundus (see Figures 7-41 and 7-44, *A-B*). The fluid-filled pylorus can appear quite round and conspicuous; this appearance is commonly misinterpreted as either an abdominal mass or a gastric foreign body (see Figures 7-44, *A*, and 7-45).

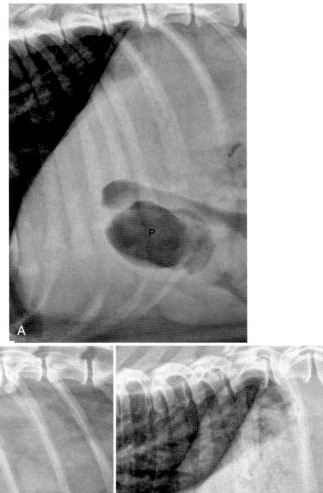
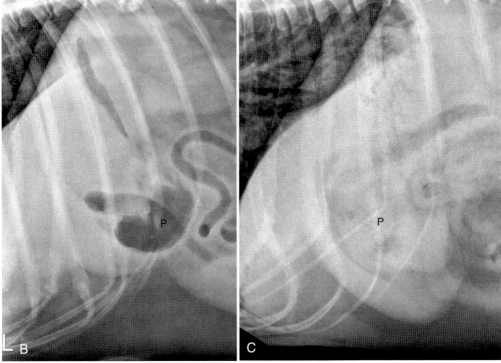

Figure 7-43. Left lateral radiographs of a 6-month-old Labrador Retriever (**A**), an 11-year-old Bluetick Coonhound (**B**), and a 9-year-old Dachshund (**C**). In **A** and **B**, there is obvious gas in the gastric pylorus *(P)*; the fundus appears different in these dogs. In **A**, the fundus is heterogeneous due to ingesta. In **B**, the fundus contains an oblong gas collection and likely some fluid, and there is heterogeneous material in the body. In **C**, a clearly demarcated gas collection is not present in the pylorus *(P)* because the stomach mostly contains ingesta and not a combination of fluid and gas.

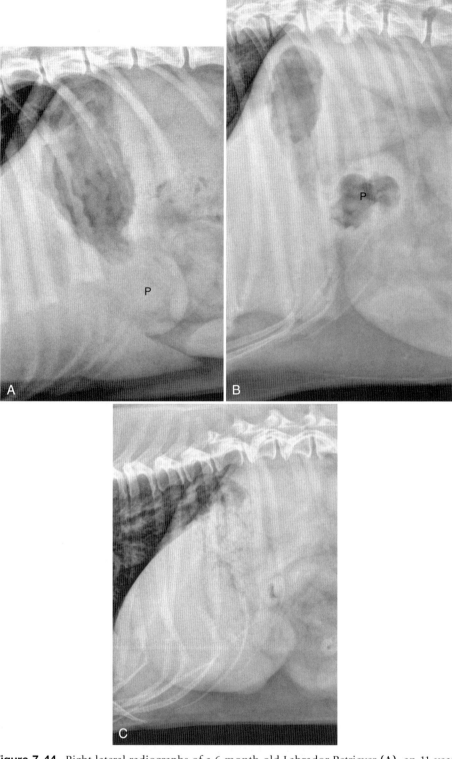

Figure 7-44. Right lateral radiographs of a 6-month-old Labrador Retriever (**A**), an 11-year-old Bluetick Coonhound (**B**), and a 9-year-old Dachshund (**C**). In **A**, fluid has filled the pylorus *(P)*, which appears as a distinct round opacity adjacent to the liver. In **B**, there is less fluid in the stomach, and the pylorus *(P)* contains gas, even in a right lateral view. In **B**, the pylorus is more dorsally positioned than expected, but there is no mass accounting for this; thus it is a normal variant. In **C**, there is mostly food in the stomach and little fluid or gas; thus the gastric lumen is heterogeneous in each lateral view, and the gravitational change in appearance is not seen in this patient. Note the gastric rugal folds in the fundus in **A** and in the body of the stomach in **B**.

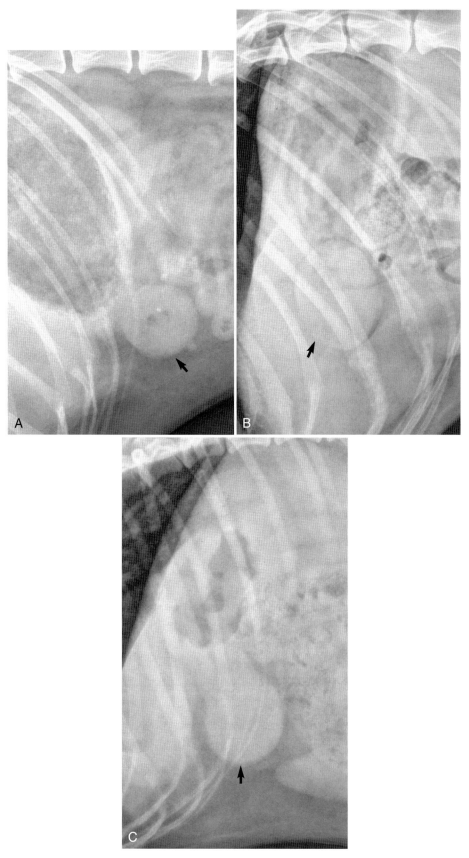

Figure 7-45. Right lateral radiographs of an 11-year-old Whippet (**A**), a 4-year-old Greyhound (**B**), and a 2-year-old Pug (**C**). In each of these patients, fluid fills the pylorus *(solid black arrow)*, creating a soft tissue mass effect that could easily be confused with an abdominal mass or a gastric foreign body. Note the gas in the fundus in each of these images, as expected in right recumbency. Also note the gastric rugal folds in the body of the stomach in **B**.

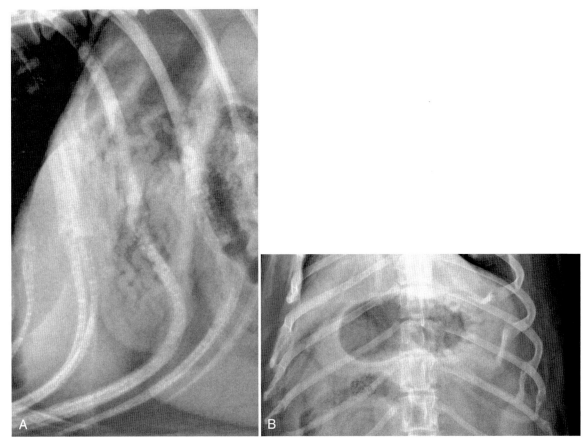

Figure 7-46. Right lateral (**A**) and ventrodorsal (**B**) radiographs of a 2-year-old Welsh Corgi. Note the conspicuous rugal folds in the body of the stomach.

For a clearly demarcated gas collection to be seen in the fundus and for fluid to gravitate to the pylorus in a right lateral radiograph, the stomach must contain both gas and fluid. If the stomach does not contain fluid and gas, a mass-like, fluid-filled pylorus will not be seen.

The mucosal surface of the stomach is characterized by folds, termed *rugal folds* or *rugae*. In survey radiographs, rugal folds can be seen only in portions of the stomach that contain gas (see Figures 7-44, *A-B*; 7-45, *B*; and 7-46). In general, rugal folds are more conspicuous in the fundus and/or body of the stomach than in the pylorus.

In the cat, rugal folds are also present in the gastric mucosa, but they are not seen as commonly as in dogs. One unique aspect of the feline stomach is the presence of submucosal fat.[16,17] This submucosal fat is present to some degree in all cats, but the amount is variable[1] (Figure 7-47). Occasionally, the submucosal fat is visible radiographically in VD radiographs. When the stomach is empty and the x-ray beam strikes a section of the fundus head-on, the fat creates striated radiolucent regions in the stomach wall (Figure 7-48).

The size and contents of the normal stomach can vary considerably. Occasionally, the amount of distension is alarming, but recent diagnostic interventional procedures, such as endoscopy or ingestion of large amounts of food, should be considered when interpreting the significance of a grossly distended stomach (see Figures 7-9 and 7-49). With physiologic or iatrogenic gastric distension, repeat radiographs every few hours should be characterized by notable change.

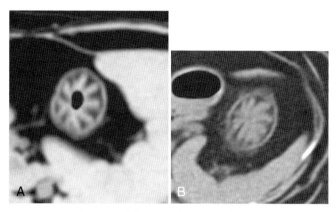

Figure 7-47. Transverse CT images of the cranial aspect of the abdomen from two cats. Note the hypoattenuating striated regions in the stomach wall created by submucosal fat. Gas is present in the fundus in **A**. The stomach is empty in **B**.

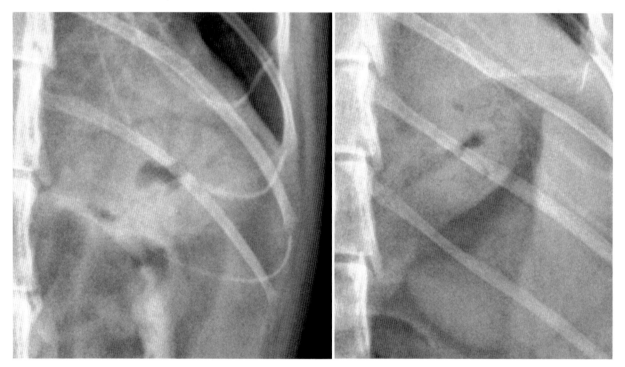

Figure 7-48. Ventrodorsal radiographs of two different cats. In each, there are striated radiolucencies in the wall of the stomach due to submucosal fat. Note that the stomach is empty in each cat; this is a requirement for visualization of the submucosal fat because the stomach must be contracted.

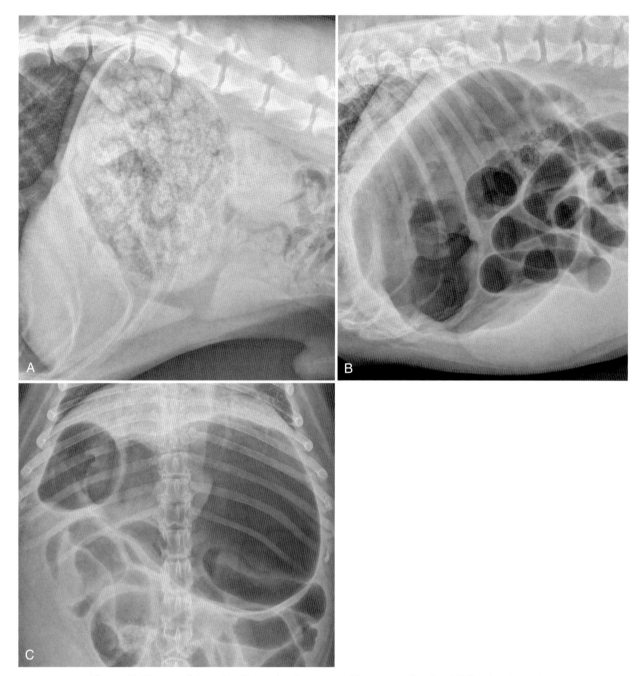

Figure 7-49. A, Left lateral radiograph of a 3-year-old German Shepherd following ingestion of a large amount of food. Right lateral (**B**) and ventrodorsal (**C**) radiographs of a 3-year-old Bulldog following endoscopy for removal of a gastric foreign body. The gas distension of the stomach and small bowel in **B** and **C** resulted from insufflation during endoscopy. In **B** and **C**, there is no evidence of gastric malpositioning; the fundus and pylorus are located in their normal positions.

SMALL INTESTINE

The position of the duodenum is relatively constant in the abdomen. The very initial part of the duodenum, the *duodenal bulb*, travels either cranially or laterally from the pylorus, depending on individual variation, before turning caudally as the descending duodenum. The descending duodenum courses caudally along the right lateral abdominal wall. Cranial to the pelvic inlet, the descending duodenum turns to the left as the caudal duodenal flexure. In the left abdomen, the duodenum then turns cranially and courses to the left of the root of the mesentery to the cranial aspect of the abdomen, where it joins the jejunum[1] (Figure 7-50). The diameter of the normal duodenum may be slightly larger than that of the normal jejunum, but this is rarely recognized radiographically.

The antimesenteric surface of the descending duodenum is characterized by mucosal depressions, called *pseudoulcers*, lying over submucosal lymphoid collections. In some dogs, pseudoulcers are large enough to be seen in upper gastrointestinal studies or when the duodenum contains gas[1,18] (Figures 7-51 and 7-52).

The jejunum continues from the duodenum and occupies the midportion of the abdomen as a mass of overlapping and folded tubular segments. The canine jejunum is rarely completely empty; typically it contains gas and fluid, but the range of normal bowel contents in the dog is wide (Figure 7-53). The finding of radiopaque material in the canine jejunum is not common but does occur in patients with dietary indiscretion or food engorgement, as well as with some radiopaque medications (see Figure 7-53, *D*).

Regardless of content, it is important to have an appreciation for the size of the normal jejunum. Various subjective rules have been proposed for assessing the diameter of normal jejunal segments.[19] (See *Textbook of Veterinary Diagnostic Radiology*[19] for these guidelines and the original

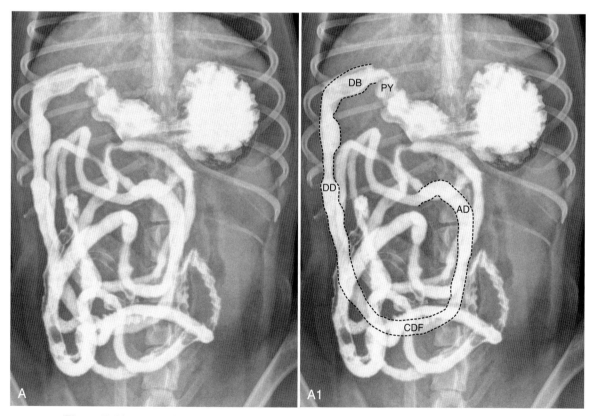

Figure 7-50. Ventrodorsal radiograph of a 7-year-old Miniature Schnauzer following administration of barium sulfate (**A**) and corresponding labeled radiograph (**A1**). Barium is present in the stomach, duodenum, and jejunum. In **A1**, the course of the duodenum is indicated by the *dotted line*. *AD*, Ascending duodenum; *CDF*, caudal duodenal flexure; *DB*, duodenal bulb; *DD*, descending duodenum; *PY*, pylorus.

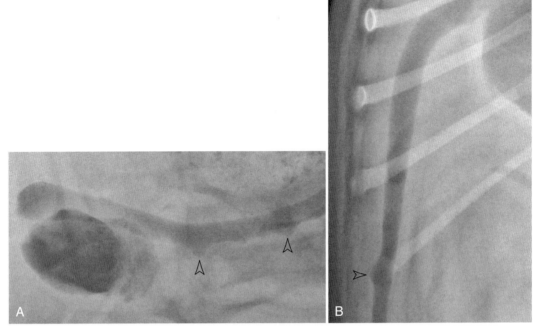

Figure 7-51. Left lateral (**A**) and ventrodorsal (**B**) radiographs of a 6-month-old Labrador Retriever. The duodenum contains gas, and outpouchings of the lumen can be seen in both views *(hollow black arrowheads)*; these are due to pseudoulcers.

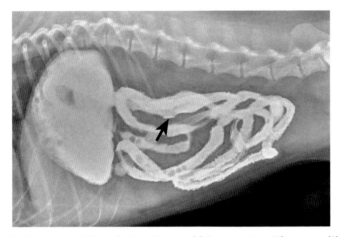

Figure 7-52. Barium upper GI study in a 6-year-old Pomeranian. The crater-like depression from the descending duodenum that is filled with barium *(black arrow)* is a pseudoulcer. The irregular, fimbriated appearance of the jejunum and duodenum is also a normal variant, commonly seen in barium upper GI studies.

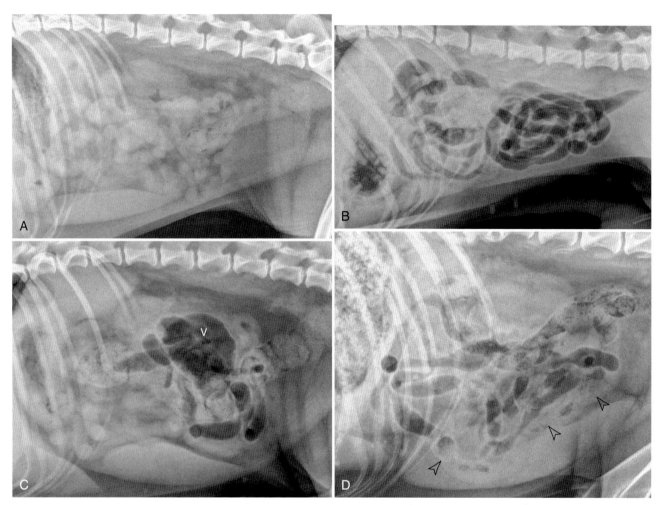

Figure 7-53. A range of normal canine jejunal radiographic appearances. **A,** A 5-year-old German Shepherd. The jejunum is mostly devoid of gas. Jejunal segments either are empty or contain a small amount of fluid. **B,** A 3-year-old Boxer. The jejunum is mainly gas-filled, but none of the segments are abnormally dilated. **C,** A 4-year-old Miniature Schnauzer. The jejunal segments contain either gas or fluid. The large gas-filled viscus *(V)* just dorsal to the fecal-containing descending colon is the normal cecum. **D,** A 13-year-old Border Collie. Occasionally, jejunal segments contain heterogeneous material rather than fluid or gas *(hollow black arrowheads)*. This can result from ingestion of nondigestible particulate matter or from failure of complete gastric digestion that might occur with food engorgement. Although this appearance can be normal, it can also signify disease and must be interpreted in concert with clinical signs and history. Repeat radiographs are often useful in such patients to assess the disposition of the jejunal contents.

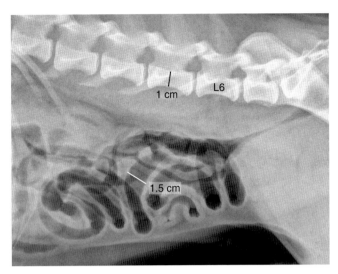

Figure 7-54. One method to quantify jejunal size is to express the diameter of the jejunal segment in question as a ratio to the height of the narrowest portion of the body of the fifth lumbar vertebra. This lateral radiograph is of a patient with jejunal gas and some modestly distended segments. One might be concerned whether the diameter is excessive. In this patient, the ratio of the jejunal diameter/vertebral height is 1.5/1 = 1.5. This is within normal limits according to published standards.[18] As noted in the text, no numeric standard is a perfect discriminator between normal and abnormal.

sources.) Most guidelines for assessing jejunal diameter are relative, comparing the size of the jejunum visually with some other structure. Based on one quantitative assessment, the diameter of a normal jejunal segment in the dog should not be larger than 1.6 times the height of the narrowest portion of the fifth lumbar vertebral body[18,20] (Figure 7-54). However, the value of making such measurements with regard to the diagnostic accuracy of suspected small intestinal obstruction has been questioned.[21] Therefore, caution must be used when employing radiographic measurement or comparison techniques to determine normal size, because none has a perfect predictive power for discriminating between normal and abnormal, and qualitative evaluation of jejunal diameter may be just as suitable. Also, jejunal size must be considered along with the clinical signs and history and radiographic features of bowel obstruction other than diameter.

Similar to the dog, the normal jejunal contents in the cat vary. However, the amount of bowel gas seen in survey abdominal radiographs may be less than commonly seen in the dog.[22,23] Although this may be true, the normal jejunum in the cat can contain gas and the presence of jejunal gas cannot be used alone as a criterion of abnormal (Figure 7-55). A feline jejunum devoid of gas may appear *bunched* or *centralized* in the midabdomen. Centralization of bowel in the midabdomen has been mentioned as a sign of a linear foreign body.[24] However, a normal empty bowel centralized in the midabdomen (see Figure 7-55, *A*) should not be confused with pathologic centralization due to plication, which is usually accompanied by other signs of a linear foreign body obstruction. The finding of radiopaque material in the feline jejunum is less common than in the dog, but ingestion of litter, foreign material, and some medications could result in the normal jejunum containing foreign material.

It is usually impossible to assess the thickness of the stomach or jejunal wall from survey radiographs, because intraluminal fluid silhouettes with the wall and creates a layer of homogeneous soft tissue opacity composed of the wall and adjacent fluid (Figure 7-56).

The jejunum terminates at the ileum in the right midabdomen. The jejunum and ileum are radiographically indistinguishable from each other.

LARGE INTESTINE

The ileum joins the ascending colon in the midaspect of the right hemiabdomen, where there is a distinct ileocolic sphincter. The dog has a distinct cecum, which is a blind-ended, coiled outpouching from the proximal part of the ascending colon. This configuration creates distinct ileocolic and cecocolic sphincters in the dog. The canine cecum is often gas-filled and appears as a coiled and/or compartmentalized gas-containing structure in the right midabdomen, usually visible in lateral and VD views (see Figures 7-53, *C*, and 7-57). There may not be an appreciable difference between the appearance of the cecum in

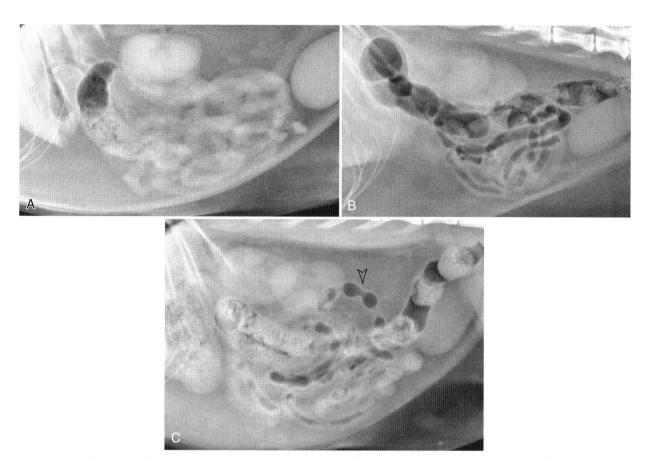

Figure 7-55. A range of normal feline jejunal radiographic appearances. **A,** A 1-year-old domestic cat. The jejunum contains no gas; a lack of jejunal gas is much more commonly encountered in the cat than in the dog. As a result of being empty, the jejunum in this cat appears centralized in the midabdomen. Centralized bowel has been associated with a linear foreign body as might occur as the bowel creeps along a linear foreign object. However, with a linear foreign body there will be abnormal gas bubbles in the bowel and not just centralization. Care must be taken not to misdiagnose a linear foreign body simply because the empty bowel appears centralized. Fecal material is present in the ascending colon in this cat. **B,** An 8-year-old domestic cat. Much of the jejunum contains gas. This is less common than in the dog, but as long as the jejunal diameter is normal, the finding of jejunal gas in a cat is not significant. Fecal material and gas are present in the descending colon in this cat. **C,** A 6-year-old domestic cat. The jejunum contains both gas and fluid. Some gas-filled portions of the jejunum are hypersegmented *(hollow black arrowhead)*. This hypersegmented appearance is due to peristalsis and has been termed the *string-of-pearls appearance.* This hypersegmentation is rarely seen in survey radiographs but is commonly seen in upper gastrointestinal examinations in cats, especially in radiographs made soon after contrast medium administration. This hypersegmented appearance should not be confused with the plication that occurs with linear foreign body. Hypersegmentation does not result in an alteration in the path of the bowel, whereas with plication, the bowel is often arranged in folds. In this cat, the stomach contains food, and fecal material is present in all portions of the colon.

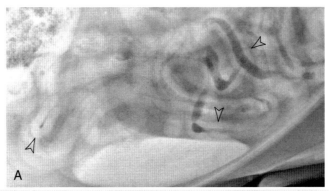
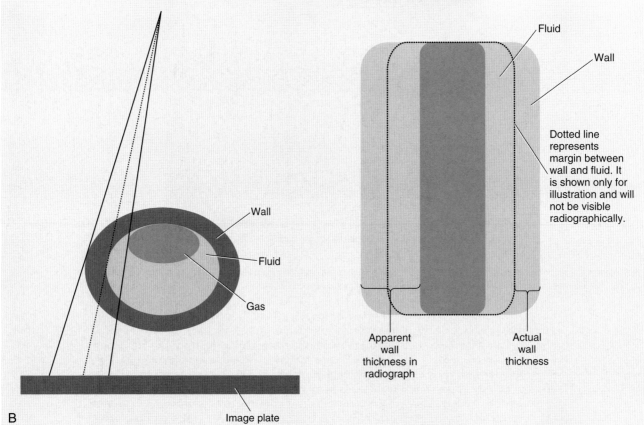

Figure 7-56. A, Radiograph of a 4-year-old Basset Hound. Some jejunal segments appear to have a thick wall *(hollow black arrowheads)*. Wall thickness can rarely be judged accurately from survey radiographs, because fluid in the lumen silhouettes with the wall, creating a layer of uniform opacity. This principle is illustrated in the drawing in **B.** A segment of intestine that contains fluid and gas is shown in the left side of the drawing; the gas will rise to the nondependent portion of the lumen. The *solid diverging black lines* in the left side of the drawing represent the trajectory of the oncoming x-ray beam that strikes the outer mucosal surface and the inner margin of the gas-fluid interface. The *dotted black line* represents the trajectory of the primary x-ray beam that strikes the mucosal-fluid interface. The wall of the jejunum and the fluid in the lumen will have the same radiographic opacity and will appear as one structure in the subsequent radiograph. The resulting radiograph is illustrated in the right side of the drawing. The wall of the bowel appears falsely thick due to the identical radiographic opacity of the bowel wall and the intraluminal fluid, which leads to border effacement between the two structures; that is, they silhouette with each other.

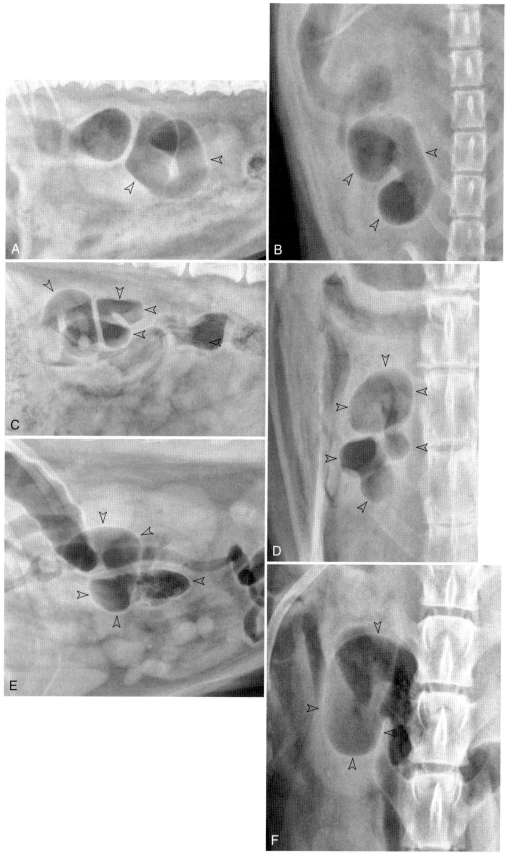

Figure 7-57. Left lateral (**A**) and ventrodorsal (**B**) radiographs of a 4-year-old Yorkshire Terrier; right lateral (**C**) and ventrodorsal (**D**) radiographs of a 9-year-old Standard Schnauzer; and right lateral (**E**) and ventrodorsal (**F**) radiographs of an 8-year-old Labrador Retriever. In each radiograph, the cecum is gas-filled and can be seen in the mid-aspect of the right abdomen *(hollow black arrowheads)*. Fecal material is present in the cecum in **E**.

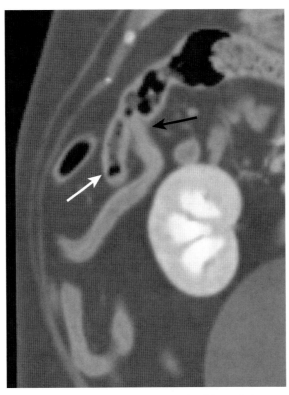

Figure 7-58. Dorsal plane post-contrast CT image of a feline abdomen at the level of the ileocolic junction *(black arrow)*. The feline cecum *(white arrow)* is a blind pouch that connects directly to the ascending colon.

left versus right lateral radiographs. The canine cecum rarely contains fecal material, but this can sometimes be seen (see Figure 7-49, E).

As in the dog, the ileum in the cat joins the ascending colon in the right midabdomen, where there is a distinct ileocolic sphincter (Figure 7-58). The cecum in the cat is not as developed as in the dog. In the cat, the cecum is a comma-shaped diverticulum from the ascending colon without a distinct cecocolic sphincter (Figure 7-58). Because there is no sphincter between the cecum and the ascending colon in the cat, the feline cecum can sometimes be identified in survey abdominal radiographs as the most proximal aspect of the colon at the ileocolic junction (see Figure 7-65).

In the dog, the colon begins at the ileocolic junction and courses cranially as the ascending colon. Just caudal to the pylorus, the colon turns to the left at the right colic flexure and then courses transversely across the abdomen, cranial to the root of the mesentery, as the transverse colon. Just medial to the spleen, the colon turns caudally at the left colic flexure and continues as the descending colon. The descending colon continues through the pelvic canal as the rectum (Figures 7-59 and 7-60). In general, the canine colon usually contains a combination

Figure 7-59. Right lateral (**A**) and ventrodorsal (**B**) radiographs of a 4-year-old Shetland Sheepdog, and corresponding labeled radiographs (**A1, B1**). In **A1**, a *question mark (?)* indicates a segment of large bowel that cannot be definitively identified due to superimposition; this may be ascending or descending colon. In **B1**, the course of the colon is outlined with a *solid black line*. It is not possible to trace the colon completely in every normal dog; visualization is a function of colon diameter and contents. In **A** and **B**, the cecum cannot be discerned; this is common. In **A1**, *solid black arrows* indicate a jejunal segment that contains heterogeneous material; finding heterogeneous material in the small bowel is not common (see the section on the small intestine). This occurrence makes small bowel difficult to differentiate from the colon in some patients. The demarcation of the cranial aspect of the pelvic canal has been designated in **A1** with a *solid black line*. Caudal to this point, the large bowel is termed the *rectum*. *AC*, Ascending colon; *DC*, descending colon; *LF*, left colic flexure; *PY*, gastric pylorus; *R*, rectum; *RF*, right colic flexure; *TC*, transverse colon.

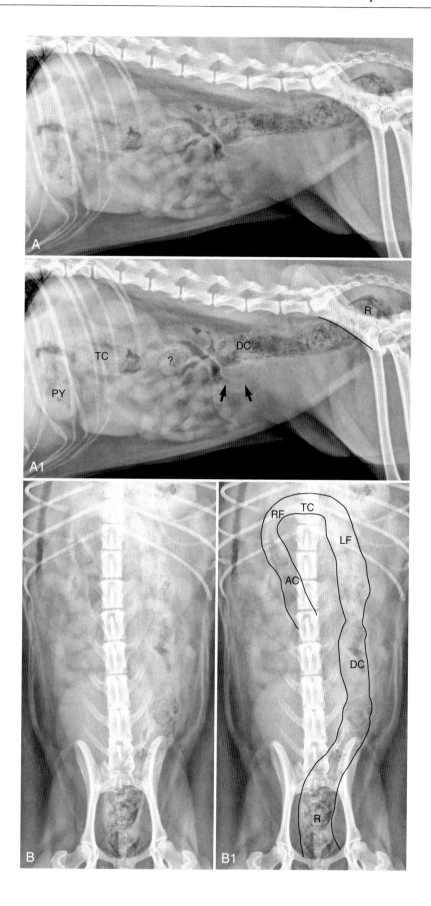

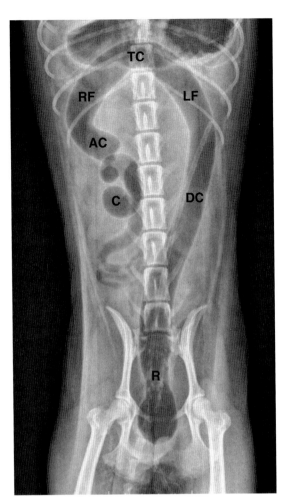

Figure 7-60. Gas has been introduced into the colon of this dog as an aid to definitively identifying colon and to distinguish between small and large intestine. A catheter is present in the descending colon. This *pneumocolonogram* has resulted in excellent depiction of the normal anatomy of the canine colon. *C*, Cecum; *AC*, ascending colon; *RF*, right flexure; *TC*, transverse colon; *LF*, left flexure; *DC*, descending colon; *R*, rectum.

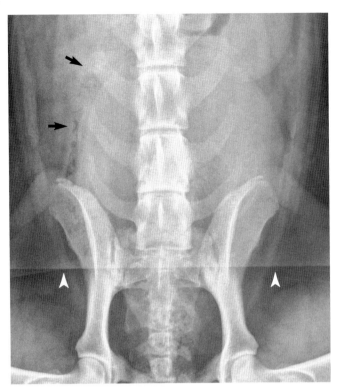

Figure 7-61. Ventrodorsal radiograph of a 7-year-old American Staffordshire Terrier. The urinary bladder is distended, and the descending colon *(solid black arrows)* is located in the right side of the abdomen; normally, the descending colon is on the left. A right-sided location of the descending colon is a commonly encountered normal variant when the bladder is distended. The *horizontal line (solid white arrowheads)* is due to the edge of the positioning trough being in the primary x-ray beam.

of gas and heterogeneously appearing fecal material (see Figure 7-59). Occasionally, the descending colon lies to the right of midline; this seems to be more commonly observed when the urinary bladder is distended. Unless an abnormal mass that is displacing the colon to the right can be identified, this is usually a normal variant (Figure 7-61).

In the dog, the colon is rarely completely filled with feces, but this is possible and can be normal, depending on the diameter. There are no quantitative guidelines for the normal size range of the colon in the dog, but the qualitative range of acceptable colon diameter is quite large. Normally, one does not encounter a fluid-filled colon in the dog. The normal canine colon can also be primarily empty, containing only gas. In these instances, peristalsis of the wall might be present and appear as multifocal regions of circumferential narrowing; this is a normal appearance and is not necessarily evidence for spasticity (Figure 7-62).

Occasionally, the normal colon in the dog is elongated such that folding occurs (Figures 7-63 and 7-64). This redundancy of the colon has no clinical significance but can complicate radiographic interpretation of the abdomen due to the increased volume of gas- and fecal-containing intestinal segments visible.

The colon in the cat is essentially the same as in the dog (Figure 7-65), except the ascending colon may be proportionally shorter. However, it is not unusual to find an ascending colon in the cat that is as comparably long as in the dog. The other difference in the feline large bowel, as noted previously, is the less developed cecum and the lack of a distinct cecocolic valve. The colon in the cat often contains fecal material and is rarely completely empty. As in the dog, finding a fluid-filled colon in the cat is not normal.

Figure 7-62. Right lateral (**A**) and ventrodorsal (**B**) radiographs of a 2-year-old Airedale Terrier. The colon mainly contains gas. Numerous regions of concentric narrowing of the colon are present. These are due to peristaltic contractions; their presence cannot be taken as evidence of hyperperistalsis or spasticity. A nonobstructive radiopaque foreign body is present in the colon near the right colic flexure. Any nonobstructive, nondigestible foreign body ultimately reaches the colon.

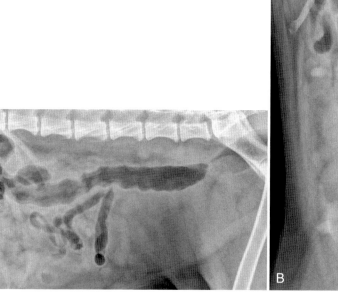
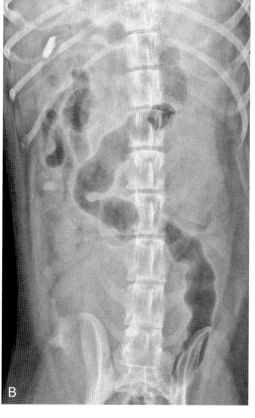

Figure 7-63. Right lateral (**A**) and ventrodorsal (**B**) radiographs of an 11-year-old Labrador Retriever. The colon is longer than typically encountered; this is a normal variant. This redundant colon leads to folds in the transverse colon that are gas-filled in this patient. The normal cecum is visible in its expected location.

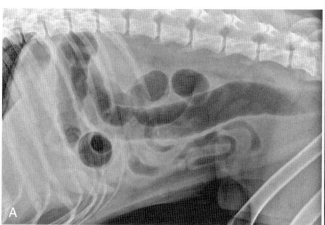
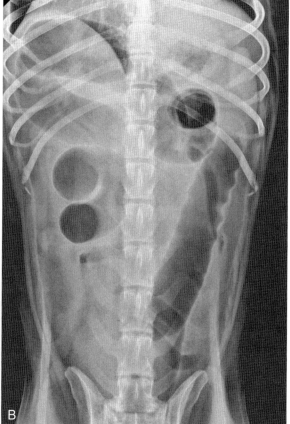

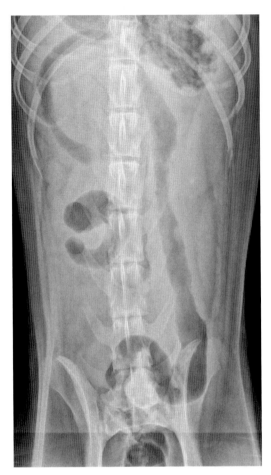

Figure 7-64. Ventrodorsal radiograph of a 2-year-old mixed breed dog. The redundancy of the descending colon has resulted in a sigmoid configuration of its distal aspect.

Figure 7-65. Right lateral (**A**) and ventrodorsal (**B**) radiographs of a 1-year-old domestic cat, and corresponding labeled radiographs (**A1, B1**). The approximate position of the cecum has been outlined with a *dotted black line* in **A1** and noted with *solid black arrows* in **B1**. This cat has a proportionally shorter ascending colon than in most dogs; this is a normal finding for cats. However, a proportionally shorter ascending colon is not always found in cats, and some cats have a longer ascending colon than shown here. In **B1**, the course of the colon is outlined with a *dotted black line*. It is not possible to trace the colon completely in every normal cat; visualization is a function of colon diameter and contents. The demarcation of the cranial aspect of the pelvic canal is designated in **A1** with a *solid black line*. Caudal to this point, the large bowel is termed the *rectum*. *AC,* Ascending colon; *DC,* descending colon; *LF,* left colic flexure; *R,* rectum; *RF,* right colic flexure; *TC,* transverse colon.

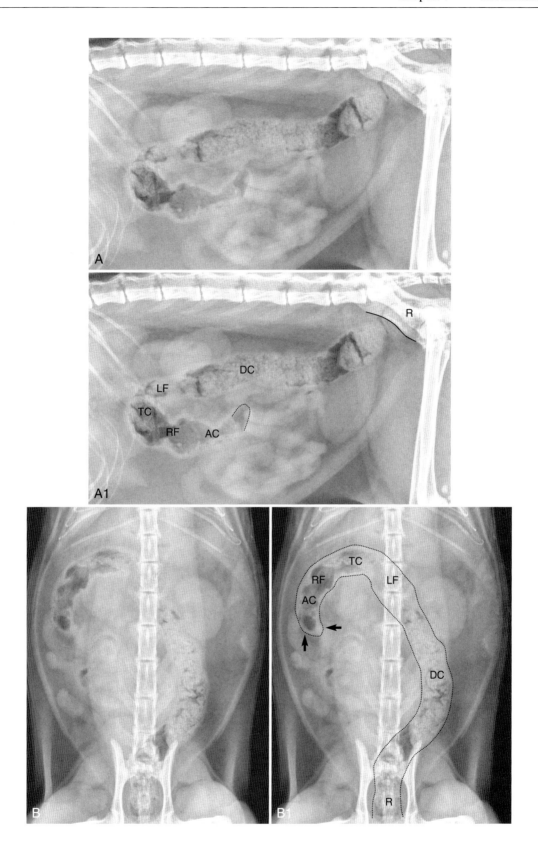

MISCELLANEOUS

The following structures, which are not mentioned elsewhere in the chapter, are usually not identifiable in radiographs of normal dogs: uterus, adrenal glands, mesenteric lymph nodes, and retroperitoneal lymph nodes. In obese cats, the colic lymph node can sometimes be visible as an oval opacity located dorsal to the descending colon (Figure 7-66). Also in cats, the adrenal glands will sometimes be mineralized and can be seen cranial to the kidneys (Figure 7-67). Although mineralization of the adrenal glands is an abnormality and not a normal variant, it typically has no clinical significance.

Many types of hemoclips applied during abdominal surgery are radiopaque and can be seen in subsequent radiographs. The most commonly encountered hemoclips are those used for ovariohysterectomy (Figure 7-68). Less commonly, hemoclips that were inserted during orchiectomy can sometimes be identified (Figure 7-69). Other types of metallic surgical implants can also be seen (Figure 7-70, *D*).

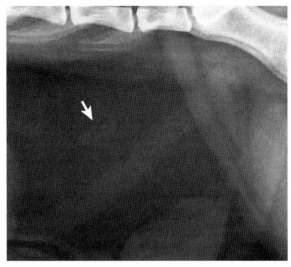

Figure 7-66. Lateral radiograph of the caudodorsal aspect of the abdomen of an overweight cat. The small nodular opacity dorsal to the descending colon is the colic lymph node. The normal colic lymph node can occasionally be seen in very obese cats.

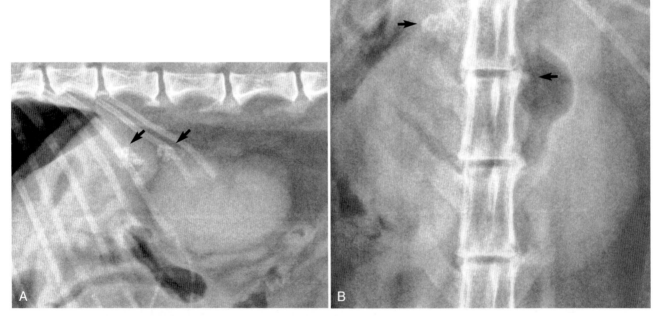

Figure 7-67. Lateral (**A**) and ventrodorsal (**B**) radiographs of an 8-year-old domestic cat. Each adrenal gland is mineralized *(solid black arrows)*. The left adrenal gland is not as conspicuous in the ventrodorsal view due to superimposition with the spine.

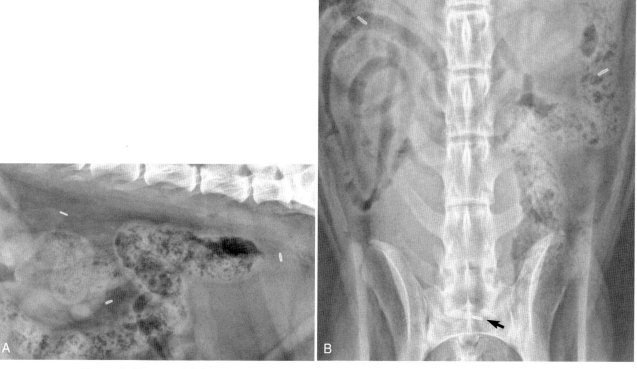

Figure 7-68. Lateral (**A**) and ventrodorsal (**B**) radiographs of a 10-year-old mixed breed dog. There are hemoclips near the caudal pole of each kidney and near the pelvic inlet that were placed by a surgeon during an ovariohysterectomy. In **B**, the hemoclip near the pelvic inlet is superimposed on the spine *(solid black arrow)*.

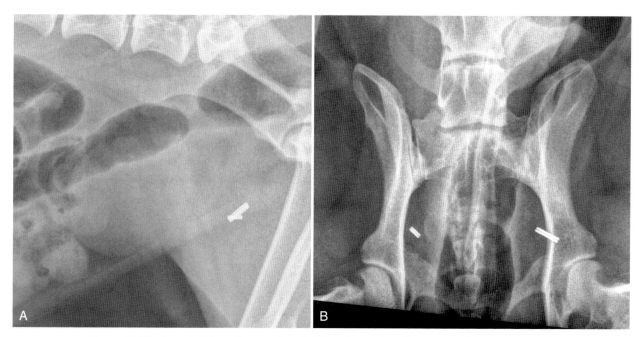

Figure 7-69. Lateral (**A**) and ventrodorsal (**B**) radiographs of an 8-year-old Basset Hound. There are two closely approximated surgical hemoclips in the left and right inguinal region that were placed around the spermatic cords during an orchiectomy procedure. The transected spermatic cords retracted through the inguinal canal, pulling the hemoclips to the level of the superficial inguinal ring.

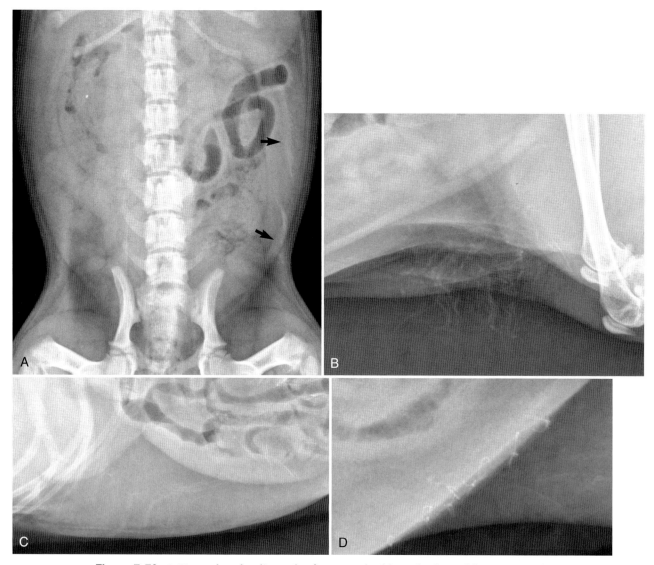

Figure 7-70. **A,** Ventrodorsal radiograph of a 4-month-old Pembroke Welsh Corgi. Residual ultrasound gel created amorphous opacities superimposed on the abdomen *(solid black arrows)*. The ultrasound-gel artifact identified by the more cranially located *arrow* is superimposed on the spleen and could be confused with a splenic lesion. **B,** Lateral radiograph of a 7-year-old Domestic Longhair cat. The streaks in the inguinal area are due to residual ultrasound gel. **C,** Lateral radiograph of a 10-year-old Golden Retriever. Residual ultrasound gel creates streaky opacities superimposed on fat in the ventral aspect of the abdomen that could easily be confused with a lesion, such as peritoneal fluid or steatitis. **D,** Same patient as in C. There are multiple small steel sutures in the ventral aspect of the abdominal wall that were placed at the time of ovariohysterectomy.

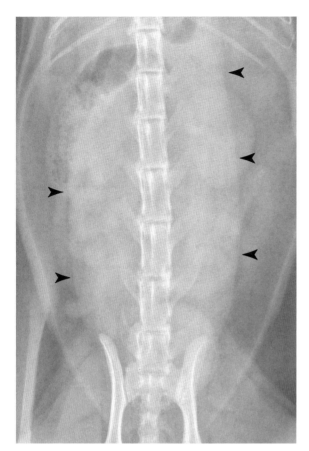

Figure 7-71. Ventrodorsal radiograph of a cat with pronounced retroperitoneal fat. The fat provides contrast for the margin of the paraspinal muscles *(solid black arrowheads)*.

Patients having abdominal ultrasonographic examinations prior to abdominal radiography are likely to have streaky radiopaque artifacts in the image from incompletely removed ultrasound gel or from wet hair following gel removal. Most patterns due to superimposed wet hair artifact are readily recognizable, but, occasionally, superimposition of the wet hair on the abdomen creates an unusual opacity that can be confused with a lesion (see Figures 7-33, *B*; 7-35, *B*; and 7-70).

In cats with moderate retroperitoneal fat, the hypaxial muscles become very conspicuous in VD radiographs due to the contrast provided by the fat (Figure 7-71). This opacity should not be confused with an abdominal mass.

REFERENCES

1. Evans HE, de Lahunta A: The digestive apparatus and abdomen. In *Miller's anatomy of the dog*, ed 4, St. Louis, 2013, Saunders.
2. Smallwood J: Digestive system. In Hudson L, Hamilton W, editors: *Atlas of feline anatomy for veterinarians*, Philadelphia, 1993, Saunders.
3. Thrall D: Intraperitoneal vs. extraperitoneal fluid. *Vet Radiol Ultrasound* 33:138-140, 1992.
4. Frank P: The peritoneal space. In Thrall D, editor: *Textbook of veterinary diagnostic radiology*, ed 6, St Louis, 2013, Saunders.
5. Bezuidenhout A: The lymphatic system. In Evans H, de Lahunta A, editors: *Miller's anatomy of the dog*, ed 4, St. Louis, 2013, Saunders.
6. Tompkins M: Lymphoid system. In Hudson L, Hamilton W, editors: *Atlas of feline anatomy for veterinarians*, Philadelphia, 1993, Saunders.
7. Valli VEO: The hematopoietic system. In Jubb KVF, Kennedy PC, Palmer N, editors: *Pathology of domestic animals*, vol 2, San Diego, 1985, Academic Press, p 216.
8. Rossi F, Rabba S, Vignoli M, et al: B-mode and contrast-enhanced sonographic assessment of accessory spleen in the dog. *Vet Radiol Ultrasound* 51:173-177, 2010.
9. Evans H, de Lahunta A, editors: *Miller's anatomy of the dog*, ed 4, St. Louis, 2013, Saunders.
10. Smith B: The urogenital system. In Hudson L, Hamilton W, editors: *Atlas of feline anatomy for veterinarians*, Philadelphia, 1993, Saunders.
11. Finco DR, Stiles NS, Kneller SK, et al: Radiologic estimation of kidney size of the dog. *J Am Vet Med Assoc* 159:995-1002, 1971.
12. Barrett RB, Kneller SK: Feline kidney mensuration. *Acta Radiol Suppl* 319:279-280, 1972.
13. Bartels JE: Feline intravenous urography. *J Am Anim Hosp Assoc* 9:349-353, 1973.
14. Shiroma J, Gabriel J, Carter T, et al: Effect of reproductive status on feline renal size. *Vet Radiol Ultrasound* 40:242-245, 1999.
15. Piola V, Posch B, Aghte P, et al: Radiographic characterization of the os penis in the cat. *Vet Radiol Ultrasound* 52:270-272, 2011.
16. Heng HG, Teoh WT, Sheikh Omar A: Gastric submucosal fat in cats. *Anat Histol Embryol* 37:362-365, 2008.
17. Heng HG, Wrigley RH, Kraft SL, et al: Fat is responsible for an intramural radiolucent band in the feline stomach wall. *Vet Radiol Ultrasound* 46:54-56, 2005.
18. O'Brien T, Morgan J, Lebel J: Pseudoulcers in the duodenum of the dog. *J Am Vet Med Assoc* 155:713-716, 1969.
19. Riedesel E: The small bowel. In Thrall D, editor: *Textbook of veterinary diagnostic radiology*, ed 6, St Louis, 2013, Saunders.
20. Graham J, Lord P, Harrison J: Quantitative estimation of intestinal dilation as a predictor of obstruction in the dog. *J Small Anim Pract* 39:521-524, 1998.
21. Ciasca TC, David FH, Lamb CR: Does measurement of small intestinal diameter increase diagnostic accuracy of radiography in dogs with suspected intestinal obstruction? *Vet Radiol Ultrasound* 54:207-211, 2013.
22. Morgan J: Upper gastrointestinal examination in the cat: normal radiographic appearance using positive contrast medium. *Vet Radiol Ultrasound* 22:159-169, 1981.
23. Weichselbaum R, Feeney D, Hayden D: Comparison of upper gastrointestinal radiographic findings to histopathologic observations: a retrospective study of 41 dogs and cats with suspected small bowel infiltrative disease (1985-1990). *Vet Radiol Ultrasound* 35:418-426, 1994.
24. Root C, Lord P: Linear radiolucent gastrointestinal foreign bodies in cats and dogs: their radiographic appearance. *Vet Radiol Ultrasound* 12:45-52, 1971.

Index

Page numbers followed by "f" indicate figures, and "t" indicate tables.

A

Abdomen, 241-295, 241f, 243f-245f
   kidneys, 257-262, 259f-264f
   large intestine, 282-288, 285f-291f
   liver, 245-248, 246f-250f
   miscellaneous, 292-295, 292f-295f
   orthogonal views of, 3t
   pancreas, 257, 258f-259f
   prostate gland, 267-268, 268f
   small intestine, 279-282, 279f-284f
   spleen, 248-257, 250f-258f
   stomach, 270-276, 271f-278f
   ureters, 265, 265f
   urethra, 268-270, 269f-270f
   urinary bladder, 265-267, 266f-267f
Abdominal aorta, 265, 265f
Accessory carpal bones, 120f, 122f-123f, 123-126
Accessory processes, of lumbar spine, 74, 75f-76f
Acetabular bone, 137, 137f-138f
Acetabular fossa, 140f, 141-142
Acetabulum, 138f, 141-142, 142f, 144f
Acromion, 91, 92f
Adrenal glands, 292, 292f
Aeration, thoracic radiography and, 184, 184f, 200, 201f
Age factors
   in antebrachium anatomy, feline, 17f
   in carpus anatomy, feline, 17f
   in crus anatomy, feline, 18f
   in elbow anatomy
      canine, 14f
      feline, 17f
   in hemipelvis anatomy
      canine, 15f
      feline, 18f
   in manus anatomy
      canine, 14f
      feline, 17f
   in ossification centers appearance, 9, 10t-11t
   in pes anatomy
      canine, 16f
      feline, 19f
   in shoulder anatomy
      canine, 13f
      feline, 16f
   in skull fusion, 13t
   in stifle anatomy
      canine, 15f
      feline, 18f

American College of Veterinary Radiology, 2-3
Amputation, of forelimb, pulmonary radiography and, 227, 229f
Analog imaging, of nasal cavity, 28-30
Anatomic directional terms, 2-3, 2f
Anconeal process, 108f, 110-111
Anesthesia, for thoracic radiography, 200, 215f
Angle, of jaw, 43-44
Antebrachiocarpal joint, 122
Antebrachium, 115-119, 118f-124f
   anatomy at specific ages in days, feline, 17f
   orthogonal views of, 3t
Anterior as directional term, 2
Anticlinal vertebra, 67, 70f
Aorta, 188f, 192f
   ascending, 206-208
   descending, 206-208
      dorsoventral view of, 196f
      ventrodorsal view of, 199f
Aortic arch, 219f
Articular process joints, 50, 51f
   caudal, 86
   cervical
      of atlas, 52
      of axis, 52
      between C5 and C6, 59f
      of C3-C5, 58-60, 59f-60f
   lumbar, 75f-76f
   thoracic, 68f-69f
Artifacts
   in pulmonary radiography, 227
   in thoracic radiography, 184, 185f
   ultrasound gel, 294f, 295
Arytenoid cartilage, 44-46, 46f
Ascending aorta, 206-208
Ascending colon, 282-286, 286f-288f, 290f-291f
Atelectasis, recumbency-associated, 184-187, 186f, 200f, 201
Atlanto-occipital articulation, 52, 53f
Atlantoaxial region, 54, 54f-55f
Atlas, 52
   age at which ossification center appears in, 10t-11t
   age when physeal closure occurs in, 12t
Atrium, 219f
   left, 216, 219f

Auricle, 219f
   left, 219f
Axis, 52
   age when physeal closure occurs in, 12t

B

Basihyoid bone, 44-46, 47f
Benign prostatic hypertrophy, 267-268, 268f
Biceps tendon, 100, 100f
Bicipital tenosynovitis, 100, 100f
Bladder, urinary, 265-267, 266f-267f
   distended, radiography and, 286-288, 288f
Blade of scapula, 91, 93f
Block vertebra, 60, 60f-61f
   sacrum as, 81
Brachium, 91-103, 103f-106f
   orthogonal views of, 3t
Brachycephalic breeds, 21
   esophagus in, 215, 216f
   frontal sinus radiography in, 30, 32f
   mediastinum in, 208, 209f-210f
   trachea in, 210, 211f
Brachygnathic mandibles, 44, 45f
Bronchial markings, 236, 236f
Bronchus (bronchi), 188f, 190, 190f-191f, 210-213, 211f-213f, 227
Bulldogs, tracheal anatomy of, 213

C

Calcaneus, 176-179, 181f
Calculi
   ureteral, misinterpreted as, 265, 265f
   urinary, 269f
      radiolucent, 265-266
Canine teeth, 21
   eruption of, 21t-22t
   retention of deciduous, 28f
Capital epiphysis, 153
Cardiomegaly, misdiagnosis of, 215-216, 217f-218f, 226
Carina, 188f, 210-211
Carnassial teeth, 21, 24f-25f
Carpal pad, 119, 124f
Carpometacarpal joint, 123-126, 126f
Carpus, 122-126, 122f-123f, 125f-128f
   age at which ossification center appears in, 10t-11t

Carpus (Continued)
  anatomy at specific ages in days, feline, 17f
Cartilage, laryngeal, 44-46, 46f
Catheters, urinary, 266, 266f-267f
Caudal articular processes, 50-51, 51f
Caudal as directional term, 2
Caudal aspect, of cranial vault, 39-43, 43f
Caudal caval reflection, 206
Caudal cervical esophagus, 211-213, 213f
Caudal lobar pulmonary artery, 219f-221f, 220
Caudal lobar pulmonary vessels, 199f
  dorsoventral view of, 196, 197f-198f
  ventrodorsal view of, 199-200, 200f
Caudal mediastinal lymph nodes, 208
Caudal spine, 86, 87f-88f
  age when physeal closure occurs in, 12t
Caudal vena cava, 187, 188f-190f, 193f, 199f, 206, 208f
  in abdominal computed tomography, 265, 265f
Caudal view of pelvis, canine, 149f
Caudocranial images, hanging of radiograph, 3
Caudoventral mediastinal reflection, 206, 207f-208f
Cecocolic sphincters, 282-286
Cecum, 282-286, 285f-291f
Centers of ossification
  age of appearance, 9, 10t-11t
  in cervical spine
    atlas, 52
    axis, 57
  of elbow, 110-111
  os penis, 268, 269f-270f
  of pelvis, 143f, 145-147
  scapular, 91, 94f
    of supraglenoid tubercle, 101, 102f
  tibial, 169
Centrum, 51-52
  of proatlas, 54f, 57, 57f
Centrum 2, 57, 57f
Ceratohyoid bones, 44-46, 47f
Cervical ribs, 64, 65f-66f
Cervical spinal nerves, 52-54
Cervical spine, 52-64, 53f-67f
  age when physeal closure occurs in, 12t
Chondrodystrophoid breeds
  antebrachial of, 118-119, 121f
  elbow of, 111, 113f
  femur of, 166, 166f
  heart of, 220, 220f
  humerus of, 103, 106f
  rib anatomy of, 203, 203f
Clavicle, 91, 95f-96f

Clavicular remnants, 91, 95f
Clock-face analogy, 216, 219f, 221f
Closure, physeal, 9, 10t-13t, 13f-19f
Cochlea, tibial, 174, 175f
Colon. see Large intestine
Computed tomography (CT)
  of abdomen, 265f, 271f
  of nasal cavity, 32f
  of pancreas, 258f
  of pelvis, 137f
  of spine, 51f
  as tomographic imaging modality, 4
Condylar process, of mandible, 36f, 37, 43-44
Condyle
  femoral, 155, 158f, 159, 166f
  flattened appearance of, 161f
  humeral, 107-109, 108f-109f
  tibial, 169
Congruency of the coxofemoral joint, 141-142, 142f, 147, 149f
Coracoid process, 91, 93f, 96f
Coronoid process
  of elbow, 108f, 111, 113f
  of mandible, 33-36, 35f, 38f, 43-44
Costal arch, 201, 203
  liver size and, 245, 247f-248f
Costal cartilage, 201, 202f
Costochondral junction, 201, 202f-203f, 229f
Coxofemoral joint, 145f, 150f, 153
  congruency of, 141-142, 142f, 147, 149f
Cranial 45-degree lateral-caudomedial view, 7t
Cranial 45-degree medial-caudolateral view, 7t
Cranial articular process, 50-51, 51f
  of C3-C5, 58-60, 59f-60f
Cranial as directional term, 2
Cranial lobe pulmonary vessels, 193-194, 194f
Cranial mediastinal arteries, 206-208
Cranial mediastinal lymph nodes, 208
Cranial vault, 30-33
  caudal aspect of, 39-43, 43f
Cranial vena cava, 206-208, 209f-210f
Craniocaudal views, 5t
  of elbow, 107-109, 108f, 111
  of femur, 154f
  hanging of, 3
Cranioventral mediastinal reflection, 205-206, 206f
Cribriform plate, 30f
Cricoid cartilage, 44-46, 46f
Crura, diaphragmatic, 187, 193-194, 193f, 237
Crus. see also Tibia and fibula
  anatomy at specific ages in days, feline, 18f

Crus (Continued)
  of diaphragm, 187, 188f-189f, 193-194, 193f, 238f
  orthogonal views of, 3t
CT. see Computed tomography (CT)
Cuboidal carpal bones, 122
Cutback zone, 114f, 118-119, 121f

D

DAR view. see Dorsal acetabular rim (DAR) view
Deciduous teeth, 21, 21t-22t, 25-28, 26f-28f
Declawed cat, 134f
Deep circumflex iliac arteries, 265, 265f
Deltoid tuberosity, 103, 104f
Dens, 54, 55f
Dental malocclusion, 44, 45f
Dentition, 21-28, 21t-22t, 24f-28f
Descending aorta, 206-208
  dorsoventral view of, 196f
  ventrodorsal view of, 199f
Descending colon, 286-288, 286f-288f, 290f
Dewclaw, 129-131, 129f, 176-179, 179f
Diaphragm, 188f-189f, 193f, 237-240, 237f-240f
  dorsoventral view, 195-196, 196f
  left lateral view of, 187
  right lateral view of, 193-194
  ventrodorsal view of, 199-200, 200f
Digital imaging
  of nasal cavity, 28-30, 31f
  of thorax, 184, 185f
Digits, 129-131, 129f
Directional terms, anatomic, 2-3, 2f
Disc. see Intervertebral disc
Divergence, of x-ray beam, in spinal radiography, 50, 50f
Dolichocephalic breeds, 21
  lungs of, 234, 235f
  temporomandibular joint and tympanic bullae radiography of, 36-37, 37f
Dorsal 45-degree lateral-palmaromedial view, 7t
Dorsal 45-degree lateral-plantaromedial view, 7t, 8f-9f
Dorsal 45-degree medial-palmarolateral view, 7t
Dorsal 45-degree medial-plantaromedial view, 7f-8f, 7t
Dorsal acetabular rim (DAR) view, 142, 144f
Dorsal as directional term, 2, 2f
Dorsal tracheal membrane, 211-213, 213f
  sagging of, 211-213
Dorsopalmar view, 5t

Dorsoplantar view, 4-6, 5f, 5t
Dorsoventral view. see also Ventrodorsal view
  of frontal sinuses, 30-33, 32f
  hanging of radiograph, 3
  of nasal cavity, 30-33, 32f
  of skull, 21, 22f-23f
  of temporomandibular joint, 36-37, 36f
  of thorax, 184, 187f, 195-196, 196f-198f
Double dewclaw, 176-179, 179f
Down pathology rises phenomenon, 193-194, 195f
Duodenal bulb, 279
Duodenum, 279, 279f-280f

E

Ectoturbinates, 28-30
Elbow, anatomy at specific ages in days
  canine, 14f
  feline, 17f
Elbow joint, 107-115, 107f-117f
End-on projection, 250, 250f
End-on pulmonary vessels, 227
  lung nodule versus, 227, 228f-229f, 231f-232f
Endotracheal tube
  removal of, before tympanic bullae radiography, 39, 42f
  superimposition of, over shoulder joint, 97-98, 98f
Endoturbinates, 28-30
Enthesopathy, radial, 118f
Epicondyle, of humerus, 107-109, 108f
Epiglottis, 44-46, 46f
Epihyoid bones, 44-46, 47f
Epiphysis
  capital, 153
  femoral, 159
  proximal humeral, 101, 101f-102f
  proximal tibial, 160f, 169
  vertebral, 51-52, 52f
  of C2, 57, 57f
Eruption, dental, 21t-22t
Esophagus, 206-208, 209f, 214-215, 214f-216f
  caudal cervical, 211-213, 213f
Eventration, diaphragmatic, 239, 239f
Exposure time, in thoracic radiography, 184
Extension, in lumbar spine, 77, 78f
Extensor fossa, 159

F

Facet, 50-51
Falciform ligament, abdominal radiography and, 242-245, 244f-245f

Fat, 209f
  abdominal radiography and, 242, 243f-245f
  in the heart, 225f
  in kidneys, 257, 259f-260f, 262
  in prostate gland radiography, 267-268, 268f
  in redundant mediastinal refection, 228, 233f
  retroperitoneal, 295, 295f
  submucosal, 276, 277f
  surrounding pancreas, 258f
Fat pad, infrapatellar, 158f, 163
Fecal material
  in cecum, 285f
  kidneys and, 262f-263f
  misdiagnosed as fracture
    in pelvic radiography, 137, 139f
    in sacral radiography, 85f
Femoral head, 153, 153f
Femur, 153-167, 153f-159f, 162f, 166f-167f
  age at which ossification center appears in, 10t-11t
  age when physeal closure occurs in, 12t
  feline, 167, 167f
Fibula. see also Tibia and fibula
  age at which ossification center appears in, 10t-11t
  age when physeal closure occurs in, 12t
Film-screen combination for thoracic radiography, 184, 185f
Fissures, pleural, 227-228, 233f-234f
Flexed lateral view, of elbow joint, 107-111, 107f-108f, 111f
Flexion
  carpal, 122
  in lumbar spine, 77, 78f
Fluid
  gastric, 270
  in jejunum, 281f, 283f
  pleural, 233f
  stomach, 271f-275f
Foramen
  cervical, 62-64
    of atlas, 52, 53f
    of axis, 56f
    of C6, 62, 63f
  lumbar, 75f-76f
  thoracic, 68f-69f
Foramen magnum, 39-43, 43f
Foreign body
  colonic, 289f
  jejunal, 282, 283f
Forelimb, amputation of, pulmonary radiography and, 227, 229f

Fossa
  acetabular, 141-142, 142f
  intercondylar, 155, 159f
  radial, 109-110
Fovea capitis, 153, 153f
Fractures, fecal material misdiagnosed as
  in pelvic radiography, 137, 139f
  in sacral radiography, 85f
Frog-leg projection, 137
Frontal bone, age at fusion of, 13t
Frontal sinus, 30-33, 30f-34f
Fundus, 270, 271f-275f
Fusion
  proximal humeral epiphysis of, 101, 102f
  of skull, 9
  of tibial tuberosity, 171f

G

Gallbladder, 248, 249f-250f
Gas
  in cecum, 282-286, 285f
  in colon, 288, 288f-289f
  in duodenum, 280f
  in esophagus, 209f, 214, 215f
  in fundus, 193f
  in gastrointestinal tract, 242
  in jejunum, 281f-283f, 282
  in pylorus, 189f
  in stomach, 270, 271f-275f
Gastric axis, 245, 246f
Gastrocnemius fabella, medial and lateral, 161-163
General anesthesia, for thoracic radiography, 200, 215f
Glenoid cavity, 91, 93f
Greater trochanter
  of femur, 153-155
  physis of, 155
Greater tubercle, of scapula, 93f

H

Hamate process, 91, 94f
Hanging film, of radiographs, 3
Hard palate, 21
Head of spleen, 248
Heart, 215-226, 217f-220f, 222f-226f
  dorsoventral view of, 195-196, 196f
  left lateral view of, 192f
  in mediastinal radiography, 206-208, 208f
  right lateral view of, 193-194, 193f-194f
  ventrodorsal view of, 199-200, 199f
Hemal arches, of caudal spine, 86, 87f
Hemipelvis, anatomy at specific ages in days
  canine, 15f
  feline, 18f
Hemivertebra, 71, 72f-73f, 86, 88f

Hemoclips, 292, 293f
Heterotopic bone, pulmonary, 234, 235f
Hiatal hernia, sliding, 237, 238f
Hiatus, of diaphragm, 237
Humeral condyle, 107-109, 108f-109f
Humeral epiphysis, 101, 101f-102f
Humeral head
  osteochondrosis in, 99-100, 99f
  in proximal humeral epiphysis, 101
  superimposition of, in scapular radiography, 97-98, 97f
  triangular region of fat at caudal aspect of, 103, 103f
Humeral shaft, 103, 103f
Humerus, 103, 103f-106f
  age at which ossification center appears in, 10t-11t
  age when physeal closure occurs in, 12t
  lucency at base of greater tubercle, 101, 101f
Hyoid bone, 53f
Hypertrophy, benign prostatic, 267-268, 268f
Hypoplasia, of xiphoid process, 203-205, 205f

I
Identification microchip, 68f-69f
Ileocolic junction, 286-288
Ileocolic sphincter, 282-286
Ileum, 282-286
Iliac crest, 75f-76f, 82f, 137-141, 141f
Ilium, 137, 138f, 141f
Imaging principles, basic, 1-19, 1f
Incisors, 21
  eruption of, 21t-22t
Infrapatellar fat pad, 158f, 163
Inspiration, thoracic radiography during peak, 184, 184f
Intercarpal joint, 123-126, 126f
Intercentrum 1, 52, 54f-55f, 57f
Intercentrum 2, 54f, 57, 57f
Intercondylar fossa, femoral, 155, 159f
Intersternebral cartilage, 203-205
Interstitial markings, in lung radiography, 227
Intervertebral disc, 50
Intervertebral disc spaces, 50, 50f
  between C5-C6, 63f
  between L2 and L3, 75f-76f
  between T4 and T5, 68f-69f
Intervertebral foramen, 51, 51f
  cervical, 62
    of atlas, 52, 53f
    of C6, 62
  lumbar, 75f-76f
  thoracic, 68f-69f

Intestine
  large, 282-288, 285f-291f
  small, 279-282, 279f-284f
Intraoral radiography, 28-30, 30f
Intrathoracic trachea, 210, 211f
Ischial arch, 148f-149f
Ischiatic tuberosity, 145-147
Ischium, 141-142, 145, 147f

J
Jaw, angle of, 43-44
Jejunum, 279, 279f-281f
Juvenile orthopedic disorders, physeal closure and, 9

K
Kidneys, 257-262
Kyphosis, 73

L
Lamina, 50, 51f
  of cervical spine, 53f
  of lumbar spine, 75f-76f
Lamina dura, 25, 27f
Large intestine, 282-288, 285f-291f
Larynx, 31f, 43-46
Lateral condyle, femoral, 155
Lateral malleolus, 175f, 176, 177f
Lateral view, 6, 6f
  of elbow joint, 107-111, 107f-108f, 111f-112f
  of frontal sinuses, 30-33, 32f
  hanging of radiograph, 3
  of kidneys, 262, 264f
  of nasal cavity, 30f
  of pelvis, 137, 138f-141f
  of scapula, 97-98, 97f
  of shoulder joint, 97-98, 98f
  of skull, 21, 22f-23f
  of spine, 50, 50f
  of spleen, 254, 254f-257f
  of temporomandibular joint, 36-37, 37f
  of thorax, 183-184, 183f-185f, 188f
    left, 187-192, 188f-190f, 192f, 195f
    right, 193-194, 193f, 195f
  of tympanic bullae, 37, 37f, 39, 39f
Lateromedial images, hanging of radiograph, 3
Left atrium, 216, 219f
Left auricle, 219f
Left caudal lobar pulmonary artery, 219f-221f, 220
Left lateral view
  of kidneys, 262, 262f-264f
  of spleen, 254, 255f-257f
  of thorax, 183-184, 183f-184f, 187-192, 188f-190f, 192f, 195f
Left-right view, 193-194

Left subclavian artery, 206-208, 209f-210f
Left ventricle, 219f
Lesser trochanter, of femur, 153-155, 155f
Lesser tubercle, of scapula, 93f, 100, 100f
Ligament, of dens, 54
Limbs, positioning of, in abdominal radiography, 242
Liver, 245-248, 246f-250f
  kidneys and, 260-262
  spleen and, 254, 254f
Lobes, lung, 210-211, 211f-212f
Lumbar spine, 74-81, 75f-81f
  age when physeal closure occurs in, 12t
Lung aeration, thoracic radiography and, 184, 184f, 200, 201f
Lung lobes, 210-211, 211f-212f
Lungs, 186f, 227-236, 228f-235f
  left lateral view of, 190-192
  right lateral view of, 193-194, 193f-194f
  ventrodorsal view of, 199-200, 200f
Lymph nodes
  mediastinal, 208, 209f
  in popliteal lymphocenter, 166f

M
Mach effect, 172-174
Magnetic resonance imaging (MRI), 4
Main pulmonary artery, 219f
Malalignment, 137
Mandible, 43-46, 43f
  age at fusion of, 13t
  condylar process of, 36f, 37, 43-44
  coronoid process of, 33-36, 35f, 38f, 43-44
  dental arcade radiography of, 25, 27f
Mandibular symphysis, 43-44, 43f
Manubrium, 203-205
Manus, 129-131, 129f-134f
  anatomy at specific ages in days
    canine, 14f
    feline, 17f
  orthogonal views of, 3t
Manx cats, lumbar and sacral region of, 84, 85f-86f
Maxilla, 25, 27f
  dentition in, 21, 24f
  intraoral radiograph of, 30f
Maxillomandibular malocclusion, 44, 45f
Maximum Intensity Projection (MIP), 137f, 145f
Medial condyle, femoral, 155
Medial malleolus, 174, 175f, 177f-178f, 180
Mediastinum, 205-210, 206f-210f

Mediolateral images, hanging of radiograph, 3
Meniscal ossicles, 169f
Meniscus, lateral, 163
Mental foramina, 43-44, 44f-45f
Mesaticephalic breeds, 21
  temporomandibular joint and tympanic bullae radiography of, 36-37, 37f
Metacarpal bones, 122, 122f-123f, 125f, 129-131
Metacarpal pad, 129f, 131
Metacarpophalangeal joints, 129-131, 129f-130f
Metacarpus
  age at which ossification center appears in, 10t-11t
  age when physeal closure occurs in, 12t
Metaphysis
  tibial, 169-172
  ulnar, 118-119, 121f
Metaplasia, pulmonary osseous, 234, 235f
Metatarsus, 179
  age at which ossification center appears in, 10t-11t
  age when physeal closure occurs in, 12t
Microchip, identification of, 68f-69f, 92f
Mineralization
  of adrenal glands, 292, 292f
  of aortic root, 226, 226f
  pattern of, in costal cartilage, 201, 202f
Molars, 21
  eruption of, 21t-22t, 24f
  tooth bud for, 25-28, 27f
MRI. *see* Magnetic resonance imaging (MRI)
Multiple radiolucent lines, 139f

N

Nasal cavity and sinuses, 28-36, 29f-36f
Nasal organ, 29f
Nasal septum, 29f-30f
Nasopharynx, 30, 31f
Nerves, spinal, 52-54
Neural arches, 50
  of atlas, 54f
  of C2, 59f
Neutral lateral view, of elbow joint, 107-109, 107f, 111, 112f
No joint, 54
Nomina Anatomica Veterinaria, 2-3, 2f
Normal images, 2
Nose up technique, in temporomandibular joint and tympanic bullae radiography, 36-37, 38f

Nutrient canal, of humerus, 103, 105f
Nutrient foramen, 118f
  of humerus, 104f
  of tibia, 172-174, 173f-174f

O

Obesity, cardiac radiography and, 215-216
Oblique projections, 3-8
  angle of, 4
  dorsoplantar view, 4-6, 5f, 5t
  lateral view, 6, 6f
  oblique views, 6-8, 7f-9f, 7t
Oblique views, 6-8, 7f-9f, 7t
  of scapula, 91-97, 97f
  of skull, 21
  of temporomandibular joint, 36-37, 38f-39f
  of tympanic bullae, 37, 39f
Obturator foramen, 145, 149f
Occipital bone, 33-36
  age at fusion of, 13t
Occipital condyles, 52, 54f
Olecranon, 107-109, 108f, 115f
Olecranon fossa, 109-110
Opacity
  kidneys, 257
  pulmonary, 228, 229f-232f
  splenic, 248, 250f-254f
  urinary bladder, 265-266, 265f
Oropharynx, 30, 31f
Orthogonal views, 3, 3t, 4f
  of scapula, 91
  of thorax, 227
Orthopedic disorders
  of elbow, 110
  juvenile, physeal closure and, 9
Os penis
  superimpose on the pelvis in ventrodorsal radiographs, 147-150
  urethral radiography and, 268, 269f-270f
Osseous tentorium, 23f
Ossification center
  age of appearance, 9, 10t-11t
  cervical spine
    atlas, 52
    axis, 57
  of elbow, 110-111
  os penis, 268, 269f-270f
  of pelvis, 143f, 145-147
  scapular, 91, 94f
    of supraglenoid tubercle, 101, 102f
  tibial, 169
Osteochondrosis
  of elbow, 110
  of shoulder, 99-100

Osteoma, pulmonary, 234, 235f
Osteophyte, 91, 94f

P

Palate
  hard, 30, 32f
  soft, 30, 32f
Palmar as directional term, 2, 2f
Pancreas, 257, 258f-259f
Parietal bone, 33-36
  age at fusion of, 13t
Patella, 161, 161f
Patellar ligament, 158f, 161
Patient positioning. *see* Positioning
Peak inspiration, thoracic radiography during, 184, 184f
Pectoral girdle, 91
Pectus excavatum, 203-205, 205f
Pedicles, 50, 51f
  cervical, 65f
  thoracic, 68f-69f
Pelvic asymmetry, 79, 80f
Pelvic limb, 136-181, 136f
  femur and stifle, 153-167, 153f-169f
  pelvis, 137-150, 137f-141f, 143f, 150f-153f
  pes, 176-179, 177f-181f
  positioning of, in abdominal radiography, 242
  tibia and fibula, 169-176, 171f-176f
Pelvis, 137-150, 137f-141f, 143f, 150f-153f
  age at which ossification center appears in, 10t-11t
  age when physeal closure occurs in, 12t
  orthogonal views of, 3t
Periodontal ligament, 27f
Permanent teeth, 21, 21t-22t, 26f
  buds of, 25-28, 27f
Pes, 176-179, 177f-181f
  anatomy at specific ages in days
    canine, 16f
    feline, 19f
  orthogonal views of, 3t
Phalanges, 129-131, 129f, 176, 180f
  age at which ossification center appears in, 10t-11t
  age when physeal closure occurs in, 12t
Pharynx, 31f
Physeal closure, 1f, 9, 10t-13t, 13f-19f
Physis
  femoral, 159, 161f
  of greater trochanter, 155
  humeral, 101, 101f-102f, 111, 114f
  metacarpal, 131, 133f
  radial, 118-119, 119f
  ulnar, 118-119, 119f-120f
  vertebral, 51-52, 52f
Plica vena cava, 206, 208f

Pneumothorax, misdiagnosis of, 203, 204f, 234, 235f
Polydactyly, 131, 134f
Popliteal sesamoid bone, 163
Positioning
　for abdominal radiography, 242
　for scapular radiography, 91
　for skull radiography, 21
　　of dental arcade, 25, 27f
　　of frontal sinuses, 31f
　　of temporomandibular joint, 36-37, 39f
　　of tympanic bullae, 39, 39f-40f
　for spinal radiography, 50, 50f
　for thoracic radiography, 183-184, 183f
　　dorsoventral view, 195-196
　　left lateral view, 187
　　of trachea, 211, 213f
　　ventrodorsal view, 199-200
Posterior as directional term, 2
Premolars, 21, 24f-25f, 25-28
　eruption of, 21t-22t
Prepuce, superimpose on the pelvis in ventrodorsal radiographs, 147-150
Prognathism, 44, 45f
Projections
　end-on, 250, 250f
　frog-leg, 137
　oblique, 3-8
　　angle of, 4
　　dorsoplantar view, 4-6, 5f, 5t
　　lateral view, 6, 6f
　　oblique views, 6-8, 7f-9f, 7t
　side-on, 250, 250f
　standard, 3
Prostate gland, 267-268, 268f
Proximal humeral epiphysis, 101, 101f-102f
Proximal tibial epiphysis, 160f, 169
Pseudoulcers, duodenal, 279, 280f
Pterygoid bone, 33-36
Pubic bone, 147
Pubic symphysis, 145, 146f
Pubis, 141-142
Pulmonary arteries
　caudal lobar, 219f-221f, 220
　dorsoventral view of, 196, 197f-198f
　left lateral view of, 187-190, 190f-192f
　ventrodorsal view of, 199-200, 200f
Pulmonary heterotopic bone, 234, 235f
Pulmonary nodule, misdiagnosis of, 227, 228f-229f, 231f-232f
Pulmonary osseous metaplasia, 234, 235f
Pulmonary osteoma, 234, 235f

Pulmonary vein
　dorsoventral view of, 196, 197f-198f
　left lateral view of, 187-190
　right lateral view of, 193-194, 194f
　ventrodorsal view of, 199-200, 200f
Pulp cavity, 25, 27f
Pylorus, 188f-189f, 270, 271f-275f

R

Radial carpal bone, 122f-123f, 123-126, 125f
Radial fossa, 109-110
Radial head, 115
Radial notch, 115
Radiographic atlas, using of, 2
Radiography
　carpal stress, 126, 128f
　hanging of images, 2
　normal, 2
　oblique projections in, 3-8
　　angle of, 4
　　dorsoplantar view, 4-6, 5f, 5t
　　lateral view, 6, 6f
　　oblique views, 6-8, 7f-9f, 7t
　standard projections in, 3
　terminology in, 2-3, 2f
　viewing images in, 3
Radiolucent urinary calculi, 265-266
Radius, 107-109, 108f-109f, 115
　age at which ossification center appears in, 10t-11t
　age when physeal closure occurs in, 12t
Range of motion
　carpal, 123-126, 126f
　of lumbar spine, 77, 78f
Recumbency-associated atelectasis, 184-187, 186f, 200f, 201
Recumbent lateral view, of thorax, 184, 186f, 189f
Redundancy
　of colon, 288, 289f-290f
　of dorsal tracheal membrane, 211-213, 213f
　esophageal, 215, 216f
Remodeling, antebrachial, 118-119, 121f
Restraint
　sedation for thoracic radiography, 184, 184f, 200, 215f
　for skull radiography, 21
Retained deciduous tooth, 25-28, 34f
Retroperitoneal fat, 295, 295f
Ribs, 188f, 201, 203f
　cervical, 64, 65f-66f
　overlapping caudal margin of heart, 226f
　thoracic, 67, 70f, 73f
Right atrium, 219f
Right caudal lobar pulmonary artery, 219f-221f, 220

Right lateral view
　of kidneys, 262, 262f-264f
　of spleen, 254, 255f-257f
　of thorax, 193-194, 193f, 195f
Right-left view, 187, 189f-190f
Right ventricle, 219f
Rostral as directional term, 2
Rostrocaudal open-mouth view, of tympanic bullae, 39, 41f
Rostrocaudal view, of frontal sinuses, 30-33, 33f
Rotation, pelvic, 147
Round ligament, 153
Rugal folds, gastric, 274f-276f, 276

S

Sacral crest, 82f
Sacral spine, 81-84, 82f-86f
Sacralization, 77, 79f-80f
Sacroiliac joint, 137, 140f
Sacroiliac junction, 137, 139f
Sacrum, age when physeal closure occurs in, 12t
Sagging, of dorsal tracheal membrane, 211-213
Scapula, 91-103, 91f-100f, 188f
　age at which ossification center appears in, 10t-11t
　age when physeal closure occurs in, 12t
Scapular notch, 91, 93f
Scapulo-humeral joint, 91
Sedation, for thoracic radiography, 184, 184f, 200, 215f
Sesamoid bone(s)
　at distal medial aspect of radial carpal bone, 123-126
　in head of gastrocnemius muscle, 158f, 161-163
　in manus, 129-131, 129f-131f
　in stifle joint, 158f, 161
　　age at which ossification center appears in, 10t-11t
　in supinator muscle, 115, 116f-117f
　in tendon of popliteus muscle, 158f, 170f
Sesamoid cartilage, 44-46, 46f
Shoulder. see also Scapula
　anatomy at specific ages in days
　　canine, 13f
　　feline, 16f
Shoulder joint, 91
Side-on projection, 250, 250f
Sinuses and nasal cavity, 28-36, 29f-36f
Size
　of heart, 216
　of jejunum, 279-282, 282f
　of kidneys, 260
　of liver, 245, 247f

Size *(Continued)*
  of spleen, 250, 251f-252f
  of stomach, 276
  of urinary bladder, 266-267, 267f
Skin folds
  abdominal radiography and, 242, 243f
  thoracic radiography and, 203, 204f, 227
Skull, 20-48, 20f
  dentition and, 21-28, 21t-22t, 24f-28f
  mandibles and larynx, 43-46, 43f-47f
  nasal cavity and sinuses, 28-36, 29f-36f
  orthogonal views of, 3t
  overview of, 21, 22f-24f
  temporomandibular joint and tympanic bullae, 36-43, 36f-43f
Sliding hiatal hernia, 237, 238f
Small intestine, 279-282, 279f-284f
Sphenoid bone, 33-36
  age at fusion of, 13t
Spinal nerves, 52-54
Spine, 49-89, 49f-50f
  age at which ossification center appears in, 10t-11t
  caudal, 86, 87f-88f
  cervical, 52-64, 53f-67f
  divergence between trachea at carina, 210-211, 211f
  lumbar, 74-81, 75f-81f
  orthogonal views of, 3t
  sacral, 81-84, 82f-86f
  of scapula, 91-97, 93f
  thoracic, 67-74, 68f-74f
Spinous process, 50, 51f
  of atlas, 53f
  of axis, 52, 56f
  of C6, 62, 62f
  of C4-C5, 59f-60f
  of lumbar spine, 75f-76f
  of thoracic spine, 67, 68f-69f
Spleen, 248-257, 250f-258f
Splenic vein, 258f
Standard projections, 3
Sternal lymph nodes, 208
Sternum, 188f, 203-205, 205f
  elevation of, for spinal radiography, 50, 50f
  positioning of, for thoracic radiography, 183-184, 183f
Stifle, 153-167, 159f-166f, 168f-169f, 173f
  anatomy at specific ages in days
    canine, 15f
    feline, 18f
  sesamoid bones of, age at which ossification center appears in, 10t-11t

Stifle joint, 155
Stifle joint space, 159
Stomach, 270-276, 271f-278f
  distended, liver size and, 245, 248f
Stress radiography, carpal, 126, 128f
Stylohyoid bones, 44-46, 47f
Styloid process
  of radius, 122f-123f
  of ulna, 115, 120f
Subclavian artery, 206-208, 209f
Subluxation, sacroiliac, 137
Submucosal fat, 276, 277f
Summation opacities, in lung, 227
Superimposition, 3-4
  in elbow radiography, 108f, 110
  of fecal material, 137
  of fused tibial tuberosity, 172f
  of phalanges, 131
  in scapular radiography, 92f
  in shoulder joint radiography, 97-98, 97f
  in skull radiography, 21
    of mandible, 46, 47f
    of nasal cavity, 28-30, 31f
  in thoracic radiography, 183-184, 183f, 194f
Supernumerary digits, 134f
Supernumerary lumbar vertebra, 74, 77f
Supraglenoid tubercle, 91, 93f, 100, 100f
Suprahamate process, 91, 94f
Supraspinatus insertionopathy, 100
Sutures
  steel, 294f
  between zygomatic process of temporal bone and temporal process of zygomatic bone, 33-36, 36f
Symphysis, mandibular, 43-44, 43f
Synchondrosis, 137
Synovial joints, 50-51

T
Talus, 174, 176-179, 177f-178f
Tarsal bone, dorsoplantar radiograph of, 5f
Tarsocrural joint, 174, 176
Tarsus, 176-179, 177f-178f, 180f-181f
  age at which ossification center appears in, 10t-11t
  age when physeal closure occurs in, 12t
Teeth, eruption of, 21t-22t. *see also* Dentition
Temporal bone, 33-36
  age at fusion of, 13t
Temporal process, 33-36, 36f
Temporomandibular joint and tympanic bullae, 36-43, 36f-43f
Tenosynovitis, bicipital, 100, 100f

Teres major tuberosity, 103, 104f
Terminology, radiographic, 2-3, 2f
Thigh, orthogonal views of, 3t
Thoracic limb, 90-135, 90f
  antebrachium of, 115-119, 118f-124f
  carpus of, 122-126, 122f-123f, 125f-128f
  elbow joint of, 107-115, 107f-117f
  manus of, 129-131, 129f-134f
  scapula and brachium of, 91-103, 92f-106f
Thoracic spine, 67-74, 68f-74f, 188f
  age when physeal closure occurs in, 12t
Thoracic wall, 201-205, 202f-205f, 227
Thorax, 182-240, 182f-183f, 185f-187f
  diaphragm, 237-240, 237f-240f
  dorsoventral view of, 195-196, 196f-198f
  esophagus, 214-215, 214f-216f
  heart, 215-226, 217f-220f, 222f-226f
  left lateral view of, 187-192, 188f-190f, 192f, 195f
  lungs, 227-236, 228f-235f
  mediastinum, 205-210, 206f-210f
  orthogonal views of, 3t
  right lateral view of, 193-194, 193f, 195f
  thoracic wall, 201-205, 202f-205f
  trachea and bronchi, 210-213, 211f-213f, 227
  ventrodorsal view of, 199-201, 199f-200f
Thymus, 210, 210f
Thyrohyoid bones, 44-46, 47f
Thyroid cartilage, 44-46, 46f
Tibia and fibula, 169-176, 171f-176f
  age at which ossification center appears in, 10t-11t
  age when physeal closure occurs in, 12t
Tibial cochleae, 174, 175f
Tibial tuberosity, 169-172
Tomographic imaging modality, 4
Tongue, radiography of soft palate and, 30, 32f
Tooth bud, 25-28, 27f
Trachea, 188f, 206-208, 210-213, 211f-213f
  intrathoracic, 210, 211f
  superimposition of humeral head on, 97-98, 98f
Tracheal membrane, dorsal, 211-213, 213f
Tracheal rings, 211-213
Tracheobronchial lymph nodes, 208
Transitional anomalies
  cervical, 65f-67f
  lumbar, 77, 79f-80f

Transitional anomalies *(Continued)*
  sacral, 81, 83f
  thoracic, 67, 71f
Transverse processes, 50, 51f
  anomalous, 67, 71f
  of atlas, 52, 53f-54f
  of C3-C5, 59f
  of lumbar spine, 75f-76f
Triadan system, of dental identification, 25, 26f
Triceps muscle group, 111-115
Tricipital line, 103, 104f
Trochanteric fossa, 153-155, 155f
Trochlea of talus, 177f-178f
Tuber ischii, 147f
Tuberosity
  deltoid, 103, 104f
  tibial, 169-172, 171f
Turbinate bones, of nasal cavity, 28-30, 29f
Tympanic bullae, and temporomandibular joint, 36-43, 36f-43f

U

Ulna, 107-111, 109f, 111f, 115
  age at which ossification center appears in, 10t-11t
  age when physeal closure occurs in, 12t
Ulnar carpal bone, 123-126, 125f
Ultrasound, 4
Ultrasound gel, artifact, 294f, 295
Ununited Anconeal Process (UAP), 110-111

Ureteral calculi, misinterpreted as, 265, 265f
Ureters, 265, 265f
Urethra, 268-270, 269f-270f
Urinary bladder, 265-267, 266f-267f
Urinary calculi, 269f
  radiolucent, 265-266
Urinary catheters, 266, 266f-267f

V

Vena cava
  caudal, 187, 188f-190f, 193f, 199f, 206, 208f
  in abdominal computed tomography, 265, 265f
  cranial, 206-208, 209f-210f
  plica, 206, 208f
Ventral cortex, of lumbar spine, 74, 75f-76f
Ventricle, 219f
Ventrodorsal view. *see also* Dorsoventral view
  of abdomen, 242, 243f
    spleen, 253f, 254, 255f-256f
    stomach, 271f-272f, 272
  of femur, 155f-157f
  hanging of radiograph, 3
  of pelvis, 137, 139f-141f, 150f-153f
  of spine, 50
  of thorax, 185f-186f, 199-201, 199f-200f
Vertebra, 50, 51f
  anticlinal, 67, 70f
  block, 60, 60f-61f
  caudal. *see* Caudal spine
  cervical. *see* Cervical spine

Vertebra *(Continued)*
  in kidney size assessment, 260
  lumbar. *see* Lumbar spine
  thoracic. *see* Thoracic spine
  transitional anomalies and, 64, 65f
Vertebral arch, 50
  of C1, 57f
  of C2, 57-58, 57f-58f
  of caudal spine, 86, 87f
Vertebral body, 50, 51f
Vertebral canal, 50, 51f, 68f-69f
Vertebral heart score, 216
Viewing images, 3
Views
  dorsoplantar, 4-6
  lateral, 6, 6f
  oblique, 6-8, 7f-9f, 7t
Vomer bone, 29f-30f

W

Wing
  of atlas, 52, 53f
  of iliac, 137-141
  of sacrum, 82f

X

Xiphoid process, hypoplasia of, 203-205, 205f

Y

Yes joint, 52

Z

Zygomatic arch, 29f, 33-36
Zygomatic process, 33-36, 36f, 38f